Essays in
Microbiology

Essays in Microbiology

Edited by

Professor J. R. Norris

*Director, Meat Research Institute,
Langford, Bristol*

and

Professor M. H. Richmond

*Head of Department of Bacteriology,
The Medical School,
University of Bristol*

John Wiley & Sons
Chichester · New York · Brisbane · Toronto

Library of Congress Cataloging in Publication Data:

Main entry under title:

Essays in microbiology.

   Includes index.
   1. Microbiology. I. Norris, John Robert.
II. Richmond, Mark H.
QR41.2.E87   576   78-2828

ISBN 0 471 99556 8

Typeset by Preface Ltd, Salisbury, Wilts. and
printed in Great Britain by Unwin Brothers Ltd, The Gresham Press,
Old Woking, Surrey

Preface

Microbiology shares with Geology and one or two other pure science courses taught in Universities and similar places the difficulty of being a subject not widely taught to children. Although it is certainly true that many young microbiologists begin their University Microbiology Courses with a sound enough knowledge of Chemistry and Biology, all too often there is only the haziest idea as to what they are actually about to undertake; this collection of sixteen Essays is therefore designed to throw some light on the problem, and to provide a solid basis on which to build an advanced knowledge of the subject from more specialized books and journals. Each author has been encouraged to take a simple introductory approach to his particular topic and the primary readership is defined as the early stage undergraduate student with some topics of direct relevance to the senior school pupil specializing in Biology. In addition, each author has been requested to try to place the particular aspect that concerns him in its context in the subject as a whole. In this way we hope that these Essays will also be of interest and value to those further on in their Microbiology Courses, and even to those specializing in other aspects of Biology — particularly the study of plants and animals, and their interactions.

The biological revolution that has been caused by the rise of Molecular Biology — an event in which the study of Microbiology has certainly played a central and crucial role — has left the teacher of Microbiology with a difficult problem. Should he tackle the subject in descriptive terms and try to teach by analysing the complex activities of microbial populations in Nature? Or should he begin with the structure of the component macromolecules that go to make up microbes and then attempt a synthesis that comes somewhere near to what we see in practice? Of course neither, if pursued without a leavening of the other, gives anything like a balanced picture; and indeed there is enormous advantage in leading students to look at the subject from these two alternative viewpoints in parallel. If this approach is well done, microbiologists are led to think about the subject in novel ways: two views of the same topic cross-illuminate one another, and from the paradoxes generated in this way the development of the subject by research is initiated.

There has been no attempt in these sixteen Essays to cover all aspects of Microbiology: clearly that would be foolhardy; and certainly no University attempts such a catholic approach. Rather we have chosen topics for the Essays to try to indicate the broad sweep of the subject. About half are broadly 'descriptive' while the others are 'molecular'. All, we hope, nevertheless help to throw more light on the question posed, and indeed partly answered, by Dr Stanier in his first Chapter: 'What is Microbiology?'

Textbooks are expensive and Microbiology is a vast subject taught in many different courses. Most teachers have had the experience of recommending students to study selected chapters from various books as being relevant to their particular courses. We therefore conceived the idea of publishing the Essays both as a complete book and as separate units available at the lowest possible price, so enabling the student economically to study those elements specifically required by his own programme of work.

In closing this brief Preface, we would like to thank a number of people: the patient contributors for writing and amending their scripts to fit the straightjacket imposed by the 32 pages allowed by our format; our secretaries, for all their help in getting the material ready for publication; and Janet Jones and John Wiley for agreeing to take part in an experiment in the form of textbook publication.

Contributors

C. Booth — Commonwealth Mycological Institute, Ferry Lane, Kew, Surrey

P. H. Clarke — Department of Biochemistry, University College, Gower Street, London WC1E 6BT

C. R. Curds — Protozoa Section, Department of Zoology, British Museum (Natural History), Cromwell Road, London SW7 5BD

D. Kay — Sir William Dunn School of Pathology, University of Oxford, South Parks Road, Oxford OX1 3RF

P. Meadow — Department of Biochemistry, University College, Gower Street, London WC1E 6BT

R. G. E. Murray — Department of Bacteriology and Immunology, Health Sciences Center, University of Western Ontario, London, Ontario, Canada N6A 5C1

C. G. Ogden — Protozoa Section, Department of Zoology, British Museum (Natural History), Cromwell Road, London SW7 5BD

M. H. Richmond — Department of Bacteriology, The Medical School, University of Bristol, University Walk, Bristol BS8 1TD

R. Slepecky — Department of Biology, Syracuse University, 130 College Place, Syracuse, NY 13210, USA

D. C. Smith — Department of Botany, University of Bristol, Woodland Road, Bristol BS8 1UG

H. Smith — Department of Microbiology, South West Campus, The University of Birmingham, P.O. Box 363, Birmingham B15 2TT

P. H. A. Sneath — MRC Microbial Systematics Unit, University of Leicester, University Road, Leicester LE1 7RH

R. Y. Stanier   *Service de Physiologie Microbienne, Institut Pasteur, 25 Rue du Docteur Roux, 75724 Paris, Cedex 15, France*

D. W. Tempest   *Laboratorium voor Microbiologie, Universiteit van Amsterdam, Plantage Muidergracht 14, Amsterdam, The Netherlands*

D. Watson   *Department of Microbiology, School of Medicine, The University of Leeds, Leeds LS2 9NL*

R. Whittenbury   *Department of Biological Sciences, School of Biology, University of Warwick, Coventry CV4 7AL*

Contents

x

What is Microbiology?

R. Y. STANIER

Institut Pasteur, 75724 Paris Cedex 15, France

I. Introduction

Unlike the physical sciences, biology can be subdivided in two planes: one organismal, the other functional. Organismal subdivision produces disciplines that deal with specific evolutionary branches of the biological world; for example, ornithology, entomology, mycology, bacteriology. Functional subdivision produces disciplines that cross organismal boundaries, such as cytology, genetics, and biochemistry.

Microbiology is seemingly an organismal discipline, since it is concerned with the properties of small forms of life, or microorganisms. However, the term 'microorganism' does not define a well-circumscribed evolutionary group. According to *Webster's New International Dictionary,* a microorganism is 'any organism of microscopic (also in a broad sense, ultramicroscopic) size; applied especially to bacteria and protozoa'. Let us follow this lexigraphic lead, and see where it takes us.

The human eye cannot resolve an object less than 1 mm in diameter, and can perceive very little structural detail in objects an order of magnitude larger. Microscopic examination is therefore essential either for the very perception, or for the determination of gross structure, of any organism that is 1 mm or less in its largest dimension. Such an organism can be reasonably described as a 'microorganism'.

With very rare exceptions, cells do not have diameters in excess of 1 mm. Consequently, unicellular organisms fall into the microbial category. These include nearly all bacteria and protozoa, as noted by *Webster.* Another class of organisms which are unquestionably microorganisms are the viruses. The only stage of viral development that can be structurally defined is the infectious particle or virion, and the size range of virions extends from the lower limits for cells (0.2–0.3 μm in diameter) to objects two orders of magnitude smaller. The resolution of most virions is thus possible only by electron microscopy, which accounts for the parenthetic proviso ('ultramicroscopic') in Webster's definition.

Some members of the fungi (e.g. yeasts) and of the algae (e.g. diatoms, photosynthetic flagellates) are unicellular and can therefore be construed without ambiguity as microorganisms. Difficulties arise, however, with other members of these two taxonomic groups. Many fungi are coenocytic mycelial organisms, and there is no fixed limit to the size which can be attained during vegetative growth of a single individual. Indeed, some basidiomycetes may produce a mycelium as much as 50 metres in diameter, from which there develop, under conditions favourable for fructification, dozens of fruiting bodies, each of macroscopic dimensions. A similar problem confronts us in the algae. Some algae are either coenocytic or multicellular, and mature individuals may attain a size considerably larger than that of many flowering plants. Strict adherence to the dictionary definition would force us to conclude that some algae and fungi fall in the domain of microbiology, whereas others do not. On the other hand, certain metazoan animals (rotifers and some nematodes) may never exceed 1 mm in their largest dimension, even though they are not commonly considered as organisms that form part of the domain of microbiology.

As this analysis shows, the concept of a 'microorganism' is highly artificial. It is at the same time remarkably broad, in the sense of covering the protozoa, the bacteria, and the viruses, three groups which differ profoundly in their biological properties; and unduly restrictive, in the sense of cutting through reasonably well-defined natural groups, such as fungi and algae. Microbiology is accordingly not comparable to most disciplines of organismal biology. Nevertheless it is without question a branch of biology that possesses both unity and coherence. These are derived from the fact that it possesses some of the attributes of a functional discipline: the diverse taxonomic groups that fall into the microbiological domain are all susceptible to analysis by a special methodology. It is the methods of microbiology that determine, in the last analysis, the kinds of organisms that constitute its biological subject matter: viruses, bacteria, protozoa, fungi, and algae.

II. The Methods of Microbiology

Microorganisms are ubiquitous in the biosphere; every natural habitat contains an extremely diverse microbial population. On very rare occasions, a population that consists predominantly of one type of microorganism may develop, but such microbial 'blooms' are typically localized in both space and time. Although the relative abundances of different microorganisms fluctuate continuously, the heterogeneity of natural populations is normally so great that any given kind of microorganism represents, at most, a very small fraction of the total. This makes it virtually impossible to study the properties of a specific microorganism in its natural habitat. The problem can be solved only by isolating it from the natural habitat, freeing it from all accompanying organisms, and propagating it as a pure strain in a suitable artificial medium which has been sterilized prior to inoculation, and protected from subsequent contamination by the other microorganisms that are omnipresent in the environment.

A. Sterilization

Fundamental to pure culture technique is the preliminary sterilization (and protection from subsequent contamination) of the media, culture vessels, and implements employed in isolating and manipulating pure cultures. Sterilization can be achieved by exposure to lethal agents, either physical or chemical; or, in the special case of solutions, by filtration. Filtration, the only means of sterilizing solutions of highly labile compounds, has an intrinsic limitation. Although filters that will retain all cellular organisms are readily available, filters sufficiently fine to retain even the smallest viruses are not, since some virions have the dimensions of large protein molecules. Filtration can be used to make a solution cell-free, without necessarily making it virus-free.

The most convenient and widely used agent of sterilization is a physical one: heat. The principles of heat sterilization are not always clearly understood and merit brief discussion. When a pure culture of a microorganism is exposed to heat (or to any other lethal agent) the kinetics of death are typically exponential: a plot of the logarithm of the number of survivors as a function of time gives a straight line, the negative slope of which expresses the rate of killing. These kinetics reflect the fact that, in a homogeneous population, probability alone determines the time of death of any given individual. The death rate defines what *fraction* of the initial population will survive any given period of exposure to the lethal agent; it does not, of itself, provide information as to the *number* of survivors at this time. This is determined by another parameter, the initial population size. The larger the initial population, the larger the number of survivors after any given period of exposure to the lethal agent. Thus, a much longer period of heating is necessary to sterilize 10 litres of a culture medium than to sterilize 10 millilitres, assuming that they both contain microbial populations of the same initial density. Long experience has shown that the endospores produced by certain bacteria are the most highly resistant of all microbial cells; bacterial spore suspensions are therefore frequently used to calibrate heat sterilization procedures.

In the light of the foregoing discussion, the goal of sterilization can be reformulated in a somewhat more sophisticated way: *the probability that the object subjected to treatment contains even one viable surviving cell should be infinitesimally small.* The sterilization procedures employed by microbiologists are designed to meet this goal, and provide a wide margin of safety.

B. Microbial Nutrition and the Design of Culture Media

In order to grow a cellular microorganism, a culture medium must be prepared which contains an adequate supply of nutrients, i.e. the various chemical substances required for the synthesis of cell materials and for the generation of ATP. The culture of viruses is a different problem which will not be further considered here; their growth is dependent on the provision of suitable conditions for the development of the cellular host.

Cellular microorganisms display an extraordinary nutritional diversity, which reflects their extreme diversity in physiological and biochemical respects. Consequently, it is impossible to prepare a universal culture medium, suitable for the growth of all kinds of microorganisms. The design of a suitable culture medium must be worked out for each particular microbial group, taking into account its special physiological and biochemical properties. For bacteria alone, literally thousands of different culture media have been proposed. Nevertheless, certain general principles govern the design of them all.

In the first place, a culture medium should contain a *balanced*

Table 1. Approximate elementary composition of the microbial cell[a]

Element	Dry weight (%)
Carbon	50
Oxygen	20
Nitrogen	14
Hydrogen	8
Phosphorus	3
Sulfur	1
Potassium	1
Sodium	1
Calcium	0.5
Magnesium	0.5
Chlorine	0.5
Iron	0.2
All others	~0.3

[a]Data for a bacterium, *Escherichia coli*, assembled by S. E. Luria, in *The Bacteria* (I. C. Gunsalus and R. Y. Stanier, eds.), Vol. I, Chap. 1 (New York: Academic Press, 1960).

mixture of the different nutrients, each being furnished in a relative amount roughly proportional to biosynthetic requirements; some nutrients are required only in traces, others in much larger amounts. This principle is of critical importance, since depletion of any one nutrient, whatever its nature, will arrest growth, and arrest is sometimes preceded by a short period of unbalanced growth which makes the population physiologically abnormal.

The chemical composition of cells, which varies little, provides a useful insight into general nutritional requirements (Table 1). Water is always the principal molecular component (80–90 per cent by weight) of the living cell, and therefore a major essential nutrient. In addition to hydrogen and oxygen (derivable metabolically from water), the dry matter of cells contains four principal non-metallic elements: carbon, nitrogen, phosphorus, and sulphur. It also contains a variety of metals, of which potassium, sodium, calcium, magnesium, and iron are quantitatively the most important. However, several additional metals, present only in traces in cells, play indispensable roles in cellular metabolism, and are therefore essential nutrients. They include manganese, cobalt, copper, molybdenum, and zinc. Although both sodium and chlorine are normally present at fairly high levels in the dry matter of cells, neither of these elements can be demonstrated to be essential for the growth of most microorganisms, with the exception of those which inhabit marine or hypersaline environments. Indigenous marine microorganisms have readily demonstrable Na^+ and Cl^- requirements, as well as quantitative requirements for Ca^{2+} and Mg^{2+} considerably higher than those of terrestrial and freshwater forms. With the partial exception of Na^+ and Cl^-, all the above mentioned elements are essential nutrients, and must be provided in any culture medium in a suitable chemical form.

All metals, together with phosphorus (as phosphate) can be provided as nutrients in the form of inorganic salts. The nutritional diversity of microorganisms largely reflects the different molecular forms in which four elements — carbon, nitrogen, sulphur, and oxygen — must be provided.

Most photosynthetic microorganisms (exception: some photo-synthetic bacteria) can use the most highly oxidized form of carbon, CO_2, as a carbon source. In these groups, ATP is derived from a physical source, by the conversion of light energy into chemical bond energy. Carbon dioxide can also be used as a carbon source by certain groups of non-photosynthetic bacteria (chemoautotrophs), which can couple the oxidation of reduced inorganic compounds (e.g. NH_3, H_2, H_2S) with ATP synthesis.

All other microorganisms belong to the nutritional category of chemoheterotrophs, which require at least one organic compound as a major nutrient, from which they derive cell carbon. This substance also has a second metabolic role, as a source of ATP; it is in part decomposed by a respiratory or fermentative pathway, the operation of which is coupled with ATP synthesis.

Many chemoheterotrophic bacteria and fungi, as well as a few protozoa, can derive both carbon and energy from the metabolism of a single organic compound. Microbial diversity with respect to the organic substances utilizable for this purpose is extreme. *Every naturally occurring organic compound can be used as a carbon and energy source by at least one type of microorganism.* Consequently the number of different organic compounds utilizable by the totality of micro-organisms runs into tens of thousands. Not surprisingly the nutritional spectrum of any given microorganism is narrow, relative to the immense total range. Nevertheless, some bacteria possess remarkably wide nutritional spectra; for example, representatives of the genus *Pseudomonas* can use as single carbon and energy sources at least 100 different organic compounds, including sugars, fatty acids, dicarboxylic acids, hydroxyacids, aminoacids, amines, benzenoid compounds, and sterols. Other bacteria have extremely limited and specialized nutritional spectra; for example, one physiological group, the obligate methylo-trophs, can use only two carbon and energy sources: methane and methanol. Methane cannot be utilized by any other microorganisms, and methanol is very rarely utilized by members of other microbial groups.

In the cell, nitrogen and sulphur occur principally in a reduced organic state, as amino ($R-NH_2$) and sulphhydryl ($R-SH$) compounds, respectively. Many microorganisms can use as sources of these elements the anions NO_3^- and SO_4^{2-}; their incorporation into organic form within the cell is preceded by a reduction to ammonia and sulphide, respectively. Microorganisms unable to perform one of these reductions must be furnished with either ammonia (or sulphide) as a nitrogen (or sulphur) source. Inability to reduce nitrate is relatively common; inability to reduce sulphate is much rarer. Many bacteria (but not other microorganisms) can use N_2 as an inorganic nitrogen source.

Simple nutrient requirements necessarily reflect a high degree of biosynthetic ability. A microorganism able to grow at the expense of a single carbon compound, nitrate, and sulphate, must be able to synthesize from these three nutrients a wide diversity of metabolic intermediates, all essential for cellular function. They include the coenzymes, as well as the monomers required for the synthesis of proteins, nucleic acids, lipids, and polysaccharides. If an organism does not possess the enzymic machinery necessary for the synthesis of any one of these coenzymes or monomers, the compound in question (or its immediate metabolic precursor) becomes an essential nutrient, and must be furnished in the growth medium. Such nutrients, individually required in quantities that are small relative to the principal carbon source, are termed *growth factors*. Growth factors can be divided by virtue of their chemical structures and biological functions into three categories: amino acids, the building blocks of proteins; purines and pyrimidines, the building blocks of nucleic acids; and vitamins, a chemically diverse array of compounds, each of which is a metabolic precursor of a particular type of coenzyme.

In microorganisms that require growth factors, the number and nature of the requirements vary widely. For many microorganisms, the requirement can be met by the provision of a single growth factor, for example, a specific amino acid or vitamin. Other groups of microorganisms have extensive growth factor requirements; this is an evolutionary expression of multiple losses of biosynthetic capacity, resulting from existence in ecological niches where these biosynthetic intermediates are readily available in the external milieu. Among bacteria, the most extreme example of growth factor dependence occur among the lactic acid bacteria. Some species of this group have absolute requirements for as many as 16 of the 20 amino acids that enter into the composition of proteins; four purines and pyrimidines, and numerous vitamins. This nutritional complexity reflects the nutrient-rich natural habitats of lactic acid bacteria, which develop in decaying plant materials, in milk, and in the body cavities of animals. Many protozoa have growth factor requirements of equivalent complexity. This often reflects their predacious mode of life (phagotrophy);

such protozoa normally use as food sources smaller microorganisms, which they ingest by phagocytosis, and digest within intracellular vacuoles.

Even when the precise growth factor requirements of such micro-organisms have been determined, the preparation of a chemically defined medium for their cultivation is rarely attempted. There is a much simpler and more expeditious solution: the preparation of a *complex* medium, which contains (in addition to the necessary minerals and a suitable organic carbon and energy source) a product of natural origin rich in growth factors, but of undefined chemical composition. Yeast extract and extracts of plant and of animal tissues, are often used for this purpose. Phagotrophic protozoa can be conveniently grown either as two-membered cultures with an appropriate microbial prey, or as pure cultures, furnished with a heat-killed suspension of the prey. This is often necessary, because many phagotrophic protozoa grow poorly or not at all at the expense of dissolved nutrients, and appear to require food materials in particulate form.

The role of oxygen in microbial nutrition requires special discussion. Although oxygen can be derived metabolically from water, all organisms that obtain energy from oxygen-linked respiration must also be furnished with another molecular form of this element, O_2, essential as a terminal electron acceptor. Organisms of this physiological type (*strict aerobes*) are widespread among bacteria, fungi, and protozoa. However, some fungi (e.g. many yeasts) and bacteria (e.g. members of the enteric group) are facultative aerobes, since they can obtain energy from either the fermentative or the respiratory dissimilation of organic compounds, and thus can grow in the absence of oxygen, provided that they are furnished with a *fermentable* organic substrate. The proviso is important: a facultative anaerobe such as the bacterium *Escherichia coli* behaves as a strict aerobe if provided with a substrate (e.g. acetate or lactate) which can be metabolized only through the respiratory pathway; on the other hand, it behaves as a facultative anaerobe if the organic substrate is a fermentable sugar.

Molecular oxygen is a very reactive compound, and all organisms that live in contact with air possess enzymic devices to prevent the accumulation in the cell of the highly toxic derivatives formed from it. The most damaging derivative is the superoxide free radical, O_2^-. Organisms that can tolerate exposure to molecular oxygen contain an enzyme, superoxide dismutase, which eliminates the free radical by the reaction:

$$2O_2^- + 2H^+ \longrightarrow O_2 + H_2O_2$$

Most organisms also contain catalase, which decomposes the much less toxic product, hydrogen peroxide, to oxygen and water:

$$2H_2O_2 \longrightarrow 2H_2O + O_2$$

However, many microorganisms (mainly bacteria, and a few protozoa) do not possess these enzymic protective devices, and as a result are very rapidly killed by exposure to molecular oxygen. The cultivation of such *strict anaerobes* requires special precautions to exclude even transient contact of the organisms with air. In the biosphere, strict anaerobes inhabit ecological niches to which oxygen never penetrates, such as the sediments of lakes and oceans and the intestinal tract of animals. Although most strict anaerobes have a fermentative mode of energy-yielding metabolism, some photosynthetic bacteria also belong to this physiological category.

The role of light in microbial growth also calls for additional comment. The mechanisms of photosynthesis and the light-harvesting pigments associated with the process in microorganisms are highly diverse. All algae share with higher plants the ability to perform *oxygenic photosynthesis*, which can be represented by the overall equation:

$$CO_2 + 2H_2O \xrightarrow{\text{light}} (CH_2O) + H_2O + O_2$$

The photochemical processes associated with this mode of photosynthesis lead, in addition to the formation of ATP, to a photochemical splitting of water:

$$2H_2O \xrightarrow{\text{light}} 4[H] + O_2$$

The photochemically derived reductant, NADPH, symbolized above as [H], serves in conjunction with ATP to mediate conversion of CO_2 to organic cell material, symbolized above as (CH_2O). The same mode of photosynthesis exists in one large group of bacteria, the cyanobacteria. All these photosynthetic microorganisms have light-harvesting pigments which absorb in the visible spectral range (roughly, 400–700 nm).

Two groups of bacteria, the purple and green bacteria, perform a less complex version of photosynthesis, termed *anoxygenic photosynthesis* because it is never accompanied by a photochemical cleavage of water, and can therefore not give rise to the formation either of oxygen or of a photochemically generated reductant. Anoxygenic photosynthesis generates only ATP, and the reducing power necessary for CO_2 assimilation must be generated enzymically, from a chemical source other than water. Many purple and green bacteria can use H_2S for this purpose. The over-all reaction for the first step of sulphide oxidation can be represented by the photosynthetic equation:

$$CO_2 + 2H_2S \xrightarrow{\text{light}} (CH_2O) + H_2O + 2S$$

Many purple bacteria and a few green bacteria can also use photochemically derived ATP to perform a direct photoassimilation of organic substrates, a reaction which can be approximately represented as:

$$\text{Organic substrate} \pm CO_2 \xrightarrow{\text{light}} (CH_2O)$$

These anoxygenic bacterial photosyntheses are anaerobic metabolic processes; many of the bacteria that perform them are, in fact, strict anaerobes. Any organism that performs oxygenic photosynthesis is of necessity an aerobe, in the sense that it must be able to tolerate the presence of molecular oxygen.

Another distinctive property of the purple and green bacteria is the possession of light-harvesting pigment systems which absorb very largely outside the visible range, in the near infrared region (750—1000 nm). This appears to be an evolutionary adaptation designed to avoid direct competition with oxygenic phototrophs for solar radiant energy. Because of their anaerobic propensities, purple and green bacteria are mostly confined to the oxygen-free, subsurface layer in natural bodies of water, and therefore have to perform photosynthesis with wavelengths of light that are transmitted through the cells of oxygenic phototrophs, present in the overlying oxygen-rich water layer. When an artificial light source is used for the growth of purple and green bacteria, it should be incandescent (i.e. with high emission in the near infrared). Fluorescent lamps, excellent energy sources for the cultivation of other photosynthetic organisms, are almost completely ineffective.

C. The Principle of the Enrichment Culture

One very valuable element of microbiological technique which falls outside the domain of pure culture methods, but is often brought into play as a preliminary to their application, is the enrichment culture. It is a device for imposing artificial selection on a heterogeneous natural microbial population, so as to favour the growth of a particular type of microorganism, possessing an ensemble of nutritional and physiological properties determined by the type of selection applied. If the selection is sufficiently rigorous, the microorganism in question, however low its initial abundance, rapidly becomes the predominant member of the population. Its subsequent isolation in pure culture is thereby greatly facilitated.

The power of this method rests on the extraordinary diversity of microorganisms in metabolic, nutritional, and physiological respects. Its development was very largely the work of Winogradsky and Beijerinck, who first systematically applied a wide range of selective conditions to mixed microbial populations, and determined the nature of the organisms that came to predominance in response to each type of selection. Once the biological outcome of a particular set of selective conditions has been determined, the same procedure can be applied with other natural source materials and (provided that organisms capable of developing under the prescribed conditions are present) the same specific enrichment will result. By making a preliminary series of dilutions of the natural material, it is possible to obtain an estimate of the numerical abundance of any type of microorganism for which there is a specific enrichment method. The enrichment culture technique is therefore also a powerful tool for the study of microbial ecology.

One example will serve to illustrate both the precision and the flexibility of the method. The ability to synthesize the enzyme nitrogenase, and hence to use N_2 as a nitrogen source, is confined to bacteria, although far from universal among them. Hence, if one inoculates the mixed microbial population present in a sample of soil or water into a liquid medium that contains all essential nutrients *except a combined nitrogen source*, and incubates this medium in contact with a gas phase containing N_2, all non-bacterial components of the population, as well as all bacteria unable to synthesize nitrogenase, will be counter-selected. Such an enrichment culture is, in principle, rigorously selective for nitrogen-fixing bacteria. In practice, there is often some carry-over of fixed nitrogen with the inoculum, but one or two further transfers in the same medium eliminate this source of non-selectivity.

Experience has shown that the ability to fix nitrogen occurs in a wide diversity of bacterial groups, which differ markedly in other biochemical and physiological respects. Accordingly, by modifying secondary parameters of the enrichment medium, it is possible to select a particular group of nitrogen-fixers to the exclusion of all others. Table 2 shows some of the specific nutritional variations which can be employed, and the biological outcome of each type of enrichment.

D. The Uses and Limits of Pure Culture Methodology

The cardinal importance of pure culture methods was recognized very early in the development of microbiology, soon after the basic techniques had been worked out by the schools of Louis Pasteur and Robert Koch. A distinguished nineteenth century mycologist, Brefeld, made the point with an aphorism: if one doesn't work with pure cultures, only nonsense and *Penicillium glaucum* (a common contaminating mould) can come out of it. Few (if any) bacteriologists or virologists would disagree with the Brefeldian dictum. However, many phycologists and protozoologists, as well as a few mycologists, still work without recourse to pure culture methodology, and would contend — no doubt justifiably — that useful information about the organisms that fall into their respective domains can be obtained by studying natural populations and even, for fungi and algae, herbarium specimens. It should be noted, however, that the methodological criterion in such

Table 2. Enrichment media and conditions of incubation selective for several different groups of nitrogen-fixing bacteria

Invariant parameters: mineral base, containing Na—K phosphate buffer, $MgSO_4$, $CaCl_2$, and the following trace elements: Mn, Co, Fe, Mo. Incubation in contact with gas phase containing N_2. Temperature $25-30\,°C$. pH $7-8$ unless otherwise stated.

A. Inorganic carbon source (CO_2): source of energy light.
1. Light of wavelengths $400-700$ nm: molecular oxygen present.
Heterocystous cyanobacteria
2. Light of wavelengths >700 nm: molecular oxygen excluded.
a. H_2S present.
Purple and green sulphur bacteria
b. Non-fermentable organic compound (e.g. ethanol or acetate) present in addition to CO_2.
Purple non-sulphur bacteria

B. Organic carbon and energy source: incubation in dark.
1. Carbon and energy source non-fermentable (e.g. ethanol or acetate): molecular oxygen present.
a. pH 7 or greater.
Azotobacter
b. pH 5 or less.
Beijerinckia
2. Carbon and energy source a fermentable sugar (e.g. glucose): molecular oxygen excluded.
a. No organic growth factors present.
Klebsiella
b. B vitamins present.
Bacillus polymyxa;
Clostridium spp.

cases clearly provides a line of demarcation between what is microbiology and what is some other kind of biology. A phycologist who confines his studies on algae to field populations and herbarium specimens is not apt to describe himself as a microbiologist; one who studies the same organisms in culture probably will.

One major restriction governs the use of pure culture methodology; *it is applicable only to organisms that can reproduce by asexual means.* Once in pure culture, such an organism can be propagated indefinitely, and on any desired scale, as a pure clone consisting of individuals of genetic near-identity (spontaneous mutations will, of course, introduce minor genetic heterogeneity into the population). *Thus the microbiologist is able to use the clonal population, rather than the individual organism, as an object of study.* The enormous experimental advantages are obvious. He can easily determine properties that are either not determinable at all, or determinable only with the greatest difficulty, on an individual organism of very small size.

Since microbiologists study populations, the principles of population biology affect almost every aspect of their work. A microbiologist cannot afford to ignore (as many macrobiologists can) the parameters of population growth; the environmental factors that influence growth rate and growth yield; or the play of mutation and selection, which inevitably becomes significant in any large population.

Although the sexual nature of their reproduction precludes the cloning of most animals, the isolation and propagation in pure culture of animal cells and tissues is now a common practice; it represents an extension into the macrobiological domain of techniques originally developed by microbiologists. However, a zoologist who undertakes cell or tissue culture at that point abandons the study of the organism, since such cultures cannot (at least so far) regenerate the animal from which they were derived; they have become permanently separated from the germ line. Cell and tissue culture of higher plants does not always suffer from the same limitation. Experience has shown that clones of plant cells often remain totipotent and can be induced by appropriate treatments to regenerate a mature plant, capable of sexual reproduction. Here, accordingly, pure culture methods can be intercalated with the study of the whole organism.

III. Microorganisms and Major Biological Categories

The notion that all living organisms can be placed in one of two kingdoms, plants and animals, was inherited by nineteenth century biology from the era of Natural History. The gradual realization that it does not accurately reflect biological realities was an outgrowth of the detailed exploration of the microbial world, which got underway in the latter half of the nineteenth century. Study of microorganisms eventually led to the recognition of three primary categories of organisms: viruses, prokaryotes (bacteria), and eukaryotes. The two former categories consist exclusively of microorganisms. The two traditional kingdoms of Natural History, plants and animals, are subgroups of eukaryotes, though it is not easy to fix their limits with precision. The various metazoan phyla can be readily defined as 'animals', and photosynthetic eukaryotes, starting at the level of complexity represented by liverworts, as 'plants'. However, many eukaryotes, including all those of interest to the micro- biologist — protozoa, algae, and fungi — cannot be construed either as plants or as animals, without introducing major qualifications into the definitions of these two assemblages, and making arbitrary assignments which lead to the splitting of some evidently natural groups. The relatively simple eukaryotes are best treated as a third major eukaryotic subgroup: the protists. This treatment is also convenient because algae, protozoa and fungi, despite their diversity, really constitute a biological continuum. In the last analysis, a rigorous dividing line cannot be drawn between protozoa and algae, or between protozoa and fungi.

A. The Properties of Viruses

Viruses exist in two different states, one extracellular, the other intracellular. The extracellular state is the infectious particle or virion, inert except when it makes contact with a cell susceptible to infection. A virion consists of one molecule (rarely more) of either DNA or RNA, enclosed by a protein coat, or capsid. The entire structure, a nucleocapsid, is of fixed form and size; and its form is definable in crystallographic terms (a polyhedron or a helix). The polyhedral nucleocapsid of some bacterial viruses bears a helically constructed tail, which plays a specific role in attachment to the host cell and in initiation of infection. The virions of some viruses are enclosed by membranous envelopes, derived from the membrane of the host cell.

The intracellular state of the virus consists of viral nucleic acid, replicated by the metabolic machinery of the host cell. The viral nucleic acid provides, either directly or indirectly, messages for the synthesis of a limited number of viral proteins. These include the subunits (capsomers) from which the capsid is assembled, the tail proteins (if formed), and some enzymes, with specific functions either in the synthesis or the release of virions. Intracellular viral development culminates in the assembly from their molecular constituents of a new population of virions, followed by their release into the external milieu. Viral development is usually but not invariably accompanied by death of the host cell.

Probably all major groups of cellular organisms, both prokaryotes and eukaryotes, can serve as hosts to viruses. Since the host ranges of viruses are relatively restricted, they can be classified in terms of the biological nature of the host, a classification which has led to the distinction of 'animal', 'plant', 'bacterial', and 'fungal' viruses. Experience has shown, however, that a more satisfactory classification can be based on the molecular properties of virions (Table 3).

B. The Common Attributes of Cellular Organisms

In all organisms except viruses, the cell serves as the ultimate unit of structure, function, and growth. The growth of a cell occurs through an orderly increase of all its molecular constituents, leading to progressive enlargement, eventually followed by division into two (or sometimes more) cells.

The genome of a cell is composed exclusively of DNA, the various forms of RNA present in the cell being produced by transcription of specific segments of the genome. Cellular RNAs are divisible into three functional classes, each of which plays a specific role in gene translation: ribosomal RNAs (rRNAs), transfer RNAs (tRNAs), and messenger RNAs (mRNAs). Part of the cellular genome consists of so-

Table 3. System for the classification of viruses

Nucleic acid	Capsid symmetry	Naked or enveloped	Size of capsid (nm)[a]	Number of capsomers	Special features	Examples		
						Bacterial	Animal	Plant
RNA	Helical	Naked	17.5 x 300					Tobacco mosaic virus
		Enveloped	9				Myxoviruses	
			18				Paramyxoviruses	
	Polyhedral	Naked	20–25	32		Coliphage f2	Picornaviruses	
			28					Bushy stunt virus
			70	92	Double-stranded RNA		Reoviruses	
DNA	Helical	Naked	5 x 800		Single-stranded DNA	Coliphage fd		
		Enveloped	9–10					
	Polyhedral	Naked	22	12	Single-stranded DNA	Coliphage φX174	Poxviruses	
			45–55	72			Papovaviruses	
			60–90	252			Adenoviruses	
		Enveloped		162			Herpesviruses	
	Binal (polyhedral 'heads', helical 'tails')	Naked	Head: 95 x 65 Tail: 17.5 x 1150			Coliphages T2, T4, T6		

[a]Diameter, in case of polyhedral virion.

Stanier, Adelberg, and Ingraham, *The Microbial World*, 4th ed., © 1976. Reprinted by permission of Prentice-Hall, Inc., Englewood Cliffs, N.J.

called structural genes, which are transcribed as specific mRNAs, and subsequently translated into the polypeptide chains of the numerous (minimum: approx. 500) proteins which enter into the composition of the cell. Most of these proteins are enzymes, endowed with specific catalytic functions, and mediate the complex network of reactions that underlie energy-yielding metabolism and biosynthesis. Metabolic integration, necessary to assure the highly ordered process of growth, is effected by low molecular weight metabolites which act at two levels: to regulate selectively the transcription of structural genes, and to regulate enzyme function.

The cell is always separated from the external milieu by a membrane. This structure, about 8 nm wide, consists of a molecular bilayer of phospholipids, into which are inserted many different proteins. The cell membrane has one universal function; regulation of the passage of solutes between the interior of the cell and the external milieu.

C. The Properties of Prokaryotes

The information necessary to assure the growth and reproduction of a prokaryotic cell is carried by a single circular molecule of double-stranded DNA, the prokaryotic chromosome (see Chapter 11); no exceptions to this rule have so far been discovered. A distinctive property of the prokaryotic genome is the infrequency of repetitive nucleotide sequences; nearly all sequences in the chromosome are unique. The only known exceptions are the genes that encode rRNA; there are several copies of each, scattered randomly over the chromosome. The DNA content of the prokaryotic chromosome ranges from about 5×10^8 daltons (some mycoplasmas) to 5×10^9 daltons (some cyanobacteria). Prokaryotes also often harbour and replicate smaller genetic elements (plasmids), similar in molecular structure to the chromosome. Plasmids rarely contain more than 10 per cent of the amount of DNA in the chromosome, and specify ancillary phenotypic traits, as shown by the fact that their elimination from the cell usually does not impair viability.

The prokaryotic genome is not segregated from the cytoplasmic region by an enclosing membrane. The replicating chromosome and plasmids are attached to the cell membrane; this attachment also determines (in a fashion not yet precisely defined) the separation of daughter genomes. Chromosomal replication can occur continuously throughout the cell cycle, and when growth is rapid, additional rounds of replication can be initiated before completion of the preceding one. Prior to division, a prokaryotic cell thus often contains multiple copies of the genome, either completed or in the course of completion.

Genetic exchange in prokaryotes occurs through unidirectional transfer of DNA from a donor to a recipient cell, effected by conjugation, by transmission in a viral vector (transduction) or by passage of free DNA through the external milieu (transformation). Only rarely (and only by conjugation) does the recipient cell acquire a complete set of chromosomal genetic determinants from the donor cell. The entry of chromosomal DNA converts the recipient into a transient (and nearly always partial) diploid; the haploid state (normal in all prokaryotes) is reestablished after recombination, and results in the elimination of unrecombined chromosomal alleles. After a transfer of plasmid DNA, the recipient cell often maintains and reproduces the plasmid indefinitely as an autonomous genetic element. However, if nucleotide sequence homologies exist, recombination between the plasmid DNA and the DNA either of the chromosome or of other plasmids already harboured by the recipient cell may occur.

Since recombination is not an obligatory sequel to plasmid transfer, plasmids can be transmitted among, and maintained in, prokaryotes that differ widely in chromosomal genetic constitution (i.e. that belong to widely diverse taxonomic groups). In this respect, the gene pools of prokaryotes are far more open than those of eukaryotes. The horizontal transmission of genetic material has probably been an important factor in prokaryotic evolution, although its role in this respect cannot yet be clearly assessed.

With one probable exception (discussed below), the cell membrane is the only unit membrane system of the prokaryotic cell. In addition to its universal biological role in regulating the transport of solutes, the bacterial cell membrane often has important functions in energy-yielding metabolism. It contains the electron transport system in all aerobic prokaryotes, since oxidative phosphorylation (respiration-linked ATP synthesis) can operate only if the components of this system are integrated into a unit membrane. In two groups of photosynthetic prokaryotes, purple and green bacteria, the photo-synthetic electron transport system and the photochemical reaction centres are integrated into the cell membrane. The light-harvesting pigments of purple bacteria are likewise incorporated into the cell membrane, which is accordingly the site of all elements of photo-synthetic function in this group of photosynthetic prokaryotes. The light-harvesting pigments of green bacteria are contained in special organelles (chlorobium vesicles), each bounded by a non-unit membrane, and attached to the inner surface of the cell membrane.

Cyanobacteria ('blue-green algae') represent one probable exception to the rule that the cell membrane is the sole unit membrane system in a prokaryotic cell. Their cells contain an extensive intracellular system of flattened membranous sacs, or thylakoids, analogous to the thylakoids of the chloroplast in photosynthetic eukaryotes. The thylakoids are the site of the cyanobacterial photosynthetic apparatus, and appear to be topologically distinct from the cell membrane.

The prokaryotic cell membrane is a barrier to the passage of objects of supramolecular dimensions. As a result, many cellular properties characteristic of eukaryotes are not possessed by any prokaryotes. These include: the ability to ingest particles or droplets by endocytosis, and to excrete cell products by the complementary process of exocytosis, and the ability to acquire cellular endosymbionts. Since prokaryotes cannot maintain water balance in a hypotonic medium by accumulation and periodic excretion of water in a contractile vacuole (a specialized form of exocytosis), protection of the prokaryotic cell from osmotic lysis in a hypotonic medium can be assured only by passive means, namely by the synthesis of a cell wall that has a tensile strength sufficient to counterbalance turgor pressure.

D. Subdivisions of Prokaryotes

Prokaryotes can be divided into three groups, distinguished by the nature of the cell envelope. The smallest group, the mycoplasmas, do not possess cell walls; the membrane is the outer boundary of the cell. The mycoplasmas are, accordingly, osmotically fragile, and can survive only in an external milieu isosmotic with the cell contents. All other prokaryotes synthesize cell walls, and two primary assemblages can be distinguished by the specific chemical and structural properties of the wall (Table 4). They are termed Gram-positive and Gram-negative bacteria, since the differences between them with respect to wall structure are as a rule correlated with the coloration ('Gram-positive') or non-coloration ('Gram-negative') of intact cells by the Gram stain. This empirical staining method was recognized to be a valuable bacteriological diagnostic tool long before its relation to wall structure was discovered.

One class of wall polymers, peptidoglycans, synthesized only by prokaryotes, is common to both assemblages. The peptidoglycans are heteropolymers, composed of two amino sugars and a small number of amino acids. The monomeric components are cross-linked; conse-

Table 4. Structural and compositional differences between the cell walls of Gram-positive and Gram-negative bacteria

	Gram-positive bacteria	Gram-negative bacteria
Fine structure:	Homogeneous, 10–50 nm wide	Inner dense layer, 2–10 nm wide; outer membrane layer, 7–8 nm wide
Location of peptidoglycan:	Throughout	Inner dense layer
Location of other wall polymers:	Throughout	Outer membrane layer
Nature of other wall polymers:	Teichoic acids; polysaccharides; in some, complex lipids	Lipopolysaccharides; lipoproteins; proteins; phospholipids

Table 5. Distribution of other major properties between the assemblages of Gram-positive and Gram-negative bacteria. None of these properties is universal within either assemblage.

	Gram-positive	Gram-negative
Performance of photosynthesis	−	+
Gliding movement	−	+
Formation of endospores	+	−
Mycelial vegetative growth	+	−

quently, peptidoglycans have the structure of a two- or three-dimensional molecular mesh. They thus confer tensile strength on the wall, and also determine the shape of the enclosed cell. In Gram-positive bacteria, peptidoglycan extends throughout the wall, intermingled with the other types of wall polymers. In Gram-negative bacteria, peptidoglycan is confined to the inner layer of the wall, and the other wall polymers are incorporated into a physically distinct outer wall layer with the fine structure of a unit membrane. This so-called 'outer membrane layer' of the Gram-negative cell wall mimics the cell membrane structurally, but differs profoundly from it in molecular composition.

Both the Gram-positive and the Gram-negative assemblages are very large and internally diverse. Nevertheless, a primary separation on the basis of wall structure results in an absolute segregation of bacterial groups distinguished by other major properties, as shown in Table 5. This suggests that the wall provides a marker to distinguish two large evolutionary subgroups among prokaryotes.

The Gram-positive subgroup includes the actinomycetes, which have a mycelial vegetative structure, associated with reproduction by the formation of spores produced from the tips of the mycelial branches; these organisms are prokaryotic structural counterparts of the mycelial fungi. Only certain Gram-positive bacteria produce the specialized resting cells termed endospores. Most endospore-forming Gram-positive bacteria are unicellular, but these resting structures are also produced by a few actinomycetes.

All photosynthetic bacteria — purple bacteria, green bacteria and cyanobacteria — are Gram-negative prokaryotes.

The property of active movement is sporadically distributed among prokaryotes, which can move by two different means, swimming and gliding. Swimming movement, displayed by some members of both the Gram-positive and Gram-negative assemblages, is effected by very thin filiform proteinaceous organelles (bacterial flagella), which extend through the cell wall (except in one Gram-negative group, the spirochaetes, where they lie between the inner and outer wall layers).

During cell movements, the flagella rotate; the motive force is derived from a structure within the cell, the flagellar motor, to which the flagellum is anchored. Gliding movement, which is confined to certain groups of Gram-negative prokaryotes (myxobacteria, flexibacteria, some green bacteria, and cyanobacteria) is not associated with the presence of visible external organelles, and contact of the cell with a solid surface is necessary for its expression. Its mechanism is still not understood.

The great internal diversity of both the Gram-positive and the Gram-negative assemblages suggests that each is composed of a large number of highly isolated evolutionary lines, which diverged early in the course of cellular evolution. This inference has been confirmed in recent years by molecular evidence. In both assemblages, there is an extreme diversity of mean base composition of the DNA; divergence with respect to this molecular parameter is a sure expression of long evolutionary separation. There are also major divergences within each assemblage with respect to rRNA nucleotide sequences, encoded by genes that are very highly conserved during evolution. The use of these and other molecular probes is gradually permitting the recognition of certain large clusters of evolutionarily related bacteria; for example, a major cluster centred around the lactic acid bacteria in the Gram-positive assemblage, and another major cluster centred around the enteric group and the vibrios in the Gram-negative assemblage. Despite these important advances in our understanding of the evolutionary structure of the prokaryotes, it is unlikely that a complete picture of their evolutionary interrelationship will ever be attainable.

E. The Properties of Eukaryotic Protists

Although the purpose of the following discussion is to summarize the cellular properties that distinguish eukaryotic protists from prokaryotes, it must be emphasized that the eukaryotic realm is indivisible at the cellular level. In terms of fundamental cellular properties, what is true collectively for fungi, protozoa, and algae is also true for elephants and flowering plants. In effect, plants and animals represent two major eukaryotic evolutionary lines distinguishable from eukaryotic protists only by their greater organismal complexity, brought about by elaborate, genetically determined developmental programmes, which unfold in the course of their ontogeny.

A basic differential property of the eukaryotic cell is its internal differentiation into two (or sometimes) three regions, each the site of part of the genetic information required to specify cellular properties, and each containing a separate system for the transcription and translation of genetic messages. These are: the nucleo-cytoplasmic region, the mitochondrion, and (in photosynthetic eukaryotes) the chloroplast.

1. Chloroplasts and Mitochondria

The mitochondrion and the chloroplast, both enclosed by the nucleo-cytoplasmic region, are separated from it by organellar membranes. Each is genetically semiautonomous, since it contains a small genome which specifies some (but not all) organellar properties. The chloroplast genome size is approximately 10^8 daltons, the mitochondrial genome, approximately 10^7 daltons; the former accordingly has an information content some 10—20 per cent of that of the smallest known prokaryotic chromosome, the latter, an order of magnitude less. The molecular structure of these organellar genomes and of the systems of organellar transcription and translation associated with them are markedly different from the corresponding functional elements located in the nucleo-cytoplasmic region.

The mitochondrion is the specific site in a eukaryotic cell of the machinery of terminal respiration; the enzymes of the tricarboxylic acid cycle and the respiratory electron transport system. Its function is to provide much of the ATP (derived from oxidative phosphorylation) required for the maintenance of cellular function. It is a well-nigh universal component of eukaryotes. A few protozoa and fungi, which have become adapted to life in anaerobic niches, do not contain mitochondria and are entirely dependent on glycolysis for ATP synthesis.

The chloroplast has a considerably more complex metabolic role, as suggested by its greater genome size. It contains the light-harvesting pigments, the photochemical reaction centres, and the photosynthetic electron transport system necessary for the generation of ATP and reductant (NADPH) through the reactions of oxygenic photosynthesis. These products of photochemical energy conversion are largely used to mediate biosynthetic reactions within the chloroplast itself; notably, the conversion of CO_2 to sugar phosphates through the reductive pentose phosphate cycle; the reduction of oxidized inorganic nitrogen compounds (nitrite, possibly also nitrate) to ammonia, and the incorporation of the latter into glutamine; the reduction of sulphate to H_2S, and its incorporation into organic sulphydryl groups. The relatively simple nutritional requirements of photosynthetic eukaryotes are thus in large measure made possible by the biosynthetic machinery located within the chloroplast.

Chloroplasts and mitochondria contain internal unit membrane systems, topologically distinct from the surrounding unit membrane of the organelle; in both, they are the sites of the machinery of electron transport and ATP synthesis, and (in chloroplasts) of the other elements of the photosynthetic apparatus. The role of the external organellar membrane is thus largely if not entirely to regulate solute transport between the interior of the organelle and the surrounding nucleo-cytoplasmic region.

2. Nucleo-cytoplasmic Region: Genetic Aspects

Most of the DNA of the eukaryotic cell is located within the membrane-bounded nucleus; it specifies both properties of the nucleo-cytoplasmic region, and some properties expressed specifically in chloroplasts and mitochondria. The nuclear DNA is dispersed over a number of separate structures, the eukaryotic chromosomes, each of which carries only part of the total genetic message. The haploid chromosome number is fixed, and is a fundamental property of each species: it ranges from 3 to over 100.

In terms of its molecular composition, the eukaryotic chromosome is a much more complex object than either the prokaryotic chromosome or the genome of a chloroplast or mitochondrion. It is an elongated structure, probably containing a continuous longitudinal strand of double helical DNA, firmly associated with a considerable number of specific proteins, both basic (histones) and acidic. This nucleoprotein complex is known as chromatin. A remarkable feature of eukaryotic chromosomal organization is the sequential clustering, on one or more chromosomes, of the numerous identical copies of the genes that encode rRNA. In the interphase (non-dividing) nucleus these particular regions are looped off from the mass of chromatin and associated with specific proteins, to form an intranuclear organelle, the nucleolus, visible by light microscopy. Another distinctive property of eukaryotic chromosomal DNA is the presence of many repeated identical or near-identical nucleotide sequences, some of which do not appear to be transcribed.

Transcription of the nuclear genome is effected by three different polymerases, each specific for the synthesis of one RNA class: rRNA, tRNA, and mRNA. Transcription of structural genes may lead to the formation of polycistronic mRNAs, since adjacent cistrons are not (as in prokaryotes) always separated by chain termination and initiation codons. The polycistronic messages pass into the cytoplasm, where they are translated on the 80S ribosomes into continuous polypeptide chains, subsequently cleaved by specific proteases into functional units (i.e. the shorter polypeptide chains of the different proteins encoded in the polycistronic message).

In eukaryotes, both chromosome replication and nuclear division are closely integrated with the cell cycle, in the sense that they normally occur only once between two succeeding cell divisions. Replication of chromosomal DNA occurs during interphase, being initiated some time after the preceding cell division, and terminated some time before the beginning of the events that lead to the next cell division. The divisional events start with division of the nucleus (mitosis), a process of great morphogenetic complexity, as well as of great diversity in different groups of protists. We shall attempt to summarize here the common denominators.

In the interphase nucleus, the chromosomes are highly extended, and hence not resolvable by light microscopy (exception: dinoflagellates). At the onset of mitosis, they shorten and thicken by coiling, and each chromosome splits longitudinally into a pair of chromatids, joined to one another at only one point, the centromere. During these events, an oriented system of proteinaceous microtubules, assembled from a pool of the monomeric precursor protein (tubulin) present in the cytoplasm, forms in the nuclear region. Some, the polar microtubules, develop from two organizing centres, located on the opposite sides of the nucleus, to form a structure termed the spindle. The organizing centres are located outside the nuclear membrane in biological groups where this structure disintegrates during mitosis; in many protists (e.g. the majority of fungi) where the nuclear membrane persists, the organizing centres are internal to it. A second set of microtubules (chromosomal microtubules) develops from the centromere of each pair of chromatids, extending to the organizing centres, and lying parallel to the polar microtubules of the spindle. The pairs of chromatids become aligned in the equatorial plane of the spindle, the centromeres split, and one copy of each chromosome is pulled towards each organizing centre, through a sliding force exerted between the chromosomal and the polar microtubules. Once chromosome separation is complete, two interphase nuclei re-form, and the microtubular system depolymerizes into tubulin.

Cell division normally begins during the terminal phase of mitosis, in a plane at right angles to the spindle, cutting across the equator. In coenocytic organisms (most fungi, many algae) mitosis is not normally followed by cell division; these organisms consequently consist of a single large, multinucleate protoplast during vegetative development. Only reproductive cells (spores, gametes) are uninucleate; and in some fungi, even spores contain several nuclei.

With exceptions among the fungi, some of which can undergo recombination resulting from the fusion of genetically distinct vegetative nuclei contained in a single coenocyte ('parasexual recombination'), genetic recombination among eukaryotes is always the sequel of a sexual process. It is preceded by the formation of haploid cells of different mating types, known as *gametes*. If they can be distinguished by structural properties (e.g. size, motility), the gametes are termed 'male' and 'female', a terminology derived from the differences in gamete structure characteristic of male and female animals. However, in many protists, the gametes are structurally indistinguishable, those of opposite mating type being termed '+' or '−'. Pairwise gametic fusion is followed by nuclear fusion, to produce a zygote with one diploid nucleus. Genetic recombination occurs subsequently, during the development of the zygote or of a diploid vegetative cell derived from it, and is effected by a special divisional process known as meiosis. Meiosis leads to the formation, from a diploid cell, of four haploid daughter cells. During meiosis, a random reassortment of the two chromosome sets contributed by the gametes occurs; and genetic recombination is further enhanced by exchange (crossing-over) of homologous regions between each diploid pair of chromosomes. Recombination is thus much more far-reaching than in prokaryotes. Moreover, whereas recombination in prokaryotes leads to the emergence of a unique haploid recombinant genome in the recipient cell, eukaryotic sexual recombination through meiosis always results in the reemergence, in the four haploid products, of all the nuclear genes contributed by both gametes, recombination is reciprocal.

Genetic recombination in eukaryotes is subject to an additional complication: the genetic information carried by mitochondria and chloroplasts is transmitted in these organelles, its inheritance thus being independent of that of the nuclear genome, and subject to different rules, alike of transfer and of recombination.

The relative importance of haplophase and diplophase in the life cycles of eukaryotic protists capable of sexual reproduction varies widely. Predominance of the diplophase, resulting from meiosis immediately preceding gamete formation, brings into operation a genetic phenomenon of profound evolutionary importance, which plays no role in the evolution of prokaryotes: *dominance.* In the diploid state, phenotypic expression of one allele carried by a pair of homologous chromosomes is often partly or wholly repressed. This permits the accumulation of a masked store of genetic variability in diploid eukaryotes.

3. Nucleocytoplasmic Region: Functional Aspects

Between the cell membrane and nuclear membrane, the cytoplasmic region is traversed by a complex membranous network termed the endoplasmic reticulum. For the most part topologically irregular, it assumes at one or two sites a more well-defined structure, the Golgi apparatus. This consists of a series of membranous vesicles in parallel array, flattened and elongated at one pole, smaller, more numerous and nearly isodiametric at the other. The organelle, continuously regenerated at the former pole and disassembled at the latter pole, is the site of the synthesis, accumulation and transport of a variety of biosynthetic products: some of the Golgi vesicles contain enzymes, others (in algal and plant cells) fragments of the cell wall fabric. Golgi vesicles that contain materials destined for export from the cell (extracellular enzymes, wall materials) migrate to the cell surface, where their enclosing membranes fuse with the cell membrane, the vesicle contents being discharged to the external milieu. This is the process of exocytosis.

The inverse process of endocytosis occurs by invagination and pinching off of a small area of the cell membrane, surrounding either a droplet of liquid (pinocytosis) or a solid particle (phagocytosis) to form a membrane-bounded vacuole which moves into the cytoplasm. Endocytosis underlies two unique and distinctive abilities of eukaryotic cells: the ability to acquire cellular endosymbionts, and the capacity for intracellular digestion. Endosymbionts, once internalized by endocytosis, proceed to grow in the cytoplasmic region, separated from the cytoplasm by the surrounding host membrane. Intracellular digestion of materials ingested as food vacuoles involves the intervention of vesicles derived from the Golgi apparatus (lysosomes and peroxysomes), which coalesce with the food vacuole, and thus introduce into it an array of enzymes, either hydrolytic (lysosomes) or oxidative (peroxysomes). The enzymic breakdown of complex food materials enclosed in the vacuole is followed by diffusion of the soluble products into the cytoplasm. This mode of nutrition, phagotrophy, is possible only in protists that are not enclosed by rigid walls. It is widespread in protozoa, relatively rare in algae, and not displayed by fungi. In fungi, as in prokaryotes, digestion of polymeric nutrients (e.g. proteins, polysaccharides) is extracellular, and is mediated by the secretion of hydrolases.

The nuclear membrane of eukaryotes is distinguished from the other membranes of the nucleocytoplasmic region by its ultrastructure: it is perforated by a regular system of pores. In many though not all protists the nuclear membrane disintegrates early in mitosis, and is reassembled (probably from the endoplasmic reticulum) at the termination of the process.

Accordingly, the different parts of the nucleocytoplasmic membrane system are largely, and perhaps completely, interchangeable. Areas of the cell membrane can be internalized, becoming vacuolar membranes, and vesicle membranes arising from the Golgi apparatus can enter the fabric of the cell membrane as a result of exocytosis. Membrane plasticity and membrane migration are fundamental properties of the nucleocytoplasmic region of a eukaryotic cell.

Implicit in these interchanges of membrane material (but also manifested in many other ways) is the internal mobility of the eukaryotic cytoplasm, often markedly vectorial (e.g. in exocytosis). Internal mobility is usually evident upon microscopic examination of a living eukaryotic cell, being manifested by marked relative displacement of its contents (cytoplasmic streaming). These continuous intracellular movements necessarily imply the existence of an internal motive force, operative even in eukaryotic cells that are immotile, in the sense of being unable to effect translatory movement through or over the external milieu. Most fungi are 'immotile' in this sense, but often display very vigorous cytoplasmic streaming. In protists not enclosed by cell walls (many protozoa and slime molds), directed cytoplasmic streaming permits slow translatory movement of the cell (so-called amoeboid locomotion).

One probable agent of intracellular mobility has been recently recognized as of wide distribution in eukaryotic cells. It is the actino-myosin system: the ensemble of proteins which (in a very highly organized state) are responsible for the contractile properties of verte-brate muscles. Elements of this system have now been identified in the cells of virtually all eukaryotes, including higher plants. Originally presumed to be synthesized only in certain highly differentiated animal tissues, the proteins of the actinomyosin system (or homologues thereof) may well prove to be universal molecular markers of eukaryotic cells.

Another protein which, in polymerized form, participates in some forms of internal eukaryotic cellular movement, and may likewise prove to be a eukaryotic molecular marker, is tubulin. The participation of microtubules composed of tubulin in chromosome movement during mitosis has already been discussed. Tubulin-based microtubular systems have additional functions in the nucleo-cytoplasmic region of eukary-otic cells. They provide the internal structural framework of the eukaryotic cilium or flagellum. This elongated microtubular complex, the *axoneme*, is surrounded by an outpocketing of the cell membrane. It is composed of nine doublet microtubules, surrounding two singlets. Ciliary movement has recently been shown to involve sliding between the two tubules of each doublet, effected by the making and breaking of bridges between them. The axonemal structure is homologous in all protists that synthesize these organelles (many protozoa and algae; aquatic fungi). Both in its machinery and its mechanism, ciliary movement is radically different from prokaryotic flagellar movement.

F. Algae, Protozoa, and Fungi

Among the eukaryotic protists, three major groups are traditionally distinguished: algae, protozoa, and fungi. At first sight, the distinctions seem clear-cut. The distinguishing property of algae is the possession of chloroplasts, which confer on them the ability to perform oxygenic photosynthesis, and to use CO_2 as a principal carbon source. Both protozoa and fungi lack these organelles, and are consequently dependent on organic compounds as sources of both carbon and energy. The traditional distinction between the two latter groups is based on cellular properties: protozoa are unicellular, and the cells are frequently not enclosed by rigid walls: fungi are for the most part coenocytic and are enclosed by rigid cell walls, which typically confer on the enclosed multinucleate coenocyte the structure of a much-branched tubular mycelium. These distinctions are satisfactory, provided that one takes into consideration only typical, specialized

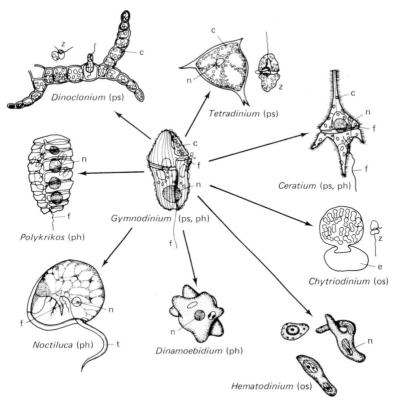

Figure 1. The different evolutionary trends that are represented among dino-flagellates. *Gymnodinium* is a relatively unspecialized photosynthetic dinoflagellate, which is both photosynthetic (ps) and phagotrophic (ph). *Ceratium* is a more specialized photosynthetic dinoflagellate, characterized by a very complex wall with spiny extensions, comprised of many plates. *Tetradinium* and *Dinoclonium* are non-motile, strictly photosynthetic organisms, which reproduce by multiple cleavage to form typical dinoflagellate zoospores. *Polykrikos*, *Noctiluca*, and *Dinamoebidium* are three free-living phagotrophic dinoflagellates. *Polykrikos* is a coenocytic, multinucleate organism, the cell of which bears a series of pairs of flagella. *Noctiluca* has one small flagellum, and bears a large and conspicuous tentacle. *Dinamoebidium* is an ameboid organism. *Chytriodinium* and *Hemato-dinium* are parasitic dinoflagellates whose nutrition is osmotrophic (os). *Chytrio-dinium* parasitizes invertebrate eggs and reproduces by cleavage of a large sac-like structure into dinoflagellate zoospores. *Hematodinium* is a blood parasite in crabs: n, nucleus; f, flagellum; c, chloroplast; z, zoospore; e, parasitized invertebrate egg; t, tentacle. (Stainer, Adelberg, and Ingraham, *The Microbial World*, 4th ed., © 1976. Reprinted by permission of Prentice-Hall, Inc., Englewood Cliffs, N.J.)

representatives of each group: a seaweed, a ciliate protozoan, and a basidiomycetous fungus conform perfectly to the differential criteria outlined above. In reality, the situation is more complex, since there are many transitions between algae and protozoa on the one hand, and between protozoa and fungi on the other. Where such transitions exist, the evidence suggests that protistan evolution has always proceeded along similar pathways, as described below.

The stem groups, evolutionarily speaking, of the eukaryotic protists appear to have been a variety of unicellular photosynthetic flagellates, from which two primary lines of divergence have occurred: firstly, maintenance of photosynthetic function (chloroplasts) accompanied by an increase of organismal complexity, leading to evolution of multicellular algae; secondly, loss of photosynthetic function (chloroplasts), leading to unicellular non-photosynthetic organisms of the protozoan type, followed by a secondary increase in organismal complexity, leading to fungi. This evolutionary scenario is revealed with particular clarity in the dinoflagellates, a large group of eukaryotic protists which display certain unique nuclear properties, notably a highly specialized form of mitosis and a failure of the chromosomes to elongate after nuclear division, as a result of which these structures remain resolvable by light microscopy in interphase. Most dinoflagellates also produce motile cells bearing two unequal flagella, one lying in an equatorial girdle, and the other directed posteriorly. These conserved properties provide evolutionary markers which make it possible to detect dinoflagellate affinities in protists that have diverged very markedly in most other respects from the unicellular photosynthetic flagellate pattern. The different evolutionary patterns displayed among dinoflagellates are schematized in Figure 1. They include: (a) typical simple immotile 'algae' (*Tetradinium, Dinoclonium*); (b) highly specialized types of 'protozoa', either free-living predators (*Polykrikos, Noctiluca, Dinamoebidium*) or parasites (*Hematodinium*), and (c) a simple parasitic 'fungus', *Chytriodinium*.

G. Possible Evolutionary Filiations Between Eukaryotes and Prokaryotes

The profound differences between eukaryotic and prokaryotic cells find their expression, at the eukaryotic level, in the nucleo-cytoplasmic region. Chloroplasts and mitochondria, even though their phenotypic properties are in considerable part determined by nuclear genes, are in many respects analogous to prokaryotic cells. In terms of energy-yielding metabolic function, their counterparts are cyanobacteria and aerobic, chemoheterotrophic bacteria, respectively. It is therefore conceivable that each class of organelle arose from free-living ancestral prokaryotes with the same mode of energy-yielding metabolism, which entered into endosymbioses with cellular organisms that already possessed most (if not all) attributes now characteristic of the nucleo-cytoplasmic region of the eukaryotic cell. If so, the initial endosymbioses were followed by far reaching evolutionary modifications *in vivo*. The most important was a reduction in the size of the endosymbiont genome, accompanied by the transfer of many genetic determinants to the genome of the host nucleus. The way in which this feat of genetic engineering might have occurred is mysterious; its adaptive function is evident. It would have assured the permanent subjection of the endosymbionts, still potentially free to escape and resume independent existence as long as they possessed full genetic autonomy. This scenario is a highly plausible one to account for the origin of chloroplasts, which still retain a relatively large genome and a very complex array of metabolic functions. It is perhaps more arguable for mitochondria, which have a far smaller genome and a more restricted array of metabolic functions.

An endosymbiotic origin for chloroplasts and mitochondria provides the most useful evolutionary working hypothesis to account for the properties of these cellular components. This hypothesis implies a prior evolution of the protoeukaryotic cell, embodying the properties of the nucleo-cytoplasmic region. Such a cell would have been dependent on fermentative mechanisms of ATP synthesis, relatively inefficient in terms of energy yield per unit of substrate metabolized. Barring the unlikely possibility of a biphyletic origin of cells, it must be assumed that the protoeukaryote evolved, in the remote past, from a cell of the prokaryotic type, and thus underwent extensive evolution prior to acquiring endosymbionts endowed with respiratory or photosynthetic function. Its divergence from the prokaryotic line could have been initiated by a change of membrane structure, which made possible a primitive type of endocytosis, opening the way to a predacious mode of life. Many of the cytoplasmic properties of the eukaryotic cell — notably the actinomyosin- and tubulin-based systems, and the Golgi apparatus — can be construed as secondary evolutionary adaptations in the context of this mode of life. The invention of tubulin was also a prerequisite for the eukaryotic type of nuclear division; microtubules are associated with nuclear division even in dinoflagellates, where the process is so atypical that it barely qualifies as 'mitotic'.

The unique nuclear properties of dinoflagellates suggest that some of the evolutionary diversity characteristic of contemporary protists may have occurred through divergence at the protoeukaryotic stage. Another possible source of very early divergence could have been the acquisition by already divergent protoeukaryotes of different types of prokaryotic endosymbionts. This possibility is suggested by the differences among chloroplasts (notably with respect to light-harvesting pigments) in the various major algal groups. Only the red algal chloroplast has a pigment system highly homologous with that of contemporary cyanobacteria. The recent discovery of a new group of prokaryotes which perform oxygenic photosynthesis but possess a pigment system of the green algal—higher plant type has provided the first solid support for the notion that contemporary chloroplast diversity could reflect an ancient divergence at the level of the prokaryotic ancestors.

IV. The Relations Between Microbiology and General Biology

Microbiology was to a very considerable extent the creation of Louis Pasteur. He was primarily responsible for developing its distinctive methodology, as well as for revealing the roles of microorganisms as agents of chemical transformations and infectious disease. During the same period, biology came of age; Darwin imposed order and meaning on the hitherto anecdotal materials of Natural History by reinterpreting them in terms of the theory of evolution based on natural selection. It might therefore have been anticipated that microbiology would take its place, together with other specialized biological disciplines, within the conceptual framework of post-Darwinian biology. This did not occur: for almost a century, the development of microbiology was directed by its own internal logic, along paths very largely separate from those followed by general biology.

The dominant interests of post-Pasteurian microbiology were the detailed analysis of the chemical activities of microorganisms, and of their roles in the turn-over of matter in the biosphere; the characterization of the microbial agents of infectious disease (which led in due course to the discovery of a new class of biological objects, the viruses), and the analysis of the interactions between agents of infectious disease and their animal or plant hosts. Exploration of the latter problem was responsible for the creation of a new discipline of functional biology, immunology. The phenomena of immunity are displayed only by animals of the vertebrate evolutionary line, and immunology is, in consequence, an area of functional biology which has no organic connection with microbiology. The historical link between the two disciplines was purely accidental, and stemmed from the fact that the basic phenomena of immunity were first recognized in vertebrate animals after an immune response had been elicited by microbial infection.

The first links between microbiology and general biology were forged at the start of the twentieth century, during the development of the new functional discipline of biochemistry. The discovery of alcoholic fermentation by cell-free preparations of yeast provided an experimental system for the chemical analysis of energy-yielding metabolic processes. Parallel studies on the mechanisms of alcoholic fermentation by yeast and of lactic fermentation by muscle tissues eventually led to the recognition of their far-reaching chemical homologies. A few years later, the analysis of the nutritional requirements of animals and microorganisms revealed a second unanticipated common denominator:

the 'vitamins', trace nutrients for animals, were shown to be chemically identical with the 'growth factors' which are trace nutrients for many microorganisms. The analysis of the metabolic role of these compounds — largely conducted, for reasons of experimental convenience, with microorganisms — showed that each vitamin is the biosynthetic precursor of a specific type of coenzyme, and that coenzymes have universal functions in cellular metabolism. Several metabolic homologies shared by all cellular organisms were thus revealed. During the 1930s, this new biological insight was epitomized in the phrase: 'the unity of biochemistry'.

However, microbiology made no contribution to the second major advance of biology in the early twentieth century: the creation of genetics. This science arose through a convergence between the Mendelian analysis of inheritance in plants and animals, and the discipline of cytology, which revealed the chromosome to be the physical vehicle of eukaryotic hereditary determinants. What little was then known about inheritance in microorganisms suggested that the laws of Mendelian inheritance did not operate at this biological level. The separation between microbiology and genetics ended only after 1940. The first junction was the isolation in 1941 by Beadle and Tatum of a series of auxotrophic mutants in the ascomycete *Neurospora*, coupled with the demonstration that the inheritance of these biochemically definable mutant traits obeys Mendelian laws. In 1943, Delbrück and Luria established a methodology (the fluctuation test) which showed that bacterial mutations are spontaneous random events, and not (as had previously been widely believed) subject to environmental determinism. This laid to rest the Lamarckian heresy, which had found a last foothold among bacteriologists. Two discoveries then opened the way to a rapid development of bacterial genetics: the demonstration of conjugation in *Escherichia coli* and of transformation in the pneumococcus. Although initially its significance was not widely recognized, the identification of the transforming principle as DNA by Avery and his collaborators first revealed the chemical nature of the gene.

During the decade between 1940 and 1950, the convergence between genetics, microbiology, and biochemistry set the scene for the second major revolution in the history of biology. Because of their relative simplicity, the bacteria and their viruses provided the principal experimental material for the elucidation of the laws of molecular biology. A century after Darwin, microbiology at last took its place in the conceptual framework of general biology.

V. Concluding Remarks

The attempt to define microbiology proves, like many problems of definition both in science and in other intellectual domains, to be a task of great complexity and difficulty. One could of course sidestep the difficulties, by an arbitrary declaration that it is the branch of biology devoted to the study of small forms of life; and , indeed, no better one-line definition can be proposed. A comparable problem faced the writer E. M. Forster when he was asked to lecture on the Novel, and this essay can be appropriately concluded by quoting his response to the definitional challenge.

'Perhaps we ought to define what a novel is before starting. This will not take a second. M. Abel Chevalley has, in his brilliant little manual (*Le Roman Anglais de Notre Temps*), provided a definition, and if a French critic cannot define the English novel, who can? It is, he says, "a fiction in prose of a certain extent". That is quite good enough for us, and we may perhaps go so far as to add that the extent should not be less than 50,000 words. . . . If this seems to you unphilosophical will you think of an alternative definition, which will include *The Pilgrim's Progress, Marius the Epicurean, The Adventures of a Younger Son, The Magic Flute, The Journal of the Plague, Zuleika Dobson, Rasselas, Ulysses* and *Green Mansion,* or else will give reasons for their exclusion?'

Aspects of the Novel (Edward Arnold, London, 1927).

VI. Further Reading

Knight, B. C. J. G. and Charles, H. P. (Eds). (1970). *Organization and Control in Prokaryotic and Eukaryotic Cells.* Cambridge: University Press.

Norris, J. R. and Ribbons, D. W. (Eds). (1969—78). *Methods in Microbiology.* 10 Vols. London: Academic Press.

Schlegel, H. G. (Eds.). (1965). *Anreicherungskultur und Mutantenauslese (Enrichment Culture and Mutant Selection).* Stuttgart: Fisher.

Stanier, R. Y., Adelberg, E. A., and Ingraham, J. L. (1976). *The Microbial World.* 4th ed. Englewood Cliffs (N.J.): Prentice-Hall; London: Macmillan.

Watson, J. D. (1976) *Molecular Biology of the Gene.* London: W. A. Benjamin, Inc.

Form and Function — I.
Bacteria

R. G. E. MURRAY

*Department of Bacteriology and Immunology, University of Western Ontario,
London, Ontario, Canada*

I. Introduction

Bacteria have developed and maintained mechanisms needed for living in dilute environments and are restricted to food of molecular dimensions. Their roles in the transformations of matter and the energy economy of the biosphere are only related to their structure in terms of physiological adaption to a particular niche or to maintaining a secure place within that environment. We identify the reactions and processes of their kind of life but our conceptual grasp is primitive unless we can put them into a structural context. We can now attempt a functional anatomy, which gives perspective to biochemistry and forms a bridge between the practicalities of functional biology and the regulatory details of molecular biology.

Structural information is essential to the effective interpretation of physiological events, the organization of biochemical processes, and the nature of subcellular fractions. Structures are the outwardly visible forms of complex associations of macromolecules and the relative stability of their genetic determinants give them taxonomic value. It should not be surprising that shared but unique structural components of cells reflect shared and unique features of molecular and metabolic biochemistry, and this implies that the broad generalizations are subject to dissection and study at much greater resolution. High resolution demands the best of preparative techniques; artefacts and uncertainty are the inevitable accompaniment and progress at the molecular/ structural level is slow.

There can be little doubt of the primacy of cell structure in biological descriptions and taxonomy, and this importance is amplified by associating microscopic cell structures, chemical constitution, and function. The bacteria can now be described as cells with unique structural features and components; they are assembled with the blue-green algae (Cyanobacteria) in a Kingdom, Procaryotae, distinct from all other kinds of living cells. This is appropriate to this ancient and honorable lineage that goes back at least 3.2 billion years.

The limited range of shapes and forms exhibited by bacteria (rods, cocci, vibrios, spirilla, and mycoplasmas) do not give an infinity of possibilities. Yet for each one of these forms there are organisms in nature displaying a fantastic range of physiological capabilities encompassing most of the known metabolic types and suggesting that the subcellular organization must exhibit important elements of distinction, some of which should involve structure. The forms may be limited but they are remarkably stable and under strict control; this consistency can be challenged by genetic and metabolic interference (e.g. mutagens or selected antibiotics), or environmental extremes. Yet these organisms in nature hold to their heritage with remarkable tenacity.

A structural description of a typical bacterium consists of a listing of components with unique features (Figure 1). They are cells enclosed, with few exceptions, by an assembly of polymers and heteropolymers to form a strong and structurally differentiated cell wall in which are included unique chemical constituents such as muramic acid and D-amino acids as well as remarkable heteropolymers. The protoplast is enclosed by a plasma membrane, which is generally without the steroids incorporated in the lipid component of eukaryotic membranes. Extension of this plasma membrane into the cytoplasm to form associated organelles is also characteristic. Some organized structures

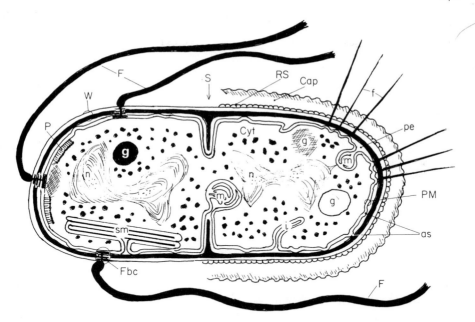

Figure 1. Diagram of a dividing rod-shaped bacterium to illustrate the anatomical relations of various structures: as, adhesion sites; cap, capsule; cyt, cytoplasm containing ribosomes; F, flagella with 'hook' attachment to Fbc, the flagellar basal complex; f, fimbriae (common pili); g, inclusion granules and ribosome-free cytoplasm; i, simple membrane intrusion; m, mesosomes; n, nucleoplasms; PM, plasma membrane; p, polar membrane; pe, periplasmic space; RS, regularly structured outer wall layer(s); S, septum; sm, stacked membranes; W, cell wall

are definitively prokaryotic (particularly the organelles of swimming motility (flagella), the small (70S) ribosomes and the form of the nucleoids or genophores) and they are essential to the circumscription of the Kingdom.

II. Nucleoplasms

It was recognized that bacteria have remarkable nuclei long before the resolving power of electron microscopy, of modern genetics, and of molecular biology gave substance to the impression. These Feulgen-positive elements divide in step with cell division but show no cyclical changes in form and staining properties; in fact, they appear to divide directly. Furthermore, staining methods designed to reveal the presence of basic proteins (such as the histones of eukaryotic chromosomes) do not stain these bodies. Operational genetics indicates no more than a single linkage group in the genome, which behaves as if it were circular.

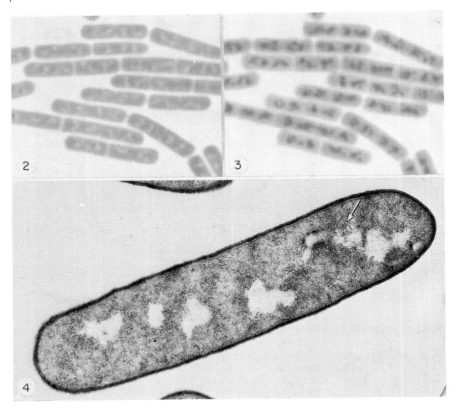

Figure 2. Light micrograph of a *Bacillus* species fixed during active growth and stained with thionine, which stains the RNA-rich cytoplasm. The nucleoplasms are unstained and very similar in shape and disposition to those of living cells observed with phase microscopy.

Figure 3. Same group of cells as in Figure 2 extracted with N-HCl at 60 °C and the nucleoplasms stained with Giemsa's stain.

Figure 4. Micrograph of a longitudinal section of *Bacillus polymyxa* showing wall, ribosome-packed cytoplasm, nucleoplasms, and a mesosome (arrow). The plasma membrane is in immediate contact with the wall

They are sometimes different and can appropriately be called geno-phores or nucleoplasms, which are terms without mitotic connotations. (Figures 2—4)

Electron microscopy added important elements to the description. The nucleoplasm is not bounded from the cytoplasm by any form of membrane and consists of a filamentous skein consistent with the concept of a wound-up DNA helix. The arrangement of these nucleoplasms in the cytoplasm is hard to discern; but they are roughly positioned in the centre and regularly spaced along the axis of the cell. The appearance of the nucleoplasm depends on the fixation medium. They behave like anionic gels within the cell because they contract and tend to fuse when the charge density is reduced by an influx of cations and disperse when the cationic environment is depleted. So position may be determined, in part, by charge and in part by some genetical component because strains of bacteria exist (e.g. of *Escherichia coli* C) in which the chromatin is peripherally distributed. Whatever the position of these chromatinic bodies the fine structure is always the same. And when released from the cell and laid out in a film of protein, the nature of the skein becomes apparent and the circularity of the genome can be inferred from the paucity of ends.

Rounds of replication of the genome require the concerted activity of a set of enzymes at the replication forks, and these are probably associated with a functionally important portion of plasma membrane. Isolation of the chromosomes generally results in a small and irreducible amount of membrane material remaining attached. Consequently, a great deal of energy has been devoted to trying to identify contact points of membrane and DNA using electron microscopy. It is not hard to see associations but it is much harder to provide more than suggestive evidence of the significance of juxtapositions. Two functions are sought in the association of membrane and nucleoplasm: one is a structure for the replicating enzymes and the other is a mechanism for separating the whole genomes that have been replicated. The latter is necessary because the microtubules and spindle structures of eukaryotes are not present.

Organisms that have mesosomes (intrusions of the plasma membrane often associated with cell division), such as *Bacillus subtilis*, usually show two or more major points of contact between mesosomal membranes and the nucleoplasm, positioned as if they could drag apart the replicated genophore as the cell and the membrane grow. This might be expected, since mesosomes are large enough to impinge on nucleoplasms and represent areas of membrane intruded close to any division sites (past, present or future). Careful microscopists have also studied organisms with few if any mesosomes and have found sections showing individual strands deriving from a nucleoplasm that comes very close to the peripheral plasma membrane. Despite this cautious interpretation, it is not unreasonable to consider that there must be functional associations between the nucleoplasm and the plasma membrane involving small but critical areas. Therefore, there is reason to believe that membrane behaviour could provide a means of genome separation.

A number of recent reports concern the association of small amounts of basic proteins with isolated DNA and the isolation of proteins that bind to DNA. This casts some doubt upon the impression that an absence of histones is one of the salient characteristics of prokaryotic nuclei. Nevertheless, the presence of small amounts might satisfy some functional requirements, e.g. the exclusion of segments of the genome from transcription, without the need for the massive associations observable in eukaryotic chromatin.

III. Cytoplasm

The functional architecture of cytoplasm is generally described in terms of membranous structures but this is only partially possible in restricted categories of prokaryotes and many show only simple intrusions of plasma membrane. They do not possess the membrane-bounded and functionally independent organelles that fill the cytoplasm of eukaryotic organisms.

The cytoplasm, lying between the nucleoplasm and the peripheral plasma membrane, is fully packed with free and randomly distributed ribosomes in functionally active bacteria (Figures 4 and 5). They are not plastered on membranes after the fashion of ergastoplasm, even when there are intruded membranes. However, a high proportion of the protein synthesis of the growing cell goes on in association with the peripheral plasma membrane and strings of ribosomes (polysomes) are found loosely attached to the inside of the plasma membrane when it is

Figure 5. Micrograph of a section through the division site of *E. coli*. The septum consists of a looped extension of the peptidoglycan layer (PG) normally on the inside of the outer membrane (OM), together forming the wall. The plasma membrane (PM) encloses the cytoplasm (C), full of densely stained ribosomes and the developing septum. This is a typical Gram-negative type of wall profile. (Reprinted with permission of American Society for Microbiology from I. D. J. Burdett and R. G. E. Murray (1974), *J. Bacteriol.*, **119**, 303–324, Figure 5c.)

Figure 6. Section of the completed septum of a *Bacillus* sp. showing a faint differentiation in the centre preparatory to delamination of the wall. This is a Gram-positive type of wall profile

Figure 7. 'Snapping division' of a *Corynebacterium* sp. showing retention of a band of peripheral wall preventing complete separation

Figure 8. Shadowed replica of a frozen-cleaved preparation of *Streptococcus faecalis* showing the plasma membrane furrow for the next division: (a) membrane cleaved showing the protein 'studs' of the outside of the inner half; (b) showing the inside of the outer half of the plasma membrane with fewer 'studs'

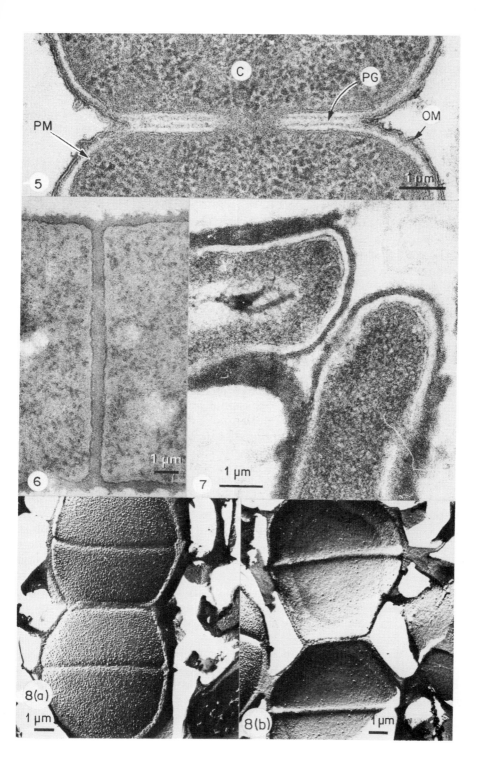

C

PG

PM

OM

1 μm

5

1 μm

6

1 μm

7

8(a)

1 μm

8(b)

1 μm

isolated (Figure 15). Thus there must be a considerable and continual traffic between the nucleoplasm and the periphery although circulation in the form of streaming is not evident.

There is no evidence of an inherent cytoplasmic architecture or cytoskeleton; the microtubule/microfilament components (actin–myosin systems) of the cytoplasm of eukaryotic cells seem to be absent in prokaryotic organisms. Microtubules have been reported several times but they have generally turned out to be polymerized components of defective phages or inactive structures (rhapidosomes) assembled from unknown cellular components, and found in bacteria during the stationary or decline phase of the life cycle. The latter do not appear to be functional elements even if they may derive from important components of cytoplasm or membrane. The only veritable microtubular structures in prokaryotic cytoplasm are those to be found running the length of the cell in the *Treponema*, in which they seem to parallel and roughly underlie the axial filaments (endoflagella). Despite such observations, no suitable functional cytoplasmic structure has been found to explain the bending movements of some bacteria, such as the *Flexibacteriaceae* and *Spirochaetaceae*.

The possibility still exists that a cytoplasmic architecture is present in prokaryotic cells but is disorganized by the preparative procedures or is not yet amenable to observation. The evidence is assumptive and indirect. Bundles of microfibres have been observed in the cytoplasm of some Cyanobacteria and the occasional patches of positively staining but ribosome-free granular material in other bacteria may represent distorted remains of some functional structure.

The cytoplasmic inclusions of prokaryotes, apart from intrusive membrane structures, are generally not of an organellar nature but represent the accumulation of metabolic by-products or reserve materials that are parked in the cytoplasm and may be recycled when metabolic events allow (Figure 1). These are usually recognizable histochemically as glycogen, lipid (poly-β-hydroxy butyrate), sulphur, polyphosphate, and other more complex molecular associations; there may be an interface with cytoplasm but they are not enclosed in a membrane. Some complex cytoplasmic inclusions of this character are to be found in the Cyanobacteria such as the polyhedral bodies and the phycobilisomes. The other types of inclusions are more properly dealt with in considering the structural interactions of the plasma membrane.

Many bacteria and Cyanobacteria contain gas vacuoles, which are presumed to be important in regulating planktonic movement in nature. They are extraordinary because they are bounded by a thin and unique membrane consisting of a single close-packed layer of macromolecules.

IV. Plasma Membrane

The plasma membrane confines the protoplast. It may be considered as the major organelle of the cell because it is the progenitor of most if not all of the functional cytoplasmic structures and it forms the organized site for the metabolic activities of the cell. The 'double track' profile in sections is remarkably uniform in appearance despite a very diverse capability (Figures 5, 12, 13). This is deceptive because the membrane is rich in proteins; some of these can be accurately located in the structure and give confidence that they are the studs seen in freeze-cleaved preparations, (Figures 8 and 9).

Differentiation is attained by either extension of the membrane or by elaboration of its structure or both. Extension of the membrane and a concomitant increase of all its components would appear to be a simple mechanism of adaptation to increased activity. Any such elaboration occurring within the relatively rigid cell wall could only fold into the cytoplasm to make intrusive structures. Such intrusions, often of lamellate form, are frequently seen in fixed preparations (especially of Gram-positive bacteria) and are called mesosomes (Figure 16). Simpler intrusions of tubes or sheets of membrane are also seen regularly in certain species. It was not difficult to accept all of these as being functional and important to that kind of bacterium, especially when other regular intrusions (e.g. those associated with photosynthesis) could be directly associated with function by identifiable constituents of the membranes (Figure 18). However, it is now clear that mesosomes, the most commonly recognized structure, are generated or at least elaborated during fixation and they are not seen in freeze-etch preparations unless pretreated with some chemical fixative. They commonly originate from the membrane in the region of septum formation but they may occur in lower frequency at any place. Despite the association with septum formation and the active synthesis of new cell wall there is no real evidence that the isolated mesosomal membranes are functionally specialized. In fact, they are relatively impoverished in protein and activity, like the outer half of the plasma membrane. If they are artifacts, they must be based upon some structural differentiation because there are preferred sites and because of a high frequency in some and an absence in other species. There is some evidence of localized distortion of membrane during the stimulation of an adaptive enzyme. Certainly it is easy enough to distort membrane; the partition of polar substances and solvents into membrane causes expansion and the rapid formation of pockets of rolled-up membrane. It is not unreasonable to consider that localized synthesis or insertion could and should do the same thing. This is

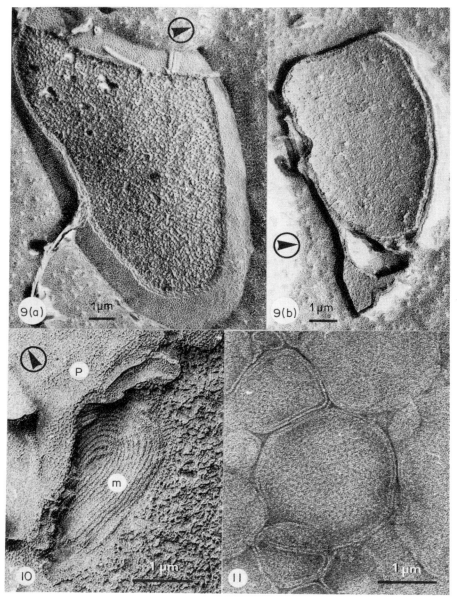

Figure 9. Replicas of a freeze-etched preparation of a *Spirillum* sp. showing the cleavage faces of the plasma membrane: (a) inner half; (b) outer half

Figure 10. Higher magnification of replica of the outer surface of the outer membrane of a *Spirillum* sp. showing pitted surface (p) and the edge of a mesosome (m)

Figure 11. Negatively stained preparation (ammonium molybdate) of isolated outer membrane vesicles of *Spirillum putridiconchylium*

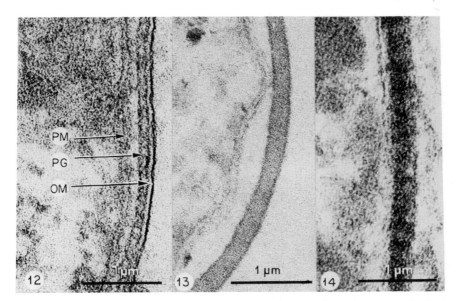

Figure 12. Profile of envelope of a Gram-negative species, *Azomonas insignis*, showing plasma membrane (PM) and the triplet wall (peptidoglycan layer, PG) and the double-track outer membrane (OM).

Figure 13. The Gram-positive profile of *Bacillus subtilis* with homogenous uptake of the 'stain' (metal salts). Other species show one or two concentric zones of differing uptake

Figure 14. Example of non-homogenous Gram-positive wall, *Micrococcus radiodurans*

represented in an extreme and almost pathological way by uncontrolled membrane production in some mutants of *E. coli*, which normally do not produce or show mesosomes. Some bacteria, e.g. *Azotobacter*, always show the profiles of tubular membrane intrusions that reach right into the central portions of the cell (Figure 17). So it may be that there are both functional and situational intrusions of membrane and these may have similar appearances.

The development of the intrusive photosynthetic membrane systems of the phototrophic bacteria, particularly the non-sulphur purple bacteria, can be studied because they disappear and only minimal amounts of photosynthetic pigments mixed with respiratory activity are found in the peripheral membrane when they are grown heterotrophically in the dark. Illumination in an appropriate anaerobic environment leads to rapid development of the membrane intrusions, an increasing content of photosynthetic pigments and a decreasing mixture of the respiratory enzymes. Eventually the internal membranes can be

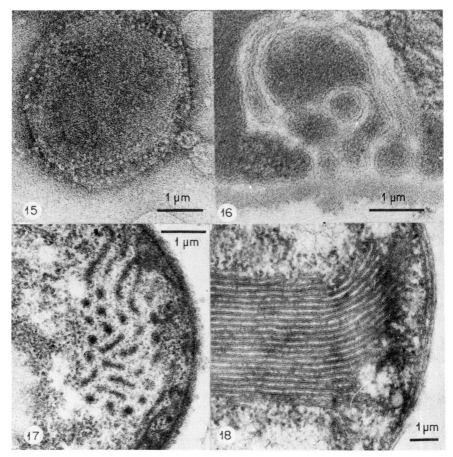

Figure 15. Isolated plasma membrane vesicle of *Spirillum putridiconchylium* showing attached strings of ribosomes (polysomes)

Figure 16. A mesosome formed of intruded plasma membrane in *Corynebacterium parvum*

Figure 17. Profiles of intruded membranous tubes in cytoplasm of *Azomonas insignis*

Figure 18. Portion of the stack of membranes in the cytoplasm of the nitrifier, *Nitrosococcus oceanus*, formed of closely appressed pairs of membranes

considered as almost entirely devoted to photosynthesis. It is most interesting to note that the arrangement of the thylakoids and the patterns of the intruded membranes are characteristic of the particular species involved, e.g. vesicular for *Rhodospirillum rubrum* but lamellar for *R. molischianum*, which indicates that the form of intrusion is controlled and probably by a genetic mechanism. In most cases, especially in the lamellar structures, continuity with the peripheral plasma membrane is easily defined in serial sections but the vesicular intrusions may be pinched off. This is less clearly understood for the Cyanobacteria, which have a more involved system including photosystems I and II.

The coherent reactions of photosynthesis are not the only energy mechanisms requiring the level of organization attainable by the stacking of membranes. A number of the chemolithotrophic organisms, especially the energy 'deficit-financing' systems exemplified by the nitrifiers, utilize very similar stacked membranes which can fill much of the available cytoplasm (Figure 18). They are functionally essential to the capture of energy, they are always present and form, in effect, a permanent organelle, but it is still connected to the periphery. In many cases the folded up membranes form 'triplet' profiles by the close fusion of pairs of double track membranes. This is further elaborated in some organisms, e.g. *Nitrobacter* by the addition of one or more layers of additional materials of unknown nature and function to the cytoplasmic side of the paired intrusion.

The structural regularity of many of the organelles derived from plasma membrane compels us to consider that they are functionally important and this is supported to a degree by a study of isolated membranes. But the experience with mesosomes leaves the suspicion that many of the simple intrusions seen in fixed preparations are illusory. It is obvious that the last word has yet to be said; we can expect development of a more functional understanding of bacterial membranes and can hope for more effective means of observation and correlation. If adhesion sites (c.f.) between membrane and wall are real, then only the areas of membrane between adhesion sites are free to intrude. Perhaps the extreme functional activity around some adhesion sites forces the focal enlargement of membrane in a closely adjacent area, thus determining the sites of intrusion.

V. Membrane Substructure

The macromolecular architecture of the plasma membrane has been an area of fruitful research ever since isolation for biochemical and biophysical study became possible. Microscopy has played a less significant role, but replicas of freeze-cleaved specimens have given something more than the unexciting profiles provided by sections and negatively stained preparations. They give some indication of the distribution of macromolecules in the outer and inner halves of the plasma membrane: the outer half is much less granular and is presumed to have much fewer proteins (and functional enzymes) than the inner half with a distribution that varies from random to organized in close-packed patches (Figures 8 and 9). The temperature from which the membrane is quickly frozen makes a great difference to the apparent distribution of the molecules and gives some feel for the remarkable plasticity and fluidity of the membrane structure. Functional associations are not discernable, so there are no specific labels to attach to the various knobs or studs visible in these interfaces. One can only imagine the substructures and micelles that might be significant in membrane function.

We can recognize that areas of membrane are structurally and functionally differentiated. An important functional differentiation can be recognized as that part of the membrane which is elaborated and invaginated for the synthesis of a septum in the process of cell division. Limited areas of membrane in motile cells synthesize and/or assemble special macromolecules, e.g. for the basal complex of flagella to which the protomers of flagellin are directed for export and assembly at the tip of the filament. The wall, through which the basal complex of rodlets and discs must emerge, is also modified to accept the structure and allow its continuity.

There must be sites on the surface of the protoplast active in the synthesis and export of each one of a number of specialized molecules, including exoenzymes and activated components for the assembly of the enveloping wall, polymers such as lipopolysaccharides, peptido-glycan, etc. Other dynamic external structures are regionally specific and it is entirely likely that they derive from functionally specialized areas of membrane (less dramatic than for flagella) as might be the case for fimbriae and F-pili. There may be, then, localized activities but there are few examples of structurally specialized areas of plasma membrane. One of these is provided by the peculiar internal elaboration called the polar membrane (see Figure 1) that is to be seen in Spirilla and organisms related to them. The function is unknown but may well be a local forest of attached macromolecules not too different from the more widely spaced 'lollipops' that seem to organize ATPase activity on the inner side of the plasma membrane of a number of bacteria.

VI. Periplasm

It is probable that we have no clear way of defining the boundaries of a structural entity that we call plasma membrane, because it is very complex in molecular and physico-chemical terms. Presumably it would include all molecules that are functionally attached on either the inside or the outside of the protein—lipid—lipid—protein bilayer, which is the formalistic first approximation to the structure of the membrane. Internal molecular associations have already been mentioned but there are also external and ill defined associations that are included in the generalized concept of a periplasmic space. In Gram-negative organisms this probably includes all loosely associated molecules between the diffusion barriers represented by the outer membrane and the plasma membrane. In Gram-positive organisms one must suppose that it represents all such loosely associated molecules within the wall and between the wall and the plasma membrane. In general these molecules are released by magnesium depletion and low ionic strength shock. There is usually a modest space in the section profile between the outer table of the plasma membrane and the inner border of the cell wall structure and this is the point at which plasmolysis occurs. Although it is spoken of as if it were a structural entity it is more in the nature of a concept for organizing enzymic functions that act outside the plasma membrane. The degree of organization is unknown. Some of the activity (e.g. phosphatase) is detectable and can be located just outside the plasma membrane by histochemical techniques involving the precipitation *in situ* of lead phosphate as a reaction product visible in the electron microscope.

VII. Cell Wall

A. Structure

Cell walls assist life in a dilute environment by acting as a corset against a high internal osmotic pressure and it is not surprising that the majority of bacteria possess them. But some free-living members of *Mollicutes,* e.g. *Acholeplasma*, are able to survive in waters without any evident cell wall. It may be, then, that some other significant components can contribute to the stability of the plasma membrane and protect against osmotic accidents.

The cell wall can be defined as that assembly of macromolecules formed into an ordered structure lying contiguous to but outside of the plasma membrane. This definition intentionally begs the question of whether or not slime, capsules, and any other loosely held materials are part of that structure. The functional definition would be that it is an encasing structure giving support to the cell and acting as the repository of shape by maintaining the macromolecular associations with a variety of cross-links. Every molecule has to be exported in some activated form of protomer or monomer to be assembled with or without enzymic help in the structural thicket of polymers forming the existing cell wall. The mechanism must be very effective and under strict control.

One of the outstanding characteristics of bacterial cell walls is that their profiles in the electron microscope show a significant layering of structures, which indicates macromolecular sorting (Figures 5 and 6, 12, 13). The recognition that the walls of Gram-positive and Gram-negative bacteria were also differentiated in terms of chemistry and molecular structure has provided one o the most powerful taxonomic generalizations that can be applied within the Procaryotae.

The murein of the wall consists of peptide cross-linked peptidoglycan that is found in all but a few (e.g. *Halobacterium*) of the walled bacteria. This component is always the innermost of the wall layers and contiguous to the external surface of the plasma membrane but separated by a 'gap' forming a part of the periplasmic space. In a few organisms (e.g. Spirochaetes and *Bdellovibrio*) the mucopeptide appears to be plastered on the outer surface of the plasma membrane so that the two appear to form one structure, but in most it is slightly separated from the plasma membrane by part of the periplasm. In any case the peptidoglycan, either alone or associated with teichoic acid polymers as in many Gram-positive bacteria, forms a strong and covalently linked structural network. It is probably the strongest element in the cell wall fabric; the degree varies according to the level of cross-linking and other associations. The effect of lysozyme is direct and dramatic as long as one can get the enzyme to the site of the peptidoglycan; with breakdown of that component of wall the cells become spherical osmometers. Therefore, the peptidoglycan contributes in large degree to shape and to strength, but all other organized components must be considered as contributory. The peptidoglycan-rich component is usually the only participant with membrane in the initiation of cell wall septa preparatory to cell division (Figure 5). So we must add this participation to the functional definition of walls when peptidoglycan is a characteristic component; other wall components have their own mechanisms and little is known about them.

B. Some Functional Discontinuities in Walls

Wall layers are usually complete and continuous; this is their strength. But we must assume that the interpolymeric spaces are sufficient for the import and export of molecules of various sizes through these concentric barriers. Major discontinuities are not often observed. In the absence of reactive chemical groupings, inappropriate charges or mere size exclusion (e.g. molecules exceeding about 600 dalton), the spaces should be sufficient for practical purposes. Some peptidoglycan layers are full of 'holes', presumably filled with other low electron scattering components, over the whole surface (e.g.

Micrococcus radiodurans (Figure 14)) or near to the division sites (e.g. in *Oscillatoria* species).

Cell walls may serve another generalized function: control of the periplasmic environment and the access of cations. Put in a simple and direct way, one can say that the bacterium lives at the bottom of a polyanionic chromatographic column which determines a form of *'milieu intérieure'* at the plasma membrane interface. It has been shown by direct titration that walls of *Staphylococcus* and *Bacillus* species provide strong buffering below pH 4.5 and above pH 8.5. It is not surprising, considering the nature of the polysaccharide, protein and lipid polymers, and heteropolymers, that there is a considerable variety of reactive groups available to interact with the solutes that pass by them in the interstices of the wall. Unfortunately, there is no equivalent data for cells that live in extremes of acid or alkaline environments to see what this may mean in circumstances of stress. The polyanions of the wall trap metal cations and may thus act as mediators for essential trace elements or provide a delaying action against toxic cations. Therefore, we must argue that these capabilities must be added to the sum of functional attributes of the wall and its components. It should be added as an aside that the electron microscopist is entirely dependent upon the metal trapping properties of the various components of biological material because most cells consist of C, H, N, and O in various combinations with much smaller amounts of heavier atoms; consequently, the scattering of electrons by biological structure is of a low order. The need for suitable contrast and much greater scattering-power is served by exposing the biological material to solutions containing relatively heavy metal salts. So the differentiation of substructure in electron micrographs represents, in a considerable degree, the varying uptake by the layers of component polymers.

Gram-negative bacterial walls present a planar discontinuity and an additional barrier to diffusion in the form of a lipopolysaccharide-rich membrane in the cell wall, the *outer membrane*. The outer aspect of this membrane can be studied by freeze-etching. In many organisms it is markedly pitted (about 5 nm in diameter and about 9 nm spacing) (Figure 10). A similar granularity, probably representing at least part of the same structure, is seen in negatively stained preparations of the outer membrane (Figure 11). This sort of discontinuity might represent columns of lipoprotein (or hydrophobic protein) molecules traversing the outer membrane and forming micelles. It seems likely that these linear molecules are assembled with their hydrophobic sides facing outwards in the lipid-rich membrane and the hydrophilic components facing inwards to form a column of useful physical properties for the conduction of aqueous solutes. It has been suggested that the lipoprotein molecules of *E. coli* wall are sufficiently long to extend through the outer membrane and, in the case of the enteric bacteria, be covalently linked through a tripeptide to the underlying peptidoglycan. Thus lipoprotein could have a function beyond the mere linking of components of the wall. There is evidence that newly synthesized lipopolysaccharide may be inserted in patches over the whole surface of the cell suggestive of the discontinuity of synthesis sites in the underlying plasma membrane.

Yet another type of discontinuity is represented by 'adhesion zones' (see Figure 1) in sections of partially plasmolysed enteric bacteria that have been fixed with great care to avoid mechanical injury during preparation. What is seen are zones 25 to 50 nm across in which the plasma membrane remains closely adherent to the overlying cell wall (peptidoglycan and outer membrane). Some of these sites can be recognized as related to flagella and fimbriae, especially F-pili, but the functions of the remaining 200 to 400 adhesion zones per growing cell are more speculative. A functional importance is suggested by the high proportion (greater than 80 per cent) of a variety of phages adsorbing preferentially to the adhesion zones. There may be a particular accessability of receptor sites in these zones. Using ferritin-labelled antibodies as a detector of a specific O-antigen, *Salmonella* have been shown to produce new antigens at a limited number of sites located over some of the adhesion zones, when the cells are converted to the synthesis of a new surface antigen by infection with a suitable lysogenic phage. So it would appear that, like plasma membrane, the cell walls of bacteria have functionally distinct areas that are important in cellular processes and events. These are not restricted to the division sites but are distributed over the whole surface of the cell. Considering that a number of functions have to go on at any given time it is not surprising that a particular function, such as the intercalation of newly synthesized lipopolysaccaride, is limited initially to 20 or 30 sites per cell.

VIII. Aspects of Growth and Division

It is hard to express all the structural implications involved in growth and fission of bacteria. But if we accept that the protoplast doubles its contents, that the replicated genomes segregate, and that membrane-associated structures increase their complement — then the most perplexing aspects of division involve the behaviour of cell walls. This is a dynamic and critical function.

The wall extends to enclose twice the original volume and also develops a crosswall so that simple fission may take place. This elaboration of the fabric involves the dimensionally specific insertion of new polymeric units so that form is preserved. Rod shaped organisms, for instance, are elongated and not widened; spherical organisms must enlarge. This implies a highly oriented substructure and, perhaps, strategic placing of effector enzymes because extension of the wall requires well regulated opening of the network for insertion of the new components.

That this happens and involves the 'autolytic' enzymes (murein hydrolases) directed at the wall is attested to by what happens to the murein and to the cell when this system gets out of control in the presence of penicillins and cephalosporins; the walls lose strength at the sites of septal synthesis, and cells burst if the medium is sufficiently dilute. Developing septa come apart and the inhibition of cross-linking assembly allows accumulation of strings of material representing, it is believed, the non-crosslinked polymers of peptidoglycan. Such materials are also found on the surface of protoplasts attempting to reform their cell walls. It is evident that the complex wall fabric is as much a functional part of the process as is the underlying membrane.

The site of new synthesis of wall preparatory to division is most generally related to the cell wall septa in process of synthesis. This is most evident in the Gram-positive cocci, e.g. *Streptococcus faecalis,* where the symmetrical outflow of wall material from the nascent septum over the growing daughter cells can be followed with the help of structural markers or of fluorescent antibody. In rod-shaped organisms the topological problem is greater and there is a degree of independence of the two wall synthesizing systems: one for growth, which operates by lengthening the cylinder, and the other for septum formation, which accomplishes cell division. Antibiotics in the β-lactam group draw attention to this distinction (varying specifically and according to dose) because some cause aseptate filamentation, others cause septal weakening and still others cause generalized relaxation of structure. Tracer studies and autoradiography indicate cyclical but major turnover in murein components (e.g. diaminopimelic acid) over the whole surface of Gram-negative and Gram-positive rods and regional bursts of activity at septal sites. This contrasts with a focus on septal synthesis in the coccal models and, probably, not all these are identical.

Many cell wall septa are laid down as thin structures within a fold of plasma membrane (Figures 1, 5, 6, 8). They consist mainly of the murein components which then differentiate within their structure (inserting a 'zipper') to allow doubling and delamination for separation (Figure 19). In some this is evident early in the synthesis, in others it is a phenomenon following completion, and in a few the septal layer is doubled from the outset as if intruded. It implies regional and dynamic effects within the wall substance and is easier to describe than to

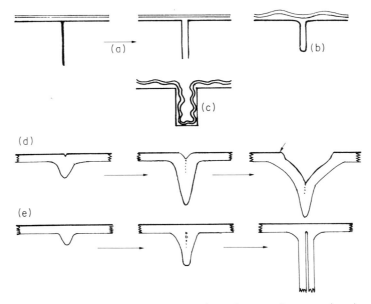

Figure 19. A diagram to show various forms of septa: (a–c) among Gram-negative bacteria; (d, e), among Gram-positive bacteria. (d) is the form seen in some cocci where the wall-band (small arrow) acts as a visible marker of synthesis and extension.

explain. In most, but not all, the septum is symmetrical and closes like an iris diaphragm.

The stimulus for septum formation and presumably for the establishment of the appropriate specificity of a ring of membrane in a specified location arises from determinants that are activated close to the end of a round of DNA replication. What is involved in the location of new sites and orientations of planes of division is unknown, but it is mediated by protein synthesis and is a phenomenon of considerable biological importance. Synchronization experiments in model systems such as *St. faecalis, Bacillus subtilis* or *E. coli* only put one in touch with the broad outline and say little about the primary events and interactions. The more complex models (budding, multiple planes of division, dimorphism) have hardly been attacked. However, budding involves asymmetrical extension, which may be closer to current studies than might be thought, because surface labelling of *E. coli* shows that extension is asymmetrical and polar. In this case the marker involves components of the outer membrane and the choice has yet to be made, definitively, for the underlying murein: diffuse, polar, septal, or a mixture of processes. Budding is also studied in those organisms that form reproductive prosthecae (*Hyphomicrobium* and *Rhodo-*

microbium) and those that form stalks or prosthecae without reproductive significance (*Caulobacter*) but the results cannot yet be incorporated into general thinking.

In any case, cell division is highly regulated as to component, time and place; much of it is translated into action by the plasma membrane and the result is reflected by the pattern of growth and behaviour of the wall. This is simply illustrated by septum formation in *E. coli*, as a representative Gram-negative model, which, when adequately fixed, shows septum formation and the differential behaviour of the major components of the wall. Even though the earliest discernable structural event may be the formation of blebs in the outer membrane the peptidoglycan septum is well advanced or even complete before the intrusion of the outer membrane starts and progresses towards separation of the sister cells (Figure 5). The same sort of thing happens in many Gram-positive organisms (Figure 19e). The separation event is under independent control because circumstances or mutations can dictate filamentous septate growth, i.e. cell division without separation, which may be a species (*Bacillus mycoides*) or a mutant characteristic (e.g. filamentous strains of *E. coli*). Interference with DNA synthesis in bacilli generally leads to aseptate filaments.

The differential synthesis of components of the cell envelope may be considered as models for processes involved in *differentiation*: directional synthesis with polarity, differential behaviour of layers of the cell envelope, the insertion of appropriate enzyme packets in an otherwise rather homogeneous cell septum to allow for splitting, buds that are smaller or bigger than the original cell, the local insertion of flagella or stalks or other special components, and the accomplishment of multiple rather than binary fission, are all the stuff of differentiation. One expression of differentiation is exemplary as a functional elaboration of cell structure and that is sporulation; this will be the subject of a later chapter. Different orders of complexity attend the formation of resting and resistant forms: Some (like endospores and arthrospores) are the result of specially directed cell divisions that may be followed by differentiation; others (like myxospores, various 'cysts', heterocysts, and akinetes) result from differentiation of the product of otherwise normal divisions.

The cell wall is so complex in chemistry and directed behaviour that it is hard to believe that it serves only as structure; a functional *credo* may help to orient the above descriptions. The wall is subject to local modification that is in part due to the regulation of enzymes incorporated *in situ*. There are the implied but uncertain functions of the periplasmic enzymes (e.g. phosphatases) that are outside of the requirements for the insertion and crosslinking of wall polymers but may be critical to the export—import trade. The wall is capable of buffering the environmental pH of the functioning plasma membrane (including periplasm) and of being selective in trapping essential rare cations or delaying the approach of toxic metals. The wall and what is attached to its external surface is the zone of interaction with the environment and has to be considered a component of the functional anatomy of the cell and may even be specific in character. This is expressed in the antigenic mosaic of pathogens, in structures such as holdfasts and, among many phenomena, the need for *Thiobacillus thiooxidans* to lie on sulphur granules in order to oxidize the sulphur.

IX. Surface Components

A number of surface components may be found outside the formal cell wall. These are usually dispensible; they can be lost by mutation without much apparent effect on the *in vitro* physiology of the cell although some interactions with the natural environment may be profoundly modified. These surface elements (Figure 1) range from the close-packing of protein macromolecules forming the many versions of the regularly structured (RS) layers, or the enclosing slime layers or capsules, to the projecting fimbriae and sex-pili. The materials for all of them have to be exported from the plasma membrane for assembly *in situ* as is the case for the formal components of wall. They are widely but not uniformly distributed among bacterial taxa.

The *RS layers* (Figures 20—22) might as well be integrated into the cell wall structure because some of them are covalently bonded to underlying components, which act as a template for their assembly. Other examples are retained only by electrostatic forces or by salt-linkages and are very easily shifted or removed. The *fimbriae* (Figure 23) and the tubular sex-pili are also highly organized because they are assemblies of protein subunits. In this case the degree of integration is uncertain but involves penetration into and probably through the cell wall as well as origination from the plasma membrane. This places them in adhesion zones and they define another area of local modification to match those induced by flagella. *Capsules* have always seemed to have a more tenuous structure but this is in part because many of them consist of long-chain polysaccharides with little or no visibility in electron microscopy. However, they must present little problem to the cell as a diffusion barrier and allow free passage to important molecules. A degree of organization becomes visible, in some, as radiating strands of material (possibly because of the presence of both polypeptides and polysaccharide) (Figure 27).

The function of the elegant but seemingly useless RS protein layers is likely to be protective for both the Gram-negative and Gram-positive

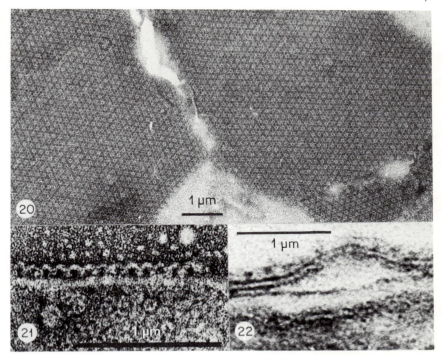

Figure 20. Negatively stained array of RS units from the external surface of *Spirillum serpens* VHA. Centre to centre spacing of units is about 15 nm

Figure 21. Folded edge of an array from *Spirillum fasciculus* in negative stain

Figure 22. Profile of *S. serpens* VHA wall showing units on the outer aspect of the outer membrane

bacteria that have them. At least for Gram-negative species it is clear that the presence and integrity of these layers protects the cell from the attachment of and predation by *Bdellovibrio,* a ubiquitous bacterial predator. Therefore, one could expect a strong selective value in favour of strains with an RS surface; but this would not be likely to apply to Gram-positive bacteria since they are not susceptible to this predation. There is no evidence that these surface layers have any role in clumping or in adherance to surfaces and there are examples to attest that neither fimbriae (e.g. *Acinetobacter*) nor capsules (e.g. *Bacillus polymyxa*) are mutally exclusive for the RS layer of any species.

However uncertain we may be about the possible functions of the RS layers, there is no doubt of the importance of the interaction of the surface of the bacterial cell envelope with a variety of solid substrates. Capsules, as is well known, are of protective value to many pathogenic organisms because they provide an interface that is inimical to phagocytosis; the same sort of result may derive in other cases from the

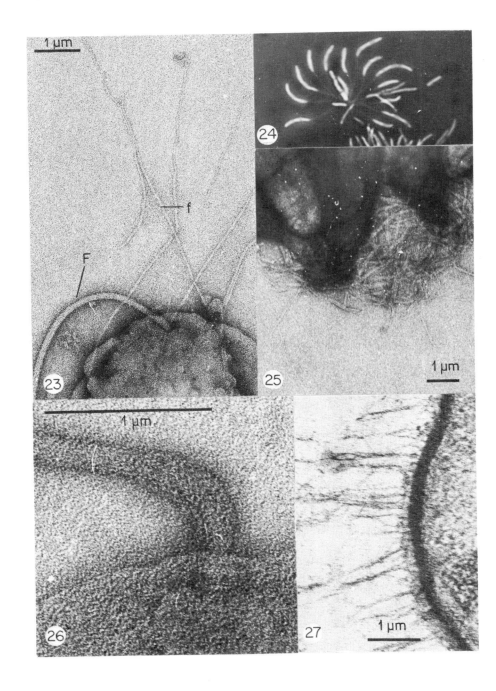

presence of specific antigenic components on the wall of a non-capsulated organism. On the other hand, adherence has a practical and selective value to organisms in many ecological niches and capsules play a part as do the fimbriae (or common pili). This often involves mutual adherence and entrapment of one or several species of bacteria by fusion of capsular material to form sessile colonies on surfaces. For instance, such colonies form the dental plaque on teeth, or the slime on the rocks in a trickling filter and the slime on rocks in a clear stream. In all such environments the selective value lies in the evasion of 'wash-out'. The formation of zoogloeal masses in slime is sometimes directed by nutritional circumstances (high C:N ratios for *Zoogloea ramigera*) whereas for others the slime forms sheaths that hold the cells in a colonial habit whether suspended or attached (*Sphaerotilus natans* and many others). The substrate interaction is probably of considerable functional importance, as explained for the cell wall, because the anionic polymers provide sites for 'contact ion-exchange' and assist in the liberation of elements essential to life. Certainly many organisms in soil enjoy a very intimate contact with the mineral particles forming microcolonies on their surfaces. The polysaccharides encrusting the colonies of many sheathed organisms accumulate oxides and hydrates of iron or manganese in great quantity. So it should not be surprising that most of the metal content of natural fresh-waters is not in solution or in organisms but bound in the 'dissolved organic carbon' fraction (presumably a variety of high molecular weight polymers that have escaped from their cells of origin). So these loose enveloping materials are not trivial in functional terms.

Adherence to a surface is a way of life and for many bacteria the sessile habit involves specific structural components. In the simplest case a holdfast area is involved instead of the whole surface and, as in the case at the non-flagellate end of the *Bdellovibrio* cell, an area of microcapsular material is visible. In the stalked bacteria such as *Caulobacter* spp., the holdfast is an even smaller area on the tip of a special prostheca where there are few very fine fimbriae in addition to a

Figure 23. Fimbriae (f) and a flagellum (F) at the pole of *Spirillum lunatum* to show relative sizes (4.5 versus 11.5 nm)

Figure 24. A rosette of *Caulobacter* sp. in nature

Figure 25. The tip of the stalk of *Caulobacter* sp. showing fimbriae and microcapsular material

Figure 26. Insertion of a flagellum into the wall of a Gram-negative organism (*Spirillum lipoferum*) in negative stain indicative of the relation of discs to wall

Figure 27. A capsule showing radiating substructure

dab of glue (Figures 24 and 25). Attachment points of one or other type are often inconspicious and are characteristic of the motile swarmer cells of *Caulobacter* or *Sphaerotilus*, or the gliding gonidia of *Leucothrix*.

The ability to attach to surfaces and interfaces becomes most specific among some parasitic and pathogenic bacteria ranging from Spirochaetes to enteric bacteria and the gonococcus, and on host membranes ranging from protozoa to termites and man. The specificity involves both partners and even the anatomical region of the host. The attachment involves a cell—cell interaction of characteristic form showing the plasma membrane of the eukaryotic cell separated from the outer part of the wall of the bacterium by a finely structured zone resembling the 'gap junctions' of tissues. A definitive recognition system involving molecules on the surface of both cells is invoked. The precise mechanism is uncertain and may be lectin-like or enzyme-like. In a great number of cases the ability to attach to surfaces is correlated with the presence of large numbers of fimbriae (common pili) and, in the case of *Neisseria gonorrhoeae*, these are also correlated with pathogenicity. Such correlations are treacherous but this one offers attractive features including the order of specificity offered by proteins and a wide distribution. There may be a further practical aspect to the fimbriate condition in that these structures project through any capsular layer and thus could still perform a function necessary to the maintenance of the species.

The greatest possible specificity attends the function of the sex-pili which recognize and attach to the appropriate partner and are involved in some mysterious way with genetic transfer in conjugation. Only a few are produced on each cell and the two types (F and I) are only distinguished by antigenic and phage specificity.

All kinds of pili and fimbriae appear to penetrate the wall. They are delicate structures and easily dislodged. When the walls of fimbriate bacteria are laid out in negative stain the electron microscope shows holes of appropriate size, which may well be available at times for the transfer of molecules that are too large to be able to diffuse through the intact wall. It is tempting to think of the tubular nature of sex-pili and the shortening on contact, which has been interpreted as contractility, as explanatory of the mechanisms of conjugation. But an explanation of transformation and competence is more difficult and discontinuities would help. To these possibilities one must add the 'leakiness' of cells approaching stationary phase and the reorganizations of wall attendant on specific regions during growth and division. The flagella insertions may offer similar opportunities (they too have a hollow section) and one should remember that the flagella-specific phages slip, within seconds, to the junction of hook and wall where the DNA is discharged.

X. Structure and Motility

Bacteria exhibit several types of motion that are a special and determinative characteristic of living members of certain groups and species. Those types of movement with structural implications include: cell flexing, post-fission movements, twitching, gliding, and swimming motility. It is fair to say that there is limited understanding of how any one of these is accomplished and how energy is transmitted for motion. There has been no shortage of hypothesis because these actions are easily observed with the light microscope but understanding is of a low order.

The flexing motion of some filamentous cells, such as species of *Flexibacter* and of Spirochaetes, is usually rapid bending from a straight to a V-form and then slow relaxation. It argues for a contractile component which has not been found in cytoplasm. The cytoplasmic microtubules in *Treponema* might be a candidate structure but these have not been found in others that perform the trick.

The Corynebacteria have a well-known habit of remaining partially attached after division and forming 'chinese character' or 'pallisade' arrangements of their rods in smears. After the completion of septa the enveloping wall snaps and the sister cells swing apart, held together by a fragment of wall (Figure 7). The suddenness of the snap may be due to turgor pressure in the delamination of the doubled septum.

Twitching is a form of intermittent, spasmodic movement of individuals in a mass of cells or in translation on an interface such as an agar surface. It is not efficient or consistent but individual cells can be found to have moved in such jerks for a number of cell lengths away from the edge of a colony. The ability to do this has now been correlated with the presence of tufts of fine (5 nm) fimbriae, but there is no idea of how it is accomplished.

Gliding is an important and widely distributed motion on moist, solid interfaces, and is characteristic of many Cyanobacteria as well as Myxobacteria and Cytophagas, and all of them can be considered to have walls of the Gram-negative type. The rates of motion are quite variable but the motion itself is normally stately and long continued. It is critical to the life cycle (involving aggregation and fruiting body formation) of the Myxobacterales and it allows them and the Cytophagas to cruise in search of adequate substrates (wood, bark, bacteria in dung, and chitinous shells) for their powerful enzymes to reduce to assimilable nutrients. All of them extend over and colonize suitable surfaces and some sessile organisms, e.g. *Leucothrix*, attain this by producing gliding gonidia. The mechanism of gliding is entirely mysterious but is reversibly damaged by proteolytic enzymes, chelators, and osmotic shock. The cell walls and surfaces are not distinctive. The

only structural characters that correlate with motility are the production of slime and the presence of polar tufts of fimbriae, and neither
of these is exclusive to gliders. Yet motile Cytophagas can mutate and
non-motile mutants have lost their fimbriae. They also have a differing
pattern of periplasmic proteins, and these antigens stimulate antisera
that inhibit gliding. Even the cell wall may not be entirely necessary to
gliding because some wall-less *Mycoplasma* (notably *M. pneumoniae*)
can glide on a glass surface in a liquid medium and at a speed
comparable to the movement of Myxobacteria ($15-30 \mu m/min$).
Unlike the Myxobacteria they move singly, the polarity is constant and
the movement is punctuated by resting periods. No aspect of the 'form'
is indicative of the mechanism and yet, as in the walled gliders, there is
reason to think that the surface of contact is important because gliding
ceases on treatment with antibody. It is a phenomenon requiring the
energies of the cell and in both cases it is inhibited by iodoacetate and
para-mercuribenzoate. The mechanism remains a mystery.

Swimming motility is much better understood and a large body of
impressive experimental data has accumulated on structure, on chemotaxis, and on genetic components of the ability to swim. There is no
doubt, from direct observation, that motion is accomplished by the
action of the unique, tubular, semi-rigid, helical assemblies of the
protein, flagellin. These are found on all swimming bacteria although
the number of them and the arrangements differ. Most flagella are
naked; others are sheathed by an extension of the outer membrane, and
others (providing a mechanistic mystery) are neatly aligned along the
axis of the helical protoplasmic cylinder (between the peptidoglycan
and outer layers) of Spirochaetes forming an axial filament. All of them
originate from the plasma membrane and have a relatively complicated
structure (the 'basal complex') for the insertion into plasma membrane
and part or all of the overlying wall (Figure 26). Undoubtedly this is
where energy is transformed into the work of the flagella. There is no
evidence of contractility in the filament, which appears to be a
semi-rigid helix, yet they move individually and in concert to provide a
potent driving force achieved, it is believed, by rotation of each
filament in its insertion complex. The concept derives from direct
observation of the spinning of singly-flagellate cells tethered by the tip
of the filament to a slide. Thus, the basal complex becomes the rotor of
a motor set in a biological bearing. The driving force is not associated
with ATP but is believed to be associated with the local proton
circulation. Other explanations have been sought.

The structure of the basal complex is simple enough: it consists of a
rod with two or four discs attached to it, which insert through and into
the wall structure (Figure 28). The distal end of the rod attaches to the

Figure 28. A diagram of one possible (plausible) set of associations for the structures of the basal complex (four discs on a central rod) of a flagellum in the wall-membrane complex of a Gram-negative bacterium (OM, outer membrane; PG, peptidoglycan layer; PM, plasma membrane)

'hook' of the filament, and the proximal end inserts in the plasma membrane with the help of a pair of discs, S on the surface and M in the inner half of the plasma membrane. This is all the Gram-positive cell needs, but Gram-negative bacteria, with projecting naked flagella filaments, have a distal pair of discs: L within and P just inside the outer membrane. There is a collar of molecules around the disc-sized (23 nm) hole in the outer membrane; the murein sacculus has a hole just big enough to accommodate the central rod (7 nm) and the cleavage plane of the plasma membrane in freeze-etched preparations has an impression in it the size of the proximal M-disc. The proponents suggest that M is a rotor, S is a stator, and that P and L are just bushings; but it seems premature to salute the concept when almost no specialized structure surrounds the complex (which is only 22.5 nm in cross section!). Some strategic component may not survive preparation, as is suggested by the outer membrane-associated concentric rings of macromolecules identified around the flagella of a number of *Spirillum* species. It has long been known that the intact wall is necessary to flagellar motility and this integrity may be needed either for the compartments required by the proton mechanism or to provide a mechanical purchase for the stator or some other pseudorotatory (or gyratory) mechanism.

XI. Assembly and Stability of Structures

It is obvious that exceedingly complex structures such as walls, flagella, and pili are put together outside the cell, and in three dimensional forms using 'building blocks' exported by the cell. It is less obvious that the same principles must apply to intracellular structures in which macromolecules are put into specific associations to form structures. The intracellular examples (e.g. plasma membrane) are almost too complex to contemplate, but, as was the case for the initiation of the study of virus capsid assembly, the RS proteins that form elaborate para-crystalline arrays on the external surface of bacteria provide relatively simple model systems.

The formation of an RS layer such as that on *Spirillum serpens* (Figure 20) involves the self-assembly of quasi-identical peptide polymers exported from the underlying plasma membrane to form units which are linked to their neighbours to make a sheet with threefold symmetry. The like charges on these polyanions prevent assembly unless a suitable charge density can be achieved, usually with monovalent cations. Assembly also requires divalent cations and a suitable nucleation surface, which is provided by the underlying lipopolysaccharide of the outer membrane. The divalent cations are a specific requirement, satisfied in this case only by Ca^{2+} or Sr^{2+} for a lifelike array; some hydrophobic bonds would then complete the construction. Some RS layers do not seem to require such specific circumstances for their particular entropy-driven assembly system. Other more complex wall assemblies (e.g. peptidoglycan) require the export and placing of suitable enzymes to cross-link the polymeric strands provided by the sub-assembly system on the membrane. One must recognize then that assemblies are natural physico-chemical experiments, each one of definitive character, that dictate the character of the nearest neighbour polymers. It is not surprising then that the components of cell walls sort themselves in layers and in consistent order. Self-assembly systems are entropy driven and the structure exists at a low energy-state equilibrium. Unless covalent bonds are introduced into assembled structures they are easily upset and subject to disassembly. It is reasonable to expect that the maintenance of assembly would require at least the presence of an appropriate concentration of the salt-link cation. This can be a very specific requirement and, as explained above, is often Ca^{2+} for walls and Mg^{2+} for membranes. A very low ionic strength environment or inappropriate cations (particularly monovalent) with or without high temperatures, appropriate chelating and other chaotropic agents can all lead to disaster. However, local control of these factors allows *in situ* addition or expansion of RS layers for growth, the export of flagellin protomer out through the tubular flagellar filament for assembly at the distal end (assembly can only take place on the distal end of a fragment of filament), or it provides for interlinking bonds to a more stable neighbouring structure.

XII. Concluding Remarks

It is well known among naturalists that you see in nature only what you are prepared to see. Cytologists are no exception and are as blind as everyone else to the significance of structures and the constituent macromolecular arrangements until serendipity or persistence brings them into focus. A general awareness, however, makes a world of difference to interpretation, and to cogitation over micrographs and the functional associations of structure. This chapter may stimulate that awareness in those who know the bacteria for other reasons.

XIII. Acknowledgements

The author is very grateful for the assistance and advice of Dr T. J. Beveridge, the expert technical work of Myrtle Hall, the drawings by Doris Murray, and the continued support of the Medical Research Council of Canada. The micrographs are from the author's laboratory and special thanks are due to Dr Beveridge for most of these micrographs.

XIV. Further Reading

Bayer, M. E. (1975). Role of adhesion zones in bacterial cell-surface function and biogenesis. In *Membrane Biogenesis*, Ed. A. Tzagoloff, pp. 393–427. New York: Plenum Press.

Begg, K. J. and W. D. Donachie. (1977). Growth of the *Escherichia coli* cell surface. *J. Bacteriol.* 129, 1524–1536.

Beveridge, T. J. and R. G. E. Murray. (1976). Uptake and retention of metals by cell walls of *Bacillus subtilis. J. Bacteriol.*, 127, 1502–1518.

Burdett, I. D. J. and R. G. E. Murray. (1974). Electron microscope study of septum formation in *E. coli* strains B and B/r during synchronous growth. *J. Bacteriol.* 119, 1039–1056.

Cell Motility: Book A. (1976). Eds. R. Goldman, T. Pollard, and J. Rosenbaum. Articles on *Primitive motile systems*, pp. 29–70. Long Island, N.Y: Cold Spring Harbor Laboratory.

Henrichsen, J. and J. Blom. (1975). Correlation between twitching motility and possession of polar fimbriae in *Acinetobacter calcoaceticus. Acta path. Microbiol. Scand. Sect. B*, 83, 103–115.

Microbiology — 1977. (1977). Ed. D. Schlessinger. Articles on *Adherence of bacteria to host tissue surfaces*, pp. 395–432. Washington, D.C.: American Society for Microbiology.

Radestock, U. and W. Bredt. (1977). Motility of *Mycoplasma pneumoniae. J. Bacteriol.*, 129, 1495–1501.

Ryter, A. (1968). Association of the nucleus and the membrane of bacteria: A morphological study. *Bacteriol. Revs.*, 32, 39–54.

Salton, M. R. J. and P. Owen. (1976). Bacterial membrane structure. *Ann. Rev. Microbiol.*, 30, 451–482.

Shockman, G. D., L. Daneo-Moore, and M. L. Higgins. (1974). Problems of cell wall and membrane growth, enlargement and division. *Ann. N.Y. Acad. Sci.*, 235, 161–197.

Silva, M. T., J. C. F. Sousa, J. J. Polonia, M. A. E. Macedo, and A. M. Parente. (1976). Bacterial mesosomes: Real structures or artifacts? *Biochim. Biophys. Acta.*, 443, 92–105.

Sleytr, U. B. (1976). Self-assembly of the hexagonally and tetragonally arranged subunits of bacterial surface layers and their reattachment to cell walls. *J. Ultrastruct. Res.*, 55, 360–377.

Form and Function — II. Fungi

C. BOOTH

Commonwealth Mycological Institute, Ferry Lane, Kew, Surrey

I. Fungi — Form and Function

Fungi form a major group of living organisms that are without chlorophyll, but because most of them have a definite cell wall they are usually placed in the plant kingdom. However, their metabolism, wall structure, and methods of reproduction are sufficiently distinct for other authorities to regard them as forming a distinct kingdom equivalent in status to the animal and plant kingdoms.

A. Myxomycota

One large group of organisms which are usually placed with the fungi but are somewhat intermediate between true fungi and protozoa are the slime moulds, the Myxomycota. In this group the vegetative plant body is a naked amoeboid mass of protoplasm which may consist of a large multinucleate body, a *plasmodium*, or be formed of an aggregation of many multinucleate protoplasts, a *pseudoplasmodium*. The division is divided into classes based on the nature of the plant body.

B. Eumycota (The True Fungi)

These fungi always have a definite cell wall. Electron microscopy shows that cell walls are laminated and formed by two to several layers of microfibrils. They are for the most part composed of hemicelluloses or chitin. True cellulose apparently is only present in the lower fungi where glucans may be a major component of the wall.

The method of sexual reproduction, which in fungi as in other organisms involves a process of nuclear fusion and meiosis, results in the formation of a fruit body and it is the nature of this fruit body which was used in the past to divide the fungi into three groups, the Phycomycetes, Ascomycetes, and Basidiomycetes. In recent years it has become apparent that the Phycomycetes consist of an assemblage that forms two distinct groups and therefore four major divisions exist within the Eumycota, these are (1) Mastigomycotina, Chytridiomycetes, Hyphochytridiomycetes, and the Oomycetes; (2) Zygomycotina which includes the Zygomycetes; (3) Ascomycotina, the Ascomycetes, and (4) Basidiomycotina, the Basidiomycetes.

A fifth group with asexual spores but in which a sexual process has not been observed is referred to as the Deuteromycotina (*fungi imperfecti*). This group is gradually being reduced as its members are linked chiefly to the Ascomycetes or in a few cases to the Basidiomycetes. There is ample evidence however that many of these imperfect fungi have lost their ability to form either asci or basidia, and comparisons with the asexual states of related species having a known perfect state suggests the division to which they belong.

1. Mastigomycotina

Both sexual and asexual reproduction processes divide this subdivision into three distinct classes.

The Chytridiomycetes: All species have a posterior uniflagellate zoospore of the whiplash type. Sexual reproduction is by fusion of similar motile gametes (isogamy) or between a small active male gamete and a larger slower female gamete (anisogamy), or by motile spermatozoids fertilizing an oogonium.

The Hyphochytridiomycetes include a small group of aquatic fungi which have an anterior uniflagellate zoospore of the so-called tinsel type which has a row of fine lateral processes along each side of the flagellum (Figure 3).

The Oomycetes have a characteristic zoospore bearing two flagella, one of the whiplash type and the other of the tinsel type. In some terrestrial species the sporangia are detached and are dispersed by wind, subsequent germination may be by zoospores or, as in downy mildews, by germ tubes which infect the host. Sexual reproduction is oogamous and results in the formation of a thick-walled oospore, or resting spore.

2. *Zygomycotina*

The Zygomycotina (Zygomycetes) reproduce asexually by aplanospores (non-motile zoospores) passively dispersed by wind, rain, or animals, although in some species such as *Pilobolus* the whole sporangium is violently discharged. Sexual reproduction is typically isogamous by fusion of two gametangia which results in the formation of a resting spore, the zygospore.

3. *Ascomycotina*

Ascomycotina (Ascomycetes) are the largest group of fungi. Asexual reproduction is by conidia or by modified hyphal structures. Sexual reproduction is by the formation of ascospores within sac-like containers (asci) which may be unprotected, as in the Hemiascomycetes, where the asci develop directly from the fusion cell, the zygote or ascogenous cell. In most other Ascomycetes the zygote or ascogonium produces binucleate ascogenous hyphae which ultimately give rise to the asci. This centrum (the ascogonium, ascogenous hyphae, and asci) is usually protected or covered by somatic tissue and the whole forms the ascocarp. The ascocarp may be an enclosed sphere, a cleistocarp, or flask-shaped with an apical pore, referred to as a perithecium and borne singly or in groups within an ascostroma, or it may be a saucer-shaped apothecium when mature.

These three ascocarpic forms were used as the basis for three divisions of the ascomycetes: The Plectomycetes with a cleistocarpic ascocarp; the Pyrenomycetes with a perithecium, and the Discomycetes with an apothecium. When serious developmental studies were undertaken it became apparent that species with these superficial characteristics in common are not necessarily related in any other way. The species with closed ascocarps within the Plectomycetes include primitive soil forms, together with highly developed obligate parasites like the Erysiphales. Within the limited range of variation possible the different lines of development have obviously each produced their own forms of cleistothecia, perithecia, and apothecia, together with the flat inverted saucer-like forms of ascocarps called thyriothecia. The superficial modifications of these ascocarps are often due to the selective pressure of the particular ecological niche which the fungus occupies. Thus true cleistocarpic species are largely soil fungi or forms which develop deeply seated in wood; in these situations violent discharge of the ascospores is not a practical method for spore dispersal. The flat thyriothecium with its protective inverted saucer-like stroma is an adaptation for fungi developing on the surface of leaves, and an ostiole may or may not be present depending upon the geneological origin of the fungus. The thick-walled cleistothecium of the powdery mildews is another adaptation to habitat by a highly specialized fungus which bears no close relationship to the primitive cleistocarpic soil or dung species which form the majority of the Plectomycetes. It is preferable therefore to include the Erysiphales in their more appropriate position related to the Sphaeriales (Pyrenomycetes) and to retain Plectomycetes (Protomycetidae) for the more primitive forms although in many cases these are obvious precursors of the Pyrenomycetes (Unitunimycetidae) or Loculoascomycetes (Bitunimycetidae). Hence we can now recognize the classes and subclasses within the Ascomycotina as described below.

1. Hemiascomycetes. Asci formed without protective tissue either directly from the fusion cell zygote as in the Endomycetales (yeasts-Schizosaccharomyces, Endomyces, and Eremascus) or from an ascogenous cell which produces a single ascus, i.e. Taphrinales (*Taphrina deformans*, the cause of peach leaf curl).

2. Euascomycetes. Asci develop from ascogenous hyphae with or without crosiers and usually with some surrounding protective somatic tissue. Three subclasses are recognized.

Protomycetidae: The protective tissue surrounding the developing asci may be primitive or highly developed. It is sparse in *Byssochlamys fulva* where it forms a loose investment of hyphae. This is more highly developed in *Gymnoascus* species where these protective hyphae are thickened and have characteristic spines. In *Thielavia* and *Cephalotheca* (Figure 1A) a distinct pseudoparenchymous wall is formed and in some species even a rudimentary ostiole is present. Asci are borne at all levels within the ascocarp and no violent discharge of ascospores occurs. Genera such as *Preussia* and *Zopfia* included here are primitive forms of the Bitunimycetidae (Loculoascomycetes) whereas *Thielavia* and *Pseudeurotium* belong to the Unitunimycetidae (Sphaeriales).

Unitunimycetidae: The subclass includes orders from the Pyrenomycetes and of the Discomycetes. The ascus wall functions as a single layer usually with an apical pore or other apparatus such as an operculum facilitating spore discharge. Protective tissue around the centrum (ascogonium, ascogenous hyphae, and asci) is usually stimulated to form by the presence of the ascogonium and the tissue

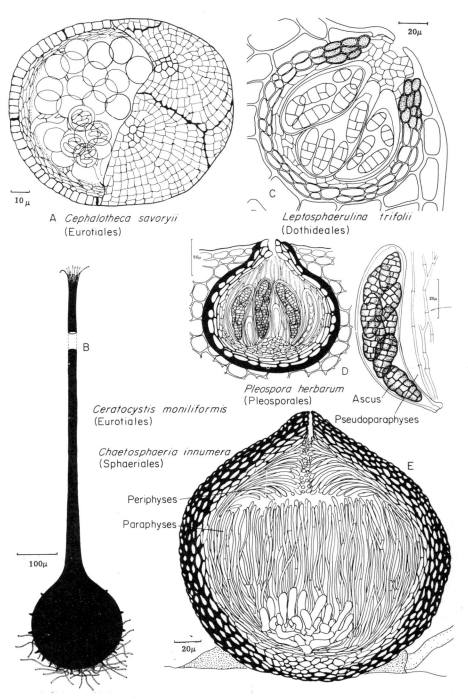

A *Cephalotheca savoryii*
(Eurotiales)

Leptosphaerulina trifolii
(Dothideales)

B

Ceratocystis moniliformis
(Eurotiales)

Pleospora herbarum
(Pleosporales)

Ascus

Pseudoparaphyses

Chaetosphaeria innumera
(Sphaeriales)

E

Periphyses

Paraphyses

Figure 1. Ascocarps (B after G. Morgan-Jones; D after K. A. Pirozynski)

develops either from the stalk of the ascogonium or from the surrounding somatic hyphae.

The mature fruit body may be a closed cleistocarp as in *Erysiphe* species, a perithecium, a flask-shaped structure with an apical pore borne singly or immersed in groups in a stroma, or an apothecium, which when mature forms an open cup or saucer-shaped fructification. The orders within this subclass show a progession from perithecial to apothecial forms. Perithecia are well developed in the Sphaeriales, often carbonaceous and with true apically free paraphyses (protective filaments which grow up between the asci from the base of the perithecium), and with periphyses lining the ostiole (Figure 1E). Whatever the final form of the apothecium the initial development around the ascogonium is similar to and resembles that found in the Sphaeriales. There are three basic methods of development; angiocarpic as found in *Ascobolus stercorarius* where the hymenium of developing asci is covered with a protective layer of tissue formed by the excipulum, the outer protective cortical tissue; the second as in *Pyronema domesticum* is gymnocarpic in which the hymenium is exposed throughout its development; and the third, hemiangiocarpic, is a somewhat intermediate condition as found in *Peziza aurantia* where hyphae initially arch over the ascogonium but then develop into cortical tissue. In passing from the perithecial form to the apothecial form of ascocarps, the Phacidiales show how these have possibly developed. *Rhytisma acerinum*, tar spot of sycamore, belongs to this order and the ascocarps are produced within the substratum and the hymenium (the spore bearing layer) forms within the fungal stroma. The upper protective layer then ruptures to expose the asci. The Phacidiales together with the Helotiales and the Lecanorales form the inoperculate discomycetes; this refers to the method of ascospore discharge which is through an apical pore. The Helotiales contain many important plant parasites such as *Sclerotinia fructigena*, brown rot of apple, *Trichoscyphella willkommii*, larch canker, and *Pseudopeziza trifolii*, leaf blotch of clover. The other major division is the operculate discomycetes represented by the order Pezizales. In this order the asci have an apical lid or operculum through which the ascospores are discharged. The Pezizales in general represent the larger more showy discomycetes that are terrestrial or coprophilous and reach their highest form of development in the morels, e.g. *Morchella esculenta*. A final highly modified group of discomycetes are the subterranean truffles in which the hymenium is completely enclosed and there is no violent discharge of spores. The young fruit body develops as a disc-like mass which invaginates to become completely enclosed.

Bitunimycetidae: This subclass also includes orders that were originally placed in the Pyrenomycetes and the Discomycetes. Here the ascus wall functionally consists of two layers; also the ascocarps are stromatic and although one or more ascogonia develop within the stromatic tissue they do not initiate its development. The ascogenous hyphae ramify through this preformed stromatic tissue in a circumscribed region and the developing asci make way for themselves by dissolving away existing tissue. In *Mycosphaerella* and *Leptosphaerulina* species the asci develop in bundles and dissolve a common cavity (Figure 1C). In the stromatic fruit bodies of *Pleospora* or *Venturia* filamentous hyphae (pseudoparaphyses) grow downwards from the apex of the stroma and destroy the thin-walled pseudoparenchymatous cells in the centre (Figure 1D). Asci which arise from the base grow upwards between these hanging filaments which serve to protect them.

In the Dothiorales the ascocarps open by longitudinal slits or by disintegration of the tissue at the apex of the lysigenously formed cavities and the ascospores are forcibly ejected. In *Pleospora* and the Pleosporales the ostiole is formed lysigenously.

4. *Basidiomycotina*

In the Basidiomycotina (Basidiomycetes) the basic organ of all members of this subdivision is the basidium which, like the ascus, is the organ in which meiosis occurs. The resulting spores are, in contrast to the ascospores, formed exogenously at the tips of special outgrowths from the basidium called sterigmata. Usually four spores are formed, one at the tip of each sterigma, but if meiosis is followed by a mitotic division, then eight basidial nuclei are formed although they rarely all develop into basidiospores. No crosiers are formed during basidial formation but clamp connections are often precursors of basidial formation (Figure 2). Many species have the basidia produced from the apical cells of dikaryotic hyphae without any clamp connections and these may produce a series of lateral buds each of which develops into a basidium. The swollen, often thick-walled, probasidium found in the Auriculariales is an outgrowth and modification of the terminal cell of the dikaryotic hyphae. In the rusts and smuts the teliospore functions as a probasidium (Figure 6).

The Basidiomycetes are divided into the following three classes:

1. Hemibasidiomycetes
2. Hymenomycetes
3. Gasteromycetes

The Hemibasidiomycetes include the rust fungi (Uredinales) and the smut fungi (Ustilaginales); both orders contain serious economic parasites. The basidia are not borne on well-developed sporophores but arise from a thick-walled resting spore. The smuts occur as parasites of Angiosperms only, whereas the rusts also parasitize ferns and may require a second, usually unrelated, host to complete their life-cycle. Further details of the life-cycles are discussed under sexual reproduction (p. 27).

The Hymenomycetes form the largest group of Basidiomycetes and they include the toadstools, bracket fungi, and jelly fungi. Basidia are usually arranged in a palisade-like hymenial layer either on the surface of a gill as in the agarics, lining a pore as in the polypores or boleti, or formed over the outer surface of resupinate species. When mature the basidiospores are violently ejected from the top of the sterigma. Various types of cystidia are formed as sterile elements in the hymenial layer. These are merely modified terminal cells of hyphae and may be conical, pyriform or cylindrical. They appear to space the gills apart in *Coprinus* species. In resupinate species cystidia may be encrusted with crystals and are more variable in structure, some being spiny and others branched.

The Gasteromycetes; this is not a natural class but an assemblage of species in which the basidiospores are not violently discharged and are generally shed into a cavity within the fruit body. Common examples are the puff-balls, earth stars, bird's nest fungi, and stinkhorns. In the puff-ball, *Lycoperdon pyriforme*, the fruit body is pyriform with an outer two-layered peridium surrounding the sponge-like gleba. This tissue has numerous small cavities which in the upper part of the fruit body are lined with a hymenial layer that produces the basidia. The hymenial chambers are made up of thick- and thin-walled hyphae. The thin-walled hyphae break down as the fruit body ripens but the thick-walled hyphae persist and form capillitium threads between which the basidiospores are contained. The outer layer of the peridium gradually cracks to form numerous warts or scales; as these are shed the gleba tissue is breaking down and the spores are left as a dusty mass inside the fruit body covered by the inner parchment-like layer of the peridium. As raindrops hit this parchment-like layer the spores are gradually puffed out through the apical pore.

In earth stars such as *Geastrum striatum* the exoperidium is more complex than in *Lycoperdon* and as the fruit body ripens it splits open in a stellate fashion from the top, the triangular flaps so formed curve outwards and make contact with the soil. In doing so they lift the inner globose glebal mass surrounded by the endoperidium into the air and spores are puffed out through the apical pore as in *Lycoperdon*. In *Cyanthus olla*, a bird's nest fungus, the peridium opens to form a cup-like structure and the gleba is differentiated into several ellipsoid peridioles which appear as eggs in the cup-shaped nest.

Sphaerobolus stellatus, which occurs on dung or on partially buried plant debris, will readily produce fruit bodies in culture. When mature the peridium opens to form a star-shaped cup and the peridial layers separate to form an inner and outer cup attached together only at the tips of the star. The inner cup contains a single brown peridiole about 1 mm in diameter containing the basidiospores. By a sudden eversion of the inner cup acting like a sling the mature peridiole is hurled into the air up to a distance of 2 m vertically or 4 m horizontally.

The stinkhorns, which also belong to the Gasteromycetes, have a well developed stipe which forms within the peridium. In *Phallus impudicus* the peridium has a well developed gelatinized middle layer called the mesoperidium. The central part of the fruit body is occupied by the cylindrical stipe with a honeycomb-like receptacle bearing the fertile part of the gleba. Basidia and basidiospores form within the honeycomb cavities and when the spores are mature the upper outer layer of the peridium splits and the stipe, using the gelatinous food reserve of the mesoperidium, elongates to become 100 to 150 mm high. The glebal honeycombed mass forms a cap on the top of the stipe and the whole outer surface is covered with a mass of basidiospores contained in a sweet, sickly smelling mass of mucus; this attracts flies, even those which normally feed on carrion, and the mucus and spore are usually removed in a few hours. The spores pass through the gut of the flies apparently unharmed.

II. Form of Somatic Structures

The assimilative or somatic part of a fungus is a thallus which in the Myxomycota consists of a plasmodium surrounded only by a hyaloplasm. All other fungi, the Eumycota, have the thallus surrounded by a true outer wall. Here the thallus may consist of a single cell as in the yeasts or more typically of hyphal filaments which intertwine or ramify to form a mycelium. Some Eumycota are dimorphic, that is they exist in two forms. Some fungal pathogens of animals and humans are unicellular and yeast-like in their host, but when growing on substrates away from the host they have a mycelial form. The opposite form occurs in some plant pathogens such as *Taphrina* (peach leaf curl) or in smuts, when the mycelial form occurs in the host and the yeast-like unicellular form occurs in culture.

In an unrestricted environment, fungal growth is centrifugal by dichotomous or lateral branching and therefore the younger more active parts are towards the periphery. The term 'fungus colony' may refer to a yeast-like mass of cells living together or to the mycelium formed by ramified hyphae which may have developed from a single spore.

New hyphae generally arise from a spore which on germination puts out a germ tube or tubes and these hyphal primordia elongate and branch to form hyphae. These microscopic fungal filaments are composed of an outer tube-like wall surrounding a cavity, the lumen, which is lined or filled with protoplasm; between the protoplasm and the cell wall is the plasmalemma, a membrane formed of two layers which surrounds the protoplasm. Growth of a hypha is distal, near the tip, and in this region the wall is thin and plastic and as elongation occurs new wall material is added. The major region of elongation takes place in a zone just behind the tip. Liquid nutrients are absorbed through the walls and suitable solids are previously digested externally by secreted enzymes. The wall consists of a single layer in a young hypha but one or more additional layers become detectable as the hypha matures. Electron microscopy shows these layers to consist of microfibrils composed for the most part of hemicelluloses or chitin, true cellulose predominating only in the lower fungi, the Phycomycetes. The young hyphae usually become divided into cells by cross walls which are formed by centripetal invagination from the existing wall material; these constrict the plasmalemma and grow inwards to form generally an incomplete septum that has a central pore large enough to allow protoplasmic streaming. In the Phycomycetes the cross walls may be absent or poorly developed with a large central pore. Even in some Ascomycetes the septal pore (Figure 2) is large enough to allow nuclear migration but in the Basidiomycetes although initially simple it often becomes much more complex. In all groups of the Basidiomycetes with the exception of the Hemibasidiomycetes an elaborate dolipore septum has been observed (Figure 2). This structure consists of a septum pierced by a narrow central pore in which the surrounding wall thickens to become dumb-bell-shaped in section, over-arched on either side with a perforated cap, the parthenosome, probably derived from the endoplasmic reticulum. The plasmalemma is continuous from cell to cell and there is good cytoplasmic continuity although nuclear migration is virtually prevented. If it does take place then the complex dolipore structure is destroyed.

As hyphae age they may become coloured and several additional wall layers may be formed. A further feature of some Basidiomycete hyphae but not all, is the presence of clamp connections. Initially Basidiomycete mycelium consists of uninucleate segments divided by cross-walls; each nucleus is derived from the original single parent nucleus in the basidiospore from which it developed and is said to be homokaryotic. Before fruiting can take place in most Basidiomycetes fusion must take place between two homokaryons of opposite mating types. Hence when two compatible homokaryotic hyphae meet in the soil or in culture, a fusion of the hyphal walls occurs and cytoplasmic continuity is achieved. This is followed by nuclear migration and plasmogamy, but no fusion of nuclei, karyogamy, occurs at this point and the mycelium develops into binucleate segments, each segment containing one nucleus from each of the compatible parent strains. Each nucleus of this dikaryotic cell undergoes conjugate division, i.e. divides at the same time. In many species of Basidiomycetes this dikaryotic mycelium has a characteristic lateral bulge at each septum termed a clamp connection. This is a device to ensure that when dikaryotic mycelium segments, each segment contains two genetically distinct nuclei (see Figure 2). Without this or a similar device dikaryotic mycelium segmenting and laying down cross walls near the tip would tend to revert to a homokaryotic state.

A. Specialized Hyphal Structures

Some fungi belonging for the most part to the Deuteromycotina, but also some Phycomycetes, develop hyphae with constricting rings or sticky knobs, which trap nematodes in the soil either by constriction or adhesion when they pass through or come in contact with such structures.

Appressoria are examples of other specialized hyphal structures which also occur. Fungi such as *Erysiphe* or *Colletotrichum* form a lobed swelling, an appressorium, on a germ tube or hypha where it comes in contact with the surface of the host (Figure 2). From the centre of this swelling on the side in contact with the host a minute infection peg develops through which the nucleus and cytoplasm enter the host. These are referred to as epiphytic appressoria. Endotrophic appressoria are formed by intracellular hyphae and these produce infection pegs which allow the hyphae to pass from cell to cell; such endotrophic appressoria are found in *Phyllachora, Pseudopeziza* and in the Ustilaginales.

In *Erysiphe graminis* (Figure 2) a small epiphytic appressorium forms and the minute hyphal threads penetrate the underlying host cell wall. Within the host cell a large digitate haustorium is formed between the cell wall and the plasmalemma, the latter being pressed inwards.

Hyphopodia may be regarded as modified forms of appressoria found in organs produced by epiphytic plant parasites such as *Meliola* and related genera, and are principal organs of attachment to and nutrition from the leaf surface. The thallus of these fungi consists of a superficial branching mycelium with characteristic lobate or rounded terminal cells borne at the apex of a one- or two-celled lateral branch (Figure 2). A second type, the mucronate hyphopodium, is also formed; these consist of a single flask-shaped cell with the tip directed away from the leaf surface. These have been regarded as performing a spermatial role but no evidence has been presented that spermatial cells are produced.

Stolons are the aerial runner hyphae which loop over the surface of a substrate infected by *Rhizopus* species. They are filamentous and anchored at intervals to the substrate by root-like branches called rhizoids; at this point, referred to as the node, sporophores usually develop.

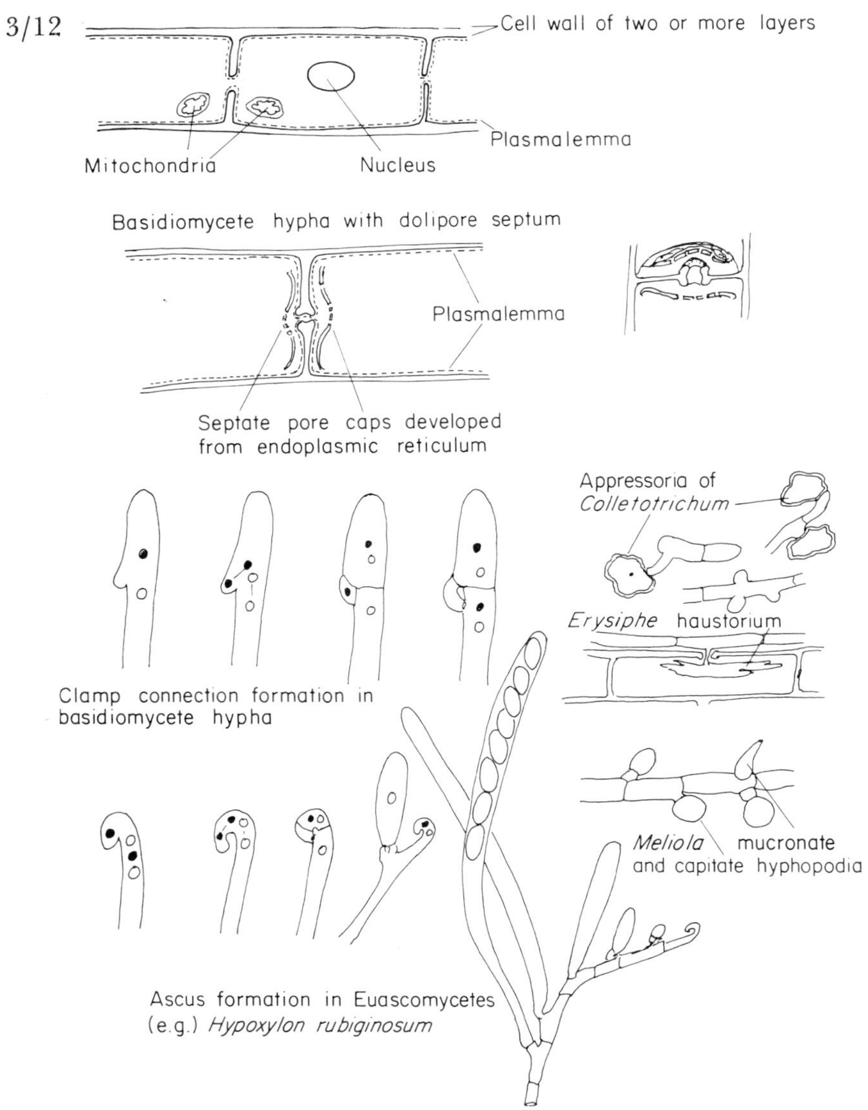

Cell wall of two or more layers

Mitochondria   Nucleus   Plasmalemma

Basidiomycete hypha with dolipore septum

Plasmalemma

Septate pore caps developed
from endoplasmic reticulum

Appressoria of
Colletotrichum

Erysiphe haustorium

Clamp connection formation in
basidiomycete hypha

Meliola mucronate
and capitate hyphopodia

Ascus formation in Euascomycetes
(e.g.) *Hypoxylon rubiginosum*

Figure 2. Ascomycete hypha with simple pores in cross septa

B. Asexual Reproduction in the Phycomycetes (Mastigomycotina and Zygomycotina)

The basic asexual organ in the Phycomycetes is the sporangium, and the cleavage of its contents into motile or non-motile asexual spores is a feature of the group. In the Oomycetes a move has occured from an aquatic habitat to a terrestrial one. With this change of habitat various modifications have taken place which facilitate air dispersal of the asexual spores. This major step has been achieved in the more advanced members of the group by the sporangia becoming deciduous and then germinating directly by a germ tube rather than producing zoospores. Although these sporangia behave like the conidia of the Ascomycetes or Basidiomycetes, they are strictly homologous to the sporangia which produce zoospores and should not be referred to as conidia. Neither should the sporangiophore be called a conidiophore. Direct germination of the sporangium may allow growth in the absence of water but it also reduces the number of potential propogules. In terrestrial species of the Oomycetes the reduction in the number of propogules has been compensated for by increasing the number of sporangia produced on each sporangiophore either by branching of the sporangiophore as in *Peronospora*, or by successive proliferation from a specialized sporogenous cell as in *Albugo*.

The zoospore is the asexual spore of the lower Phycomycetes although after release in the more primitive species they may fuse in pairs and thus act as gametes. The zoospore is of characteristic form in the orders of the Mastigomycotina, e.g. Chytridiomycetes, Hyphochytridiomycetes, and Oomycetes. The most primitive of the Chytridiomycetes have a single posterior whiplash-type flagellum. The Hyphochytridiomycetes have a single anterior tinsel-type flagellum, and the Oomycetes have anterior or sublateral biflagellate zoospores, one flagellum of the whiplash type and the other of the tinsel type (Figure 3).

In the more primitive Phycomycetes no clear distinction is present between the thallus and the sporangium. In *Synchytrium endobioticum*, the cause of wart disease of potatoes, or in *Olpidium brassicae*, club root of brassicas, the mature thallus is spherical or cylindrical and is referred to as holocarpic because the entire contents become transformed into zoospores or gametes (Figure 3).

In eucarpic species the thallus is differentiated into a vegetative part and a reproductive part. If the whole thallus is within the host cell it is referred to as endobiotic, as in *Diplophlyctis,* whereas if rhizoids only penetrate the host cell and the sporangia or gametangia are outside, it is epibiotic as in *Catenochytridium* (Figure 3).

There is an obvious transition between species in which zoospores give rise to a rhizoidal system which produces a single sporangium, to those with a well developed thallus producing numerous sporangia. *Saprolegnia* species are common in fresh water and some such as *S. parasitica* are parasites of fish. In this species sporangia develop by modification of a pointed hyphal tip (Figure 3). Gradually as the tip swells it becomes rounded and the sporangium as a whole becomes obovoid in shape. The cytoplasm becomes denser around the still-visible central vacuole and a septum is laid down across the base of the swollen part. The young sporangium contains many nuclei and cleavage furrows separate the cytoplasm into uninucleate pieces which differentiate into the apically biflagellate zoospores. When the sporangium is mature a pore forms at the tip and the zoospores escape backwards one at a time through this pore. After discharge of the zoospores growth is frequently renewed from the base of the effete sporangium and a further sporangium develops within the old sporangial wall, a process which may be repeated several times. Released zoospores in *Saprolegnia* species usually encyst after a period of motility which in most species is up to 1 hour. Germination may then occur or a further bean-shaped zoospore may be formed.

In *Pythium*, as in *Saprolegnia*, sporangiophores differ little from the assimilative hyphae and the sporangia may either be filamentous, as in *P. gracile* where any part of the hyphae may have the cytoplasm differentiated into zoospores, or well defined and globose as in *P. debaryeanum*. In *Pythium* zoospores are not differentiated within the sporangium but form in a thin walled vesicle which develops at the tip of a fine tube formed from the wall of the sporangium; the undifferentiated protoplasm passes into this vesicle before forming zoospores. After release the spores swim for a time and then encyst before germinating by means of a germ tube. However, under dry conditions sporangia may germinate direct without formation of zoospores.

In *Phytophthora infestans*, the cause of the destructive late blight of potatoes, and in other terrestrial species of the genus, zoospore formation is often suppressed. The usually pear-shaped sporangia arise on well differentiated simple or branched sporangiophores and the whole sporangium is detached and dispersed by wind before the differentiation of its contents (Figure 3). Subsequent germination, under moist conditions, is either by means of zoospores produced in the sporangium or, under dryer conditions, by sporangia germinating direct by means of a germ tube. In aquatic species of *Phytophthora* the sporangia usually form zoospores within the sporangium whilst still attached to the sporangiophore.

In the terrestrial *Peronosporaceae*, collectively known as the downy mildews, sporangial germination is usually by germ tube but zoospore formation has not been entirely suppressed. In *Peronospora parasitica*, a parasite frequently found on shepherd's purse (*Capsella bursa-pastoris*), the sporangiophores are well differentiated and emerge singly or in groups from the stomata of the host. They are extensively branched towards the apex and the incurved tip of each branch bears an oval sporangium. These are dispersed by wind and germination is by formation of a germ tube.

The Family *Albuginaceae* has only one genus, *Albugo*, the cause of white rust or white blister rust of flowering plants; here the globose sporangia develop in chains from club-shaped sporangiophores which form a palisade below the epidermis of the host. This sorus appears as a

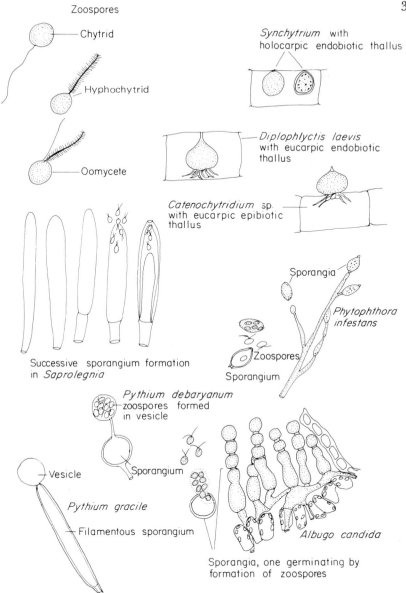

Figure 3. Asexual reproduction in Mastigomycotina (Phycomycetes)

white blister as the sporangia develop and before the epidermis is finally ruptured. When rupture occurs the sporangia are dispersed by wind and if they alight on a susceptible host under suitably moist conditions, they will produce usually about eight biflagellate zoospores (Figure 3). On release from the sporangium these swim in the surface moisture of the leaf and then encyst before forming a germ tube which penetrates the epidermis of the host.

In the Mucorales (Zygomycotina) zoospores are absent and asexual reproduction is by non-motile aplanospores carried by wind, by insects, or dispersed by rain splash.

The characteristic sporangium as found in *Mucor* or *Rhizopus* (Figure 4), whose species are common on organic matter or in soil, has a long sporangiophore bearing an apical globose sporangium with a columella. During development the tip of the sporangiophore expands to form the globose sporangial initial. This contains numerous nuclei which continue to divide, as a dome shaped septum is laid down cutting off the swollen distal portion in which the spores form and it is this curved septum which forms the columella. The spores are formed by cleavage of the cytoplasm which separates the individual nuclei. The number of spores formed is variable and often related to the amount of nutriment present. Rupture of the sporangial wall is either mechanical by touch or, as in *Mucor hiemalis*, it deliquesces leaving the spores in a drop of mucilage at the tip of the sporangiophore adhering to the columella.

There is also a trend in the Mucorales to reduce the number of spores in a sporangium to one and for the sporangia to behave like conidia as found in the Ascomycetes. If the wall of the aplanospore fuses with the wall of the sporangium then it is difficult to distinguish such a spore from a conidium having a two layered wall.

Reduced globose sporangia without a columella are known as sporangioles and in *Thamnidium elegans* both normal apical columellate sporangia occur together with sporangioles often formed from the same sporangiophore (Figure 4). The sporangiophore produces lateral branches from a given region and the sporangiola form on the ultimate branchlets of these dichotomously branched lateral conidiophores. Each sporangiola contains only a few aplanospores. In some *Choanephora* species the sporangiola have one spore only.

In *Cunninghamella monosporeae* sporangiola only are produced. They form on lateral branches with swollen ends, the multispored columellate sporangium is completely suppressed; these sporangioles behave and act like conidia (Figure 4).

A further developmental trend in the Mucorales is the formation of merosporangia. In genera such as *Syncephalastrum* and *Piptocephalis* the swollen apices of the sporangiophore branches bear dense heads of linear merosporangia on small denticles. In *Syncephalastrum racemosum* the cytoplasm in the cylindrical merosporangium cleaves to form a single row of five to ten sporangiospores (merospores) (Figure 4). At maturity the merosporangial wall splits and contracts leaving the spores which appear to be in chains, reminiscent of *Aspergillus*.

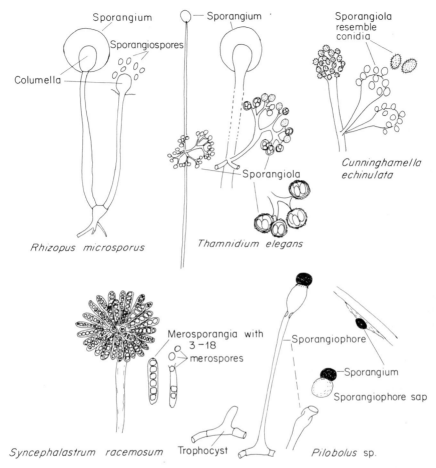

Figure 4. Asexual reproduction in Zygomycotina (Phycomycetes)

In *Pilobolus* species which occur on the dung of herbivorous animals, the sporangia are violently discharged into the air and sporangial production follows a daily sequence. During the afternoon the sporangiophore grows away from the trophocyst, which is a swollen yellow coloured segment in the hypha, and during the early part of the night the apical sporangium develops. After midnight the subsporangial vesicle forms and in this modification of the sporangiophore pressure builds up in the internal liquid which produces the violent discharge of the sporangium usually in late morning but depending somewhat upon the prevailing temperature.

The sporangiophores are extremely phototrophic and direct the sporangium towards the light (Figure 4). On discharge the sporangia which alight on the leaves of adjacent vegetation stick in position until the leaves are eaten by animals and the sporangial contents are then released into the gut.

In the Entomophthorales most species are parasites of insects or animals, or occur on animal excreta. In this order violent discharge of the sporangium, which is often modified to act as a single conidium-like propagule, is a common feature. (Although homologous to a mono-spored sporangium, it is, for practical purposes, a conidium and is often referred to as such.)

C. Asexual Spore Formation in the Ascomycetes and Basidiomycetes

As the asexual spores of all other fungi are not enclosed in a sporangium they may be referred to as conidia. It should be realized that this term covers a wide range of different methods of spore production. In the Ascomycetes and the Basidiomycetes conidia are usually produced from specialized cells called sporogenous cells and are formed by the modification or elaboration of one or both of the two wall layers of these cells. Two basic methods of conidial ontogeny have been defined: (1) *Thallic*, in which the spore (conidium) differentiates from the whole cell. If there is any enlargement of the conidial initial it takes place after the initial has been delimited by a true septum. (2) *Blastic*, in which the conidium differentiates from part of a cell wall and undergoes marked enlargement before the initial is delimited by a septum.

These sporogenous cells occasionally develop directly from the hyphae but more frequently they are produced by specialized hyphal structures called conidiophores. In the *Fungi Imperfecti* these are extremely variable and originally formed the basis of the classification within the group. If the conidiophores occur free either amongst aerial mycelium or arise from superficial or immersed hyphae, they are placed in the Hyphomycetales. If either the conidiophores or the conidia are dark they are placed in the *Dematiaceae* and if hyaline in the *Moniliaceae*. Two further orders are also recognized for species with aerial conidiophores. If they are aggregated into a fascicle to form a synnema they are placed in the Stilbellales or if aggregated into a cushion-like mass, a sporodochium, they are placed in the *Tuberculariaceae*. If the mass of conidiophores is effuse without the pulvinate cushion-like stroma they are referred to as pionnotal sporodochia, a form many of the *Tuberculariaceae* assume when grown in culture.

Two other divisions of the *Fungi Imperfecti* based on the gross morphology of the asexual fructification are the Melanconiales and Sphaeropsidales. In the Melanconiales the aggregation of hyphae or stroma producing the conidiophores is initially subcuticular or subepidermal but with the conidiophores or spore mass becoming exposed at maturity. The other order is the Sphaeropsidales in which the conidiophores are borne in a pycnidium, a flask-shaped structure with an apical pore. Within these orders further subdivisions are made depending on the nature of the associated stromatic tissue and whether this is superficial or immersed. Genera and species are distinguished on spore form, colour, and size, but with the recognition of the importance of the sporogenous cell some genera have had to be split or revised.

III. Sexual Reproduction Structures in Fungi

Sexual reproduction is the process involving nuclear fusion followed by meiosis, a nuclear division in which the chromosomes are reduced from the diploid to the haploid number. The process begins with the anastomosis of two cells and fusion of their protoplasts (plasmogamy) thus bringing together two haploid nuclei of opposite sexual potential (mating types) into one cell.

The fusion of these compatible haploid nuclei to form a diploid nucleus is karyogamy. Such fusion may occur immediately after plasmogamy in the lower more primitive fungi but in the higher fungi it is generally delayed. The anastomosed cell which contains the two nuclei (a dikaryon or dikaryotic cell) may grow and divide, the conjugate nuclei undergoing simultaneous division.

As mentioned, this process and the resulting fruit body form the basis for the division of the true fungi into four major groups: Mastigomycotina and Zygomycotina (Phycomycetes), Ascomycotina (Ascomycetes) and Basidiomycotina (Basidiomycetes).

A. Mastigomycotina

Sexual reproduction in the Mastigomycotina ranges from fusion of undifferentiated zoospores in the primitive Chytridiales to the fertilization of an oogonium by an antheridium in the Oomycetes. Confirmation of the detailed life cycles is still needed for some species and there is evidence in the Oomycetes that some species have a diploid, vegetative thallus.

In the Chytridiomycetes, if sexual reproduction is by fusion of morphologically indistinguishable gametes it is isogamous. If it occurs by fusion of a smaller motile gamete, designated male, and a larger, more sluggish but still motile, female gamete, it is anisogamous.

Even in the most simple holocarpic Chytrids such as *Olpidium brassicae* or *Synchytrium endobioticum*, where the whole contents of the globose or cylindrical thallus may be converted into zoospores, there is strong evidence of sexual reproduction. The uniflagellate zoospores are the primary means of infection and they penetrate the root hairs or epidermal cells to produce a further thallus and ultimately a further sporangium. However, there is evidence that the zoospores exist in two distinct strains and when two opposite mating types meet, they fuse to form a biflagellate zoospore which on contact with the host will penetrate the epidermal cells and form a binucleate, thick-walled, resting sporangium. Before the contents undergo differentiation into zoospores it is believed that the two nuclei fuse and undergo meiosis before the further mitotic divisions which produce the many uninucleate zoospores.

Many Chytrids have a eucarpic thallus which means it is differentiated into vegetative and reproductive parts. In the Blastocladiales as exemplified by *Allomyces arbuscula* (Figure 5) there are both haploid gametophytic and diploid sporophytic generations. Diploid sporophytes produce asexually thin-walled zoosporangia whose zoospores give rise to a similar diploid generation and thick-walled, brown resting sporangia. On germination the nuclei of the resting sporangium undergo a meiotic division to form 48 haploid zoospores, sometimes called meiospores to distinguish them from the mitospores formed from the thin-walled zoosporangia. The meiospores have a single posterior flagellum and swim for some time before settling down to form a branched haploid eucarpic thallus. After a period of growth the tips of the branches swell to form paired swellings, an apical female and a subapical male gametangia. The contents of the male differentiate into small orange gametes and the female produces larger but fewer colourless female gametes. After swimming for some time the male and female gametes fuse in pairs and the resulting zygote with two flagella may be seen before it finally comes to rest and encysts. Nuclear fusion occurs and a diploid asexual plant begins to develop often immediately after fusion.

In *Monoblepharis polymorpha* only the male gametes are motile and are often referred to as spermatozooids. The antheridia are formed at the tip and are cut off by a cross-wall. Below this the hypha swells asymmetrically to form an oogonium. Although both types of sexual organs are produced on one thallus, the terminal antheridium has often discharged all its spermatozooids before the subterminal oosphere is mature. The oogonium consists of a single spherical uninucleate oosphere and when this is mature an apical receptive papilla on the oogonium becomes gelatinous, an approaching spermatozoid is caught up in this mucilage and plasmogamy occurs. The oospore then secretes a golden-brown membrane around its outer wall and nuclear fusion occurs later.

Apart from the distinctive biflagellate, vegetatively produced zoospores, the Oomyces are also distinguished by their oogamous reproduction, whereas in *Saprolegnia*, a large, usually spherical, oogonium contains one to many eggs which are fertilized by antheridial branches which apply themselves to the oogonial wall (Figure 5).

Fertilization is by the penetration of the oogonial wall by a fertilization tube through which a single male nucleus passes into each egg. The majority of species are homothallic (monoecious) in that cultures arising from a single zoospore give rise to both oogonia and antheridia.

In the Peronosporales, each oogonium contains in most cases only a single egg (*Pythium multisporum* has several), and it is believed that meiosis occurs during the development of the gametangia, a feature that suggests the vegetative thallus is diploid.

Germination of the oospore throughout the Oomycetes is by germ tube which produces further hyphae or which gives rise to a sporangium and zoospores.

B. Zygomycotina

The major class in the Zygomycotina is the Zygomycetes, represented by the Entomophthorales and the Mucorales. As the name suggests sexual reproduction results in the formation of a zygospore.

In the *Mucoraceae* some species are homothallic, zygospores being formed in cultures derived from a single sporangiospore as in *Rhizopus sexualis*, or heterothallic, where zygospores are produced only when two compatible mating strains meet as in *Mucor mucedo* (Figure 5).

When two compatible strains approach each other over a culture substrate, aerial club-shaped zygophores develop and grow towards each other; these cannot be differentiated as male and female and are therefore referred to as '+' and '−' strains. Only opposite strains attract each other. When compatible zygophores make contact, the tip of each becomes cut off by a septum to form a progametangium and a suspensor.

The walls between the two progametangia break down so that the numerous nuclei from each cell become mingled and surrounded by a common cytoplasm. This fusion cell, or zygote, swells and becomes globose or it may appear somewhat compressed between the two suspensors and generally develops a thick, warted, outer wall. On germination usually an unbranched sporangiophore develops and produces an apical sporangium.

C. Ascomycetes

Ascomycetes may be homo- or heterothallic. Homothallic strains of *Neurospora terricola* or *Gibberella zeae* will produce ascocarps without the intervention of any other strain of the same species and no trichogyne or other structure relating to the male element is produced. Heterothallic strains require the presence of a compatible mating strain before ascocarps are produced. Many ascomycetes exhibit physiological heterothallism because although they produce both male and female

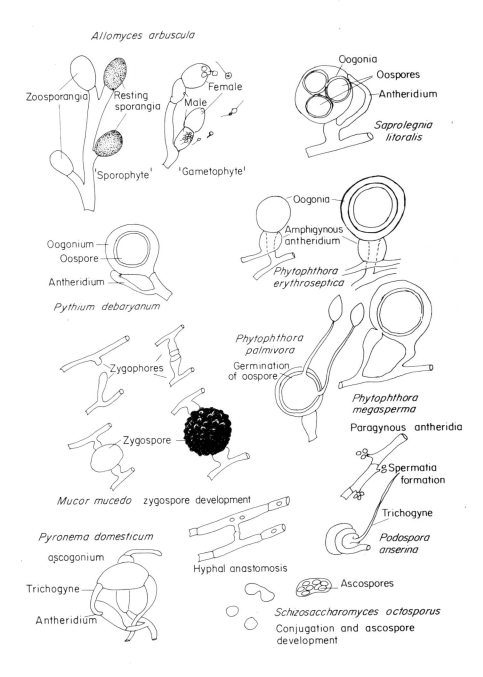

Figure 5. Sexual reproduction in the Phycomycetes and Ascomycetes

sex organs, these are self-incompatible and will not produce ascocarps without contact with a compatible mating strain. True heterothallic strains are either male or female; the latter will often develop ascocarp primordia which do not complete their development without proto-plasmic and nuclear fusion with the male strain.

The simplest form of sexual reproduction occurs in yeasts such as *Schizosaccharomyces* (Hemiascomycetes) and related species. Here the sexual phase or ascus formation follows fusion of two cells which may have developed from the same parent. As the two cells come together a pore is formed at the point of contact and this widens to form a common cell, the ascus, from the two mating cells (Figure 5). The cell contents intermingle (plasmogamy), and the two nuclei fuse immediately (karyogamy), with no intervening dikaryophase. Nuclear fusion is followed by meiosis to produce four ascospores. In species with eight ascospores meiosis is followed by a further mitotic division.

As there is a tendency throughout the class for a reduction from a filamentous thallus as in *Dipodascus* to one consisting of separate cells as found in *Saccharomyces*, there is a corresponding reduction in the morphological characteristics upon which a classification can be based. In these yeasts or yeast-like forms, therefore, the classification is based largely on the physiological characteristics, especially the fermentative capabilities.

1. Euascomycetes

The Euascomycetes are characterized by sexual nuclear pairing without immediate fusion of the two nuclei. This process of copulation and pairing is plasmogamy and the resulting binucleate cell is the first cell of the dikaryon stage. The paired nuclei tend to remain close together and divide simultaneously to produce the dikaryotic cells of the ascogenous hyphae. This dikaryotic stage is usually dependent upon the monokaryotic mycelium for its nutritional requirements.

Many highly developed ascomycetes have no recognizable sex organs and, as in *Taphrina deformans* or *Sclerotinia sclerotiorum*, fusion takes place between undifferentiated hyphae. The process of hyphal fusion is a common phenomenon in fungal hyphae but it has sexual significance only when it brings together compatible nuclei into one cell to form the dikaryotic cell. Other ascomycetes produce more definite sexual organs. Species in genera such as *Cochliobolus, Gelasinospora, Mycosphaerella, Pleurage,* and *Podospora* produce minute uninucleate spermatia which fertilize a receptive trichogyne produced by the ascogonium. Spermatia are incapable of germination but fertilization of the trichogyne by microconidia has been observed and these have the additional capacity of germinating freely and producing new homokaryotic fungal colonies (Figure 5).

Other species may produce one or more multinucleate antheridia which grow towards and make contact with the female ascogonium. These may be homothallic as in *Pyronema domesticum* where both organs occur on the same thallus, or heterothallic as in *Venturia inaequalis*, the apple scab fungus, where the male and female gametangia occur on different thalli. A pore develops at the point of contact and plasmogamy follows. Thus in the next stage of development after plasmogamy the ascogonium usually produces numerous dikaryotic ascogenous hyphae which ultimately bear the asci and which are recognizable by their curved crozier-like tips. The dikaryotic tissue, or ascogenous hyphae, together with the ascogonium, if definable, forms the centrum of the ascocarp and fusion of the nuclei (karyogamy) ultimately takes place in the penultimate cell of the ascogenous hyphae. It is from this cell that the ascus develops, as described below. Asci usually form a palisade known as a hymenium which assumes various forms depending upon the species. It is usually protected by monokaryotic (somatic) hyphal structures although in *Taphrina deformans* an extensive hymenium of naked asci is formed.

At the onset of ascus formation the pair of nuclei near the tip undergo conjugate division and two septa are formed which cut off the binucleate penultimate cell to form the ascus initial (Figure 2). The two nuclei in this cell fuse to form the diploid phase and this nucleus will later undergo meiotic division to form the ascospores. The bent over uninucleate terminal cell fuses with the ascogenous hypha in a similar manner to that in clamp formation in the Basidiomycetes. As the ascus enlarges, successive new crozier formation usually takes place from the supporting ascogenous hyphae and each crozier gives rise to an ascus. Subsequent development of the ascocarp depends upon the group to which the species belongs. In most classifications ascomycetes with protective tissue around the asci were referred to as Euascomycetes (Euascomycotina) and were divided into three classes: the Plectomycetes, the Pyrenomycetes, and Discomycetes.

Much evidence has now accumulated which suggests the superficial structure of the ascocarp is as much a response to the particular habitat the species occupies as to any close relationship to species showing a superficially similar structure. In fact homoplasy, the production of similar morphological forms along several unrelated phyletic lines, is a common feature of the ascomycetes.

The three basic divisions of the Euascomycetidae are now based upon the nature of the ascus and the structure of the centrum. These divisions are given the rank of subclass and are: the Protomycetidae; the Unitunimycetidae, and the Bitunimycetidae.

a. The Protomycetidae. The primitive protunicate ascus is the major feature of the modern concept of the Plectomycetes. This type of ascus is formed without the intervention of croziers and is produced singly or in chains from branched ascogenous hyphae which ramify through the centre of the ascocarp. The wall is thin, undifferentiated and often breaks down extremely early, releasing the ascospores into the cavity at the centre of the ascocarp; no spore discharge mechanism is present. Families with this type of ascus structure include the *Gymnoascaceae* in which there is no true peridial wall. The ascogenous hyphae and asci rise from an ascogonium which coils round the antheridium. They form a cluster which may be surrounded by an enmeshed network of hyphae which may or may not be differentiated. In *Arachniotus ruber* and *Amauroascus verrucosus* the peridium is sparse and formed of undifferentiated hyphae, whereas *Gymnoascus reesii*, which forms yellowish-brown ascocarps on jute or hemp ropes or fabric, has well developed peridial hyphae. These develop from the vegetative hyphae in the region of the gametangia and are distinguished by their short, bent, blunt spines. The morphology of these peridial spines is characteristic for the species.

In the family *Eurotiaceae* the ascocarps vary from the poorly defined ascocarps of *Byssochlamys fulva* to the thick-walled sclerotioid fructifications of *Eupenicillium* and *Dichlaena*. In the *Cephalothecaceae* the outer wall has a number of meristematic zones which produce plates of cells. On the lines where these plates meet, the ascocarp ultimately ruptures as in *Cephalotheca savoryi* (Figure 3). A more advanced family is the *Ophiostomatiaceae* whose members possess an ostiole often at the apex of a long neck as in *Cerotocystis ulmi*, the cause of Dutch Elm disease. In this species ascospores are released into the cavity of the ascocarp and extruded passively in mucous. They accumulate at the tip of the neck where a fringe of setae tends to hold them in place, ideally positioned for insect dispersal.

b. The Unitunimycetidae. The Unitunimycetidae have asci which arise from croziers and have a functionally differentiated wall and usually a mechanism for active spore discharge. The ascus wall appears as a single layer and an apical pore structure is often visible. The group includes species without ascocarps as in the Taphrinales, *Taphrina deformans;* species with closed ascocarps e.g. *Erysiphe graminis*, powdery mildew of Gramineae; species with true flask-shaped perithecia, the Pyrenomycetes, and species with cup-shaped apothecia, the Discomycetes.

The Pyrenomycetes originally referred to all fungi which had asci produced in a locule with an apical aperture or pore. The term is now restricted to those fungi in which the asci are surrounded by a true perithecial wall derived from the base of the ascogonium or from neighbouring vegetative hyphae. The locule is usually flask-shaped with the apical ostiole lined by periphyses (sterile protective filaments). In the Sphaeriales paraphyses grow upwards from the wall tissues at the base of the locule and form protective filaments between which the asci develop. In the Hypocreales where the perithecia are bright coloured and fleshy, similar protective filaments grow downwards in many species from the roof of the locule and between the asci which are developing from ascogenous hyphae at the base. The perithecia in either order may occur singly or aggregated, seated on a stroma or completely immersed in stromatic tissue. These stromatic or non-stromatic perithecia may be superficial, or immersed in the tissue of the host. Such modifications of the perithecia and their associated structures together with the form of the ascospores and asci are the basis of the families and genera within the orders.

In the Discomycetes a further major subdivision is made based on the form of ascus. Species in which the asci have a small lid, an operculum, are placed in the Operculatae; many of the showy coloured cup fungi such as *Peziza aurantia*, the orange peel fungus, are included here. The inoperculate species of Discomycetes have asci with an apical pore or canal. Many lichenized fungi and some important plant parasites such as *Rhytisma acerinum*, tar spot of sycamore and *Lophodermium pinastri*, the cause of die-back in pines, belong to this subdivision.

   c. *The Bitunimycetidae*.   Asci of the Bitunimycetidae (Loculo-ascomycetes) also arise from croziers but the primary character of this group is the bitunicate ascus, the wall of which consists of two layers which are separable under conditions of normal spore discharge. The outer layer or ectoascus is a thin inextensible layer which at maturity ruptures at the apex, or an apical cap is shed, and this allows the thick extensible inner layer, the endoascus, to form a tube-like extension two to three times the length of the original ascus which extends towards the ostiole. The ascospores move up this tube and the uppermost spore is trapped momentarily, usually at its widest point in the elastic pore at the apex of the tube, pressure is built up in the tube below it and this results in the spore being shot into the air. The second ascospore then moves into its place, again it is momentarily trapped whilst pressure builds up before being shot. After all the spores have been thus discharged the extensible ascus tube collapses into a collar round the top of the ectoascus. In some species, such as *Botryosphaeria*, the endoascus is not completely functional; it tends to swell and dissolve as soon as it emerges from the ectoascus. When this happens ascospores tend to ooze out in a gelatinous sheath.

   The bitunicate ascus is always associated with an ascostroma rather than a perithecium. Karyogamy takes place within the young ascostroma; it does not initiate its formation. The asci may either make a space for themselves as in the Myriangiales, e.g. *Elsinoe fawcettii*, citrus scab, or the asci form in a cluster from the ascogonium as in the Dothideales and either dissolve a common cavity or grow up between the pseudoparenchyma cells. If a unilocular ascostroma, as found in *Mycosphaerella* spp., is squashed on a microscope slide the centrum with its palmate group of asci will often be released as a unit from the ascocarp. Within this class the ascocarps may be separate, grouped on a common basal stroma or completely immersed in the stroma and this stroma may be superficial, erumpent or immersed on or in the host tissue.

In the order Pleosporales the ascocarps more nearly resemble a perithecium and are often referred to as pseudothecia. The lysigenous locule forms before the asci develop and protective filaments, pseudo-paraphyses, grow downwards from the top of the cavity and fuse with the tissues at the base. Asci arise from ascogenous hyphae in the basal region and grow up between the protective pseudoparaphyses. As they mature an ostiole forms lysigenously or schizogenously at the apex of the pseudothecium. This is a very large order with a similar range of ascocarp form to that found in the Dothideales. In *Letendraea* spp. the usually bright coloured ascocarps are formed superficially on a basal stroma. In *Stigmatea* spp. the ascocarps are erumpent and in *Didymella* and *Venturia* species, wholly immersed. In other genera they may be completely immersed in a stroma which develops either superficially on the host tissue or immersed in it. The ascospores range from 1-septate hyaline spores of *Venturia* to large dark coloured dictyospores found in *Clathrospora* and *Cucurbitaria*. Dehiscence is commonly through an apical pore but in the *Lophiostomataceae* the ostiole has the form of a longitudinal slit.

In the Hysteriales the ascocarp has a linear folded appearance and at maturity the fold opens to become apothecioid and exposes the tips of the asci and pseudoparaphysoidal tissue.

D. Basidiomycetes

The basic difference between the Ascomycetes and the Basidiomycetes is in the way the spores are produced from the meiosis following karyogamy. In the Ascomycetes they are produced endo-genously within a sac-like structure, the ascus, and referred to as ascospores. In the Basidiomycetes, although meiosis takes place within a cell called the basidium, the resulting nuclei migrate through stalk-like, often tiny, outgrowths called sterigmata and the resulting cells, the basidiospores, are formed exogenously. These may or may not be forcibly discharged.

Morphologically differentiated sex organs are lacking in the Basidio-mycetes, apart from the rusts, and sexual reproduction is initiated by hyphal anastomosis and by the transfer of a nucleus and cytoplasm from one haploid cell to another; this dikaryotization may occur at any time, even between two germinating basidiospores, and does not immediately precede basidial formation. Whereas in the Ascomycetes the dikaryotic mycelium arising after fertilization is still nutritionally dependent upon the haploid thallus, in the Basidiomycetes the dikaryotic phase is independent of the haplophase, and in many species most vegetative growth takes place in the diplophase.

The presence of a basidium, and there are many morphological types, is the only common feature in the Basidiomycetes. Generally three major divisions are made: the Hemibasidiomycetes in which are placed the rusts and smuts, the Hymenomycetes which include the agarics (toadstools), polypores (bracket fungi), fairy clubs and jelly fungi, and the Gasteromycetes which include the truffles, bird's nest fungi, puff-balls, earth stars, and stinkhorns.

1. Hemibasidiomycetes

This important group contains the plant pathogens referred to as rusts (Uredinales) and smuts (Ustilaginales). These are probably not closely related but have in common the fact that the basidia arise from a thick-walled teliospore. In some literature this is referred to as a teleutospore or, in the smuts, as a chlamydospore or brand spore.

a. Uredinales. The rusts are obligate parasites. Most are confined to one host, and are autoecious. However, there are quite a large number in which the life cycle alternates between two hosts, the heteroecious rusts. Rusts have a pleomorphic life cycle which in its most complex form contains five distinct spore forms. These are the pycnia, aecia, uredinia, telia, and basidia, and the life cycle may include heteroecism. All the variations known today appear to be derived from this five-state cycle by elimination or reduction of its parts. Thus, in microcyclic species of *Coleosporium* the pycnia, aecia, and uredinia are no longer present.

The full five-state cycle is found in *Puccinia graminis*, the black rust of wheat and other graminaceous hosts.

Biological specialization is highly developed in the rusts and has been extensively studied in *Puccinia graminis* where a number of subspecies have been defined.

b. Ustilaginales. In this order the teliospores which are also known as chlamydospores or brand spores are produced in a sorus but are not strictly homologous with those of the rusts. They are formed from dikaryotic mycelium and the young teliospore, which is dikaryotic acts as a probasidium, in which nuclei fuse so that the mature teliospore has a single diploid nucleus. On germination the teliospore gives rise to variable metabasidia. Meiosis takes place in the metabasidial cells which may then branch and bud off small sessile basidiospores. There are over 100 species of smuts which attack angiosperms in over 75 families and many are important plant pathogens, especially of cereals. Many are organ specific as in *Ustilago violaceae* (Figure 6) and many will grow on laboratory media to produce a yeast-like colony. *Ustilago violacea* produces teliospores in the anthers of certain *Caryophyllaceae* and the systemically infected host supplies the nutriment to produce a mass of violet-brown spores.

2. *Hymenomycetes*

Most Hymenomycetes are heterothallic and the dikaryotic mycelium arises only after fusion of two compatible opposite mating strains. A few, possibly less than 10 per cent, are homothallic and, as in *Coprinus sterquilinus*, mycelium from a single basidiospore soon becomes binucleate and can, after a suitable period of growth, produce fruit bodies and basidia. The structure and development of the basidium can be observed from a study of the gill bearing fungi in the Hymenomycetes. Basidia arise as terminal cells of the hyphae forming the gill tissue and almost the whole surface area becomes covered with a palisade of basidia forming the hymenium. The young basidia are binucleate with dense cytoplasm. As the basidium elongates and becomes clavate a vacuole develops in the basal region and the cytoplasm is pushed towards the top. Nuclear fusion occurs and is immediately followed by meiosis. Typically four haploid nuclei are formed and one ultimately passes into each basidiospore. At the exposed tip of the basidium four sterigmata develop, more or less one at each corner (Figure 6); these are minute pegs and one of the haploid nuclei migrates through each sterigmata into the spore which develops from the tip. In some basidia only two basidiospores are formed and two nuclei may pass into each spore. In others a further mitotic division in the basidium will produce eight nuclei but seldom are eight basidiospores produced. Observations have shown that basidiospores in the Hymenomycetes are projected violently into the space between the gills or into the pore cavity in the *Polyporaceae* before they fall into the wind currents below. Thus the basidium in the Hymenomycetes may be divided into the probasidium, the part where karyogamy occurs, and the metabasidium which arises from the probasidium by elongation and is the part where meiosis occurs. Sterigmata form as outgrowths from the metabasidium. In the Hymenomycetes basidia are most commonly club shaped with two, four or more apical sterigma bearing the basidiospores.

Here the metabasidium is merely a later stage of development of probasidium. This is also the case in the Dacrymycetales and Tulasnellales where the basidia have strongly enlarged or inflated sterigmata

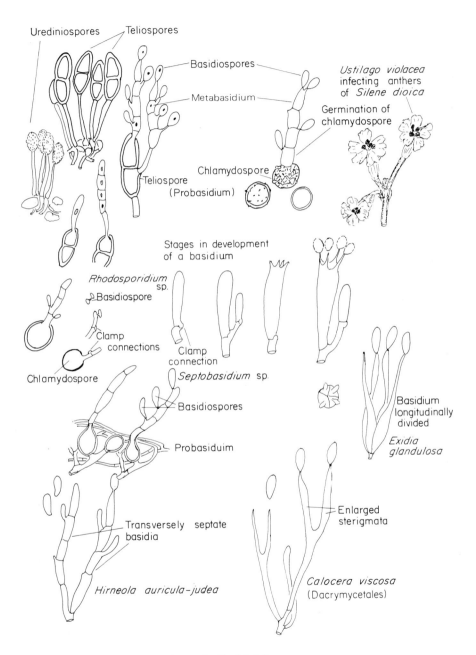

Figure 6. Basidial forms

and in *Exidia* or *Tremella* where the basidium is also longitudinally divided. In *Septobasidium*, whose species are often associated with scale insects, the probasidium is a thick-walled, oval cell (Figure 6). The metabasidium develops from this and becomes four-celled, each cell producing a sterigmata and basidiospore. Similar transversely septate basidia occur in the fungus *Hirneola auricula-judea*, the jew's ear fungus common on elder throughout the year. In this species the probasidium is poorly developed.

3. *Gasteromycetes*

Observations on *Lycoperdon* and other members of the Gastero-mycetes suggest that young basidia are binucleate and that nuclear fusion and meiosis occur in a similar manner to that found in the Hymenomycetes. Usually basidia surround cavities within an enclosed fruit-body and the basidiospores are not discharged violently into those cavities. The ultimate liberation of the basidiospores is achieved by a quite remarkable series of adaptations as mentioned in Section I.

IV. A General Purpose Classification of the Fungi

Note: Divisions end in '-mycota'; Sub-divisions in '-mycotina'; Classes in '-mycetes'; Subclasses in '-mycetidae'; Orders in '-ales'; Families in '-aceae'.

Families are excluded. The orders cited, which should be regarded as only representative, are listed alphabetically by classes.

A few common names and alternative names have been inserted.

FUNGI (Mycota)
I. MYXOMYCOTA (Myxobionta)
    (1) Acrasiomycetes
        Acrasiales
    (2) Hydromyxomycetes
        Hydromyxales
        Labyrinthulales
    (3) Myxomycetes (slime moulds)
    (4) Plasmodiophoromycetes
        Plasmodiophorales
II. EUMYCOTA (Mycobionta; eumycetes)
    1. Mastigomycotina (Phycomycetes)
        (1) Chytridiomycetes
            Blastocladiales
            Chytridiales (chytrids)
            Monoblepharidales
        (2) Hyphochytridiomycetes
            Hyphochytriales

        (3) Oomycetes
            Lagenidiales
            Leptomitales
            Peronosporales (downy mildews, etc.)
            Saprolegniales
    2. Zygomycotina (Phycomycetes)
        (1) Zygomycetes
            Entomophthorales
            Mucorales
        (2) Trichomycetes
    3. Ascomycotina (ascomycetes)
        (1) Hemiascomycetes
            Endomycetales
            Taphrinales
        (2) Euascomycetes
            Protomycetidae (Plecto-mycetes)
            Eurotiales

Unitunimycetidae
Erysiphales
Sphaeriales —
Hypocreales
Helotiales ∥inoperculate
Phacidiales ∫discomycetes
Pezizales — operculate
            discomycetes
Tuberales
Bitunimycetidae
(Loculoascomycetes)
Myriangiales
Capnodiales
Dothideales
Hysteriales
Microthyriales
(Hemisphaeriales)
Pleosporales
Laboulbeniomycetes
4. Basidiomycotina
   (basidiomycetes)
   (1) Hemibasidiomycetes
       Uredinales (rusts)
       Ustilaginales (smuts)

(2) Hymenomycetes
    Agaricales (agarics,
    boleti)
    Aphyllophorales
    (polypores, etc.)
    Tulasnellales
(3) Gasteromycetes
    Hymenogastrales
    Lycoperdales
    Nidulariales
    Phallales
    Sclerodermatales
5. Deuteromycotina (*fungi
   imperfecti*)
   (1) Coelomycetes
       Melanconiales
       Sphaeropsidales
   (2) Hyphomycetes
       Hyphales (syn.
       Moniliales)
   (3) Agonomycetes (mycelia
       sterilia)
       Agonomycetales (syn.
       Myceliales)
(amended from G. C. Ainsworth)

V. Further Reading

Ainsworth, G. C. (1971). *Dictionary of the Fungi*, 6th ed. Kew: Commonwealth Mycological Institute.

Ainsworth, G. C. and Austwick, P. K. C. (1973). *Fungal Diseases of Animals*, 2nd ed. Review Series of the Commonwealth Bureau of Animal Health 6. London: Commonwealth Agricultural Bureaux.

Alexopoulos, C. J. (1962). *Introductory Mycology*, 2nd ed. New York: John Wiley.

Arx, J. A. Von. (1974). *The Genera of Fungi Sporulating in Pure Culture*, 2nd ed. Germany: J. Cramer.

Booth, C. (1971). *Methods in Microbiology*, Vol. 4. London: Academic Press.

Emmons, E. W., Binford, C. H., Utz, J. P. and Kwon-Chung, K. J. (1977). *Medical Mycology*, 3rd ed. Philadelphia and London: Lea & Febiger.

Hawksworth, D. L. (1974). *Mycologist's Handbook*. Kew: Commonwealth Mycological Institute.

Kendrick, B. (1971). *Taxonomy of Fungi Imperfecti*. Toronto: University of Toronto Press.

Martin, G. W. and Alexopoulos, C. J. (1969). *The Myxomycetes*. Iowa: University of Iowa Press.

Müller, E. and Loeffler, W. (1976). *Mykologie*. Stuttgart: Georg Thieme. (English ed.)

Smith, G. (1969). *An Introduction to Industrial Mycology*, 6th ed. London: Edward Arnold.

Talbot, P. H. B. (1971). *Principles of Fungal Taxonomy*. London: Macmillan.

Watling, R. (1973). *Identification of the Larger Fungi*. Amersham: Hulton Educational Publications.

Webster, J. (1970). *Introduction to Fungi*. London: Cambridge University Press.

Form and Function — III. Viruses

D. H. WATSON

Department of Microbiology, The University of Leeds, Leeds LS2 9JT

I. The Nature of Viruses

It is plainly necessary as a preliminary to discussing form and function of viruses to attempt to define a virus. This topic has been mentioned by Stanier in the first essay in this series of Essays in Microbiology, but it is appropriate to review this point again here in slightly more detail.

The question 'What is a virus?' has aroused considerable controversy over the years and the fact that there is no absolutely correct answer has made the question an examiners' favourite. Perhaps the only faultless answer is that of André Lwoff who, with impeccable Gallic logic, has postulated that 'Viruses are viruses'. Following earlier attempts at precise definition, organisms which did not 'fit' (e.g. the agent causing psittacosis, an infection of parrots sometimes passed on to those who keep them as pets) were deprived of the right to the title of virus. Nowadays a more cautious attitude prevails and virologists, perhaps fearing that pursuit of the exclusive policy might leave them with no viruses at all, have tended to bend the definition whenever anomalies or exceptions have been uncovered.

Accordingly it is convenient to consider the question 'What is a virus?' under four headings 'What is a virus (a) physically, (b) chemically, (c) biologically, and (d) clinically?' This has the further advantage in that proceeding in this way, we can see how the recognition of what we consider to be viruses has evolved historically, as well as allowing the somewhat hazy interface between viruses and other microorganisms to be delineated as clearly as possible.

(a) What is a virus physically? During the latter half of the nineteenth century bacterial filters had been produced which were capable of yielding filtrates free of bacteria. In 1892 Iwanowski in Russia and in 1898 Beijerinck, working independently in Holland, had found that the pathological condition of tobacco plants known as tobacco mosaic could be produced by sap from infected plants even after it had been passed through a bacterial filter. Similar observations were made on the agent of foot-and-mouth disease by Loeffler and Frosch in 1909. The responsible organisms were therefore defined as filterable viruses which were recognized as being smaller than bacteria. Indeed their size was such as to place them beyond the limit of resolution of conventional light microscopy, except in the cases of the very largest; it is clear that Buist in Kirkintilloch became the first man to see a virus by light microscopy of vesicle fluid from a smallpox patient. However detection of most viruses only became possible with the development of the electron microscope and viruses became a favourite test sample for all early practitioners of electron microscopy, including Ruska who is credited with the construction of the first electron microscope. From the results of electron microscopy we can say that viruses are ten to one hundred times smaller than bacteria with a size range of 20 to 300 nm approximately.

(b) What is a virus chemically? Chemically the simplest viruses consist of protein surrounding a nucleic acid core, which is remarkable in being predominantly of one kind — either DNA or RNA. This is a characteristic of viruses which marks them out from all other organisms. More complex viruses contain, for example, lipid, carbo-hydrate, polyamines, in addition to nucleic acid and protein.

(c) What are viruses biologically? Viruses are all obligate intracellular parasites. That is they are incapable of independent growth in artificial media such as nutrient agar, glucose broth, etc. They can only grow in animal or plant cells or in other microorganisms. Obligate intracellular

parasitism is, of course, not a property unique to viruses since it is shared by other microorganisms such as chlamydia or rickettsiae. However the means by which viruses grow is quite characteristic. Their growth is characterized not by the fission of a cell into daughter cells which themselves later divide in similar fashion, but by a process termed *replication*. In this process many copies or replicas are made of each component (i.e. nucleic acid and protein for the simplest viruses) of the infecting virus particle and then these are assembled to produce progeny virus.

d) What is a virus clinically? Clinically viruses are responsible for diseases which are generally insensitive to the broad range of antibiotics (penicillin, streptomycin, etc.) to which bacteria are normally sensitive.

II. Biological Reality of Virus Particles

Since we are concerned in this chapter with form *and* function it is important before considering virus structure to be sure that we shall be considering structures which are still capable of exercising the basic function of virus particles — that is they are still infective. Not only must we hope that we are not examining artefactual structures generated by the preparative procedures but we must also be satisfied that virus particles *per se* have a biological reality. We might, for example, consider that virus infectivity was a cooperative property of virus particles, or was due to some component of a virus preparation other than the characteristic 'virus' particles — that is the particles would be a contaminant.

The second proposition seems unlikely since even the most rigorous purification procedures fail to separate virus infectivity from visible virus particles. Although this is to a certain extent a negative criterion there is no evidence that the ratio between numbers of particles visible in the electron microscope and the numbers of infective particles increases as a preparation is purified.

The first objection may be answered by quantitative considerations of virus infectivity assays. Although such assays may require use of susceptible animals or plants where the only 'scoring' can be quantal (i.e. the host is alive or dead, infected or uninfected) in many cases *focal response* assays are possible. Thus a virus preparation may be allowed to infect a monolayer of confluent tissue culture cells or a lawn of bacteria such that the cells infected directly by the virus inoculum infect, by contact, neighbouring cells which themselves similarly infect their neighbours. The number of infected foci (called *plaques*) may be subsequently identified, for example, as clear regions in a bacterial lawn. From the number of such plaques we can draw conclusions about the number of infective viruses in the original inoculum. For most viruses it turns out that the number of plaques is linearly related to the dilution of the original virus inoculum. That is, a twofold dilution of the inoculum halves the number of plaques produced. Such a dilution response suggests one-hit kinetics — i.e. the primary infection is produced by 'hitting' the susceptible cell with one particle. We may confirm this suggestion by a more detailed statistical analysis. If we have V virus particles and N cells in our assay, then each cell receives on average V/N virus particles. Obviously V/N may be non-integral and indeed, under conditions of a virus assay will be much less than unity since we should not wish to infect all the cells in the assay monolayer. However each cell can obviously only receive $0, 1, 2, \ldots$, virus particles. The proportion of cells receiving $0, 1, 2, \ldots$, particles may be predicted by the Poisson distribution which tells us that a proportion $e^{-V/N}$ receives 0 particles, a proportion $(V/N)e^{-V/N}$ receives 1 particle,

$$\text{a proportion} \quad e^{-V/N} \quad \left(\frac{V}{N}\right)^2 \Big/ 2! \quad \text{receives 2 particles,}$$

$$\text{a proportion} \quad e^{-V/N} \quad \left(\frac{V}{N}\right)^3 \Big/ 3! \quad \text{receives 3 particles,}$$

$$\text{a proportion} \quad e^{-V/N} \quad \left(\frac{V}{N}\right)^n \Big/ n! \quad \text{receives } n \text{ particles}$$

It will be noted that the exponential $e^{-V/N}$ appears in each term and that the multipliers $1, V/N, (V/N)^2/2!$ etc. represent the expansion of the series $e^{V/N}$ so that the sum to infinity of the proportion of cells receiving $0, 1, 2, \ldots$, particles will be $e^{V/N} \times e^{-V/N} = 1$. Fortunately for the less mathematically minded virologist, the first term of this series is the most important in most experimental situations.

If we make the hypothesis that only one virus particle is necessary for infection (i.e. no cooperative phenomena between two or more particles), then we may say that *all* cells other than those which are hit by 0 particles will be infected, i.e.

Proportion infected = 1 — proportion 'hit' by 0 particles
$$= 1 - e^{-V/N}$$

Now V/N is very much less than zero and examination of exponential tables will show that under these circumstances $e^{-V/N} \doteqdot 1 - V/N$

Accordingly in our assay

Number of plaques = No. of cells per assay × proportion of cells infected

$$= N \times (1 - e^{-V/N}) = N \times \left[1 - \left(1 - \frac{V}{N}\right)\right] = V$$

Therefore a situation where the number of plaques varies with dilution (i.e. with V) is consistent with a one hit hypothesis. Plainly a two-hit hypothesis (i.e. that infection of a cell requires two particles) would mean that only those cells receiving $\geqslant 1$ particle would be infected so that

Proportion infected = 1 $-$ Proportion uninfected

= 1 $-$ (Proportion of cells receiving 0 particles

+ proportion of cells receiving 1 particle)

The reader should be able to verify easily what would be intuitively expected, i.e. that in this circumstance the number of plaques will be proportional to the square of the number of inoculum virus particles.

The experimental results with a number of viruses show that one particle is sufficient to initiate infection although there are situations where this is not so; some viruses require simultaneous infection with a 'helper' virus for successful infection (e.g. the parvovirus adeno-associated virus requires helper adenovirus) and as we shall see some 'divided genome' viruses (e.g. tobacco rattle viruses) divide their genomes between two particles, each of which is necessary for infection. However, it should be noted that it does not necessarily follow that all particles visible in the electron microscope are capable of initiating infection. Methods for deriving the absolute count of physical particles seen in the electron microscope show that for some bacteriophages the ratio physical/infective particles is about unity. For many animal viruses it is significantly greater than this — ratios from just over 1 to over 10^6 have been recorded. This is partly due to assay inefficiency — infectivity assays for animal viruses are inherently less efficient than those for bacterial viruses. Another contributory factor is the presence of 'empty' particles devoid of nucleic acid or damaged particles which cannot be identified as such in the electron microscope.

III. Chemical Analysis of Virus Particles — The Coding Defect

The previous section showed how quantitative consideration of virus infectivity assays led to an appreciation of the biological reality of virus particles as seen in the electron microscope. In this section we shall see how studies on the chemical composition of viruses led to some theoretical ideas on virus structure, which we shall use as a basis for our detailed discussion of this topic.

If we examine figures for the chemical analysis of a number of small RNA plant viruses such as tomato bushy stunt virus or turnip yellow mosaic viruses, we find that the number of amino acids of coat protein per nucelotide of virus RNA is usually much greater than unity and indeed is often greater than ten. Thus a typical small spherical plant virus may consist of 70 per cent protein and 30 per cent RNA of molecular weight 2×10^6. Assuming that the molecular weight of a codon of three nucleotides is approximately 10^3, the virus would possess enough information to code for about 2000 amino acids — whose total molecular weight would be just over 2×10^5. Plainly the total molecular weight of the coat protein must be over 4×10^6 from the results of the chemical analysis. This coding defect demands that the coat protein must be composed of identical subunits as was first pointed out by Crick and Watson (1957) — interestingly enough long before they had come to a definite idea on the number of nucleotides in one codon. These ideas led them to suggest that the coats of spherical viruses would be composed of repeating protein subunits arranged in symmetrical fashion. X-Ray diffraction analysis of crystals of tomato bushy stunt virus showed that this virus indeed possessed the so-called 5:3:2 symmetry typical of the icosahedron (Figure 1). This Platonic solid possesses fivefold, threefold, and twofold axes of symmetry (see Figure 1). This structure is capable of accommodating 60 identical units — as can be seen by reference to Figure 1. Suppose we place one unit in the vertex of one of the five triangular faces surrounding a fivefold axis of symmetry. To maintain the symmetry we have to place a similar unit on the other four faces. Further, since each of the five faces has an axis of threefold symmetry passing through its centroid, maintenance of symmetry will require units to be placed in *each* vertex. Proceeding in this way we see that maintenance of symmetry requires 60 or a multiple of 60 units to be placed on the icosahedron. The advantage of icosahedral symmetry over simpler forms of cubic symmetry is that it allows the packing of the greatest possible number (i.e. 60) of identical units to build a spherical or approximately spherical shell. If a protective protein shell of a given size is made of the greatest possible number of units, it follows that they must be of the smallest possible size and icosahedral symmetry,

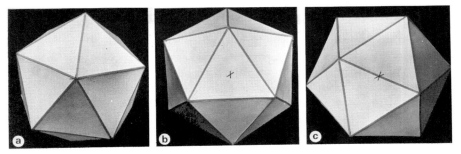

Figure 1. Icosahedron viewed down (a) fivefold, (b) threefold, (c) twofold axes of symmetry. The appropriate axes emerge perpendicular to the page from the icosahedron at the points marked X. Rotation of the icosahedron about, for example, a fivefold axis produces the same appearance five times within a complete revolution of $360°$

therefore, allows the most economical construction of a protein shell.

As already noted, X-ray diffraction studies provided confirmation of the existence of icosahedral symmetry in a spherical virus. Thereafter it was shown that the *Tipula* iridescent virus particle had an icosahedral shape. This does not itself show that the particle possesses icosahedral symmetry and as we shall see icosahedral symmetry does not require icosahedral shape. In following sections we shall discuss detailed electron microscopic evidence for icosahedral symmetry in spherical viruses. First, however, it is necessary to consider briefly methods used to study the form of viruses in the electron microscope.

Crick and Watson were, of course, aware that some viruses were rod shaped rather than spherical and suggested that these viruses would consist of a helical array of protein subunits surrounding the nucleic acid.

IV. Methods for Electron Microscopy of Viruses

A. Shadowcasting

Many earlier electron microscopic studies on the form of viruses used the method of shadowcasting in which a pseudo three-dimensional effect was achieved by evaporating a beam of heavy metal at an angle on to virus particles supported on a thin support film. The consequential differential deposition of metal and in particular the existence of 'shadow' regions devoid of deposited metal led to differential contrast in the electron beam and gave information in the third or vertical dimension, not normally available since in the electron microscope the whole depth of the specimen is simultaneously in focus in contrast to the light microscope where the depth of field is less than the specimen thickness, allowing differential focusing on upper and lower planes of the specimen. An ingenious variation to shadowcasting was to shadow in two different directions inclined at an angle of 60°. The form of resulting shadows cast by particles of *Tipula* iridescent virus allowed the deduction of its icosahedral shape, the particular shadows observed being uniquely correlated with this shape. In general, however, the amount of information derived from this technique was somewhat limited. Indeed one prominent virologist commented in 1959 on the rather small return up to that time from the great amount of money, time, and effort, expended on the electron microscopy of viruses. Paradoxically, 1959 saw a revolutionary advance with the introduction of negative staining for electron microscopy.

B. Negative Staining

The principle of negative staining had, of course, been used in light microscopy for decades and should be familiar to microbiologists through its use for example to reveal capsular material of bacteria. If encapsulated bacteria are mixed with nigrosin or India ink, the capsule being impervious to the black dye results in the whole extent of the organism being revealed in negative contrast, i.e. it appears white against a black background, while components within the capsule may be revealed by conventional positive staining with coloured dyes.

Many electron microscopic methods are derived by analogy from preparative methods in light microscopy so it is perhaps surprising that negative staining was not introduced earlier into electron microscopy. In fact most effort was devoted to positive staining with salts of heavy metals and negative staining was only 'discovered' as an accidental by-product from unsatisfactory attempts to use such positive stains in electron microscopy of viruses. When viruses were 'stained' with heavy metal salt solutions, it was often difficult to completely wash away excess stain. Two groups of workers noted that this often led to virus particles being seen in negative contrast where they were left in pools of excess stain. They both commented that the resolution of structure was often better in this situation than on 'clean' stained particles. It was left to Brenner and Horne in 1959, on obtaining the same effect, to proceed to negative stain on purpose, rather than by accident. The term 'negative staining' is of course self-contradictory, and the technique should be more correctly termed 'negative contrast', but the earlier term has perhaps been justified by usage. The technique has two outstanding characteristics which were important in studies of virus structure. First it is unaffected by impurities of small molecular size which fade into the background. Secondly, as we shall see, it reveals characteristic structural patterns on the particles of many viruses which may then be clearly differentiated from larger impurities. This has not only made it possible to examine structure of particles in crude preparations but has allowed exploitation of the electron microscope as a diagnostic and quantitative tool in virology. Quantitative methods are of course mandatory in any attempt to relate form with function. Particle counting methods were already available before negative contrast techniques were introduced. They relied either on mixing virus preparations with reference particles (e.g. polystyrene latex) of known count and deriving the virus count from the observed ratio of virus and reference particle, or on methods in which virus particles were sedimented on to specimen grids with the count being derived from the number of particles observed in a given area of the grid. The main problem in such methods was described by an early practitioner as being not how to count but what to count, since in shadowcast preparations it was often a matter of guesswork to distinguish which 'blob' was the virus. The characteristic structures revealed by negative staining solved this problem in many cases. Using a quantitative approach it is then possible to relate differential particle morphologies to biological activities; for example, if two morphologically different particles A and B, are seen in a preparation of a particular virus then it may be possible to draw conclusions on the biological properties of A and B by comparison of particle count data with infectivity assays. Hence if the number of infective particles exceeds the number of A particles then it is plain that the B particles must be infective with A particles being non-infective or at best only accounting for some of the observed number of infective particles.

Additional information can be obtained as we shall see by combining immunological methods with this approach. Hence, by observing whether particles are agglutinated by an antiserum to a particular component we may determine whether this component is represented on the surface of some or all of the particles. Once again use of particle counting methods allows these observations to be placed on a quantitative basis.

C. Thin Sectioning

While negative contrast techniques provide a wealth of information on the exterior structure of virus particles they are usually less informative about internal structure. Thin-sectioning techniques are more suited to this purpose. Virus particles, either in suspension or in infected cells, are fixed, embedded and sectioned for electron microscopy. Staining with heavy metal salt solutions after fixation or on mounted sections is frequently used to increase contrast. Thin-section methods have been used in combination with enzyme digestion or other cytochemical methods for revealing the chemical nature of nucleic acid cores, site of virus particle enzymes, etc. The use of ferritin tagged antibodies allows antigenic components to be located in virus particles.

D. Freeze Etching

This technique is claimed to show surface structure of virus particles by methods avoiding drying artefacts which may wrongly suggest pleomorphism. The virus preparation is frozen in Freon followed by liquid nitrogen. Surface relief is heightened by etching or sublimation under vacuum under controlled temperature conditions. A shadowed replica of the surface is made by coating with platinum + carbon. The replica is then cleaned with concentrated sulphuric acid before examination in the electron microscope.

V. Virus Groups

Viruses may be classified into groups on the basis of a number of properties, but it is sufficient to note here that this can be done almost

Table 1. Classification of viruses

Family	Vernacular group name	Typical genera within family	Typical members of genus	Host	Associated disease[a]	Morphology/symmetry[b]
DNA viruses						
Iridoviridae	Icosahedral cytoplasmic deoxyvirus	*Iridovirus*	*Tipula* iridescent	*Tipula* (arthropod)	–	Icosahedral (?1472)
Poxviridae	Poxvirus	*Orthopoxvirus*	Variola	Man	Smallpox	Brick shaped
			Vaccinia	Man	(smallpox vaccine)	Brick shaped
		Leporipoxvirus	Myxoma	Rabbit	Myxomatosis	Brick shaped
		Parapoxvirus	Orf	Sheep	Contagious pustular dermatitis	Cigar shaped
			Pseudocowpox	Man	Milkers' nodes	Cigar shaped
Adenoviridae	Adenovirus	*Mastadenovirus*	Adenovirus (many types)	Man	Respiratory infection	Icosahedral (252)
Herpetoviridae	Herpesvirus	*Herpesvirus*	Herpes simplex	Man	Cold sore	Icosahedral (162) + envelope
		?	Varicella	Man	Chickenpox	Icosahedral (162) + envelope
		?	EB virus	Man	Infectious mononucleosis (glandular fever)	Icosahedral (162) + envelope
		?	Marek's disease	Chicken	Lymphoproliferative disease of nervous tissue	Icosahedral (162) + envelope
Papovaviridae	Papovavirus	*Papillomavirus*	Papilloma viruses of many species	Man, rabbit, cow, etc.	Warts	Icosahedral (72)
		Polyomavirus	Polyoma	Mouse	(Tumours in hamsters)	Icosahedral (72)
Parvoviridae	Parvovirus	Adenoassociated virus group	Adeno-associated virus	–	–	Icosahedral (32)
Microviridae	φX phage group	*Moralavirus*	φX174	*E. coli*	–	Icosahedral (12)
Myoviridae	Complex tail phage group	T-even phage group	T2	*E. coli*	–	Bipyramidal hexagonal prism head + tail
?	?	?	Hepatitis B	Man	Serum hepatitis	Spherical double shelled particles

Family/group		Genus	Virus[a]	Host	Symptoms/disease	Structure[b]
Tymovirus group			Turnip yellow mosaic, Eggplant mosaic	Eggplant	Mosaic/mottle symptoms	Icosahedral (32)
Reoviridae		*Reovirus*	Reovirus			Icosahedral double shell (?92)
		?	Rotaviruses	Man, Cow	Infantile diarrhoea, Calf diarrhoea	Icosahedral double shell (?92)
Picornaviridae		*Enterovirus*	Poliovirus, Swine vesicular, Encephalomyocarditis	Man, Pigs, Mouse	Polio, Vesicular disease	Icosahedral, Icosahedral, Icosahedral
		Rhinovirus	Human rhinovirus	Man	Common cold	Icosahedral
		Calicivirus	Foot-and-mouth, Vesicular exanthema	Sheep, cattle, Pigs		Icosahedral, Icosahedral
Tombusvirus group			Tomato bushy stunt	Tomatoes	Mottle/distorted growth	Icosahedral (90)
Ribophage group			Ribophage R17, MS2, f2	E. coli		Icosahedral
Tobamovirus group			Tobacco mosaic virus	Tobacco	Mosaic/mottle symptoms	Helical
Tobravirus group			Tobacco rattle virus	Tobacco	Necrosis	Helical (divided genome)
Potyvirus group			Potato virus Y	Potato	Mosaic symptoms	Helical (flexuous)
Orthomyxoviridae	Myxovirus	*Influenzavirus*	Influenza	Man	'Flu'	Helical + envelope
Paramyxoviridae	Paramyxovirus	*Paramyxovirus*	Newcastle disease	Chicken		Helical + envelope
Rhabdoviridae	Rhabdovirus	*Morbillivirus*, *Vesiculovirus*	Measles, Vesicular stomatitis	Man, Cattle		Helical + envelope (bullet shape)
		Lyssavirus	Rabies	Dog, fox, man		Helical + envelope (bullet shape)
Retroviridae Subfamily: Oncovirinae	RNA tumour viruses	Type C oncovirus, Type B oncovirus	Moloney sarcoma, Rous sarcoma, Mouse mammary tumour	Mouse, Chickens, Mouse		C type particles, C type particles, B type particles

[a] Not shown where obvious from virus name. [b] Numbers of capsomers for icosahedral viruses shown in brackets

entirely on the basis of morphology. It is therefore desirable to indicate the nature and composition of the virus groups as a preliminary to describing the characteristic structures displayed by members of each group. While so far the taxonomically neutral word 'group' has been used, virologists no less than other biologists have been consumed with a desire to devise a Linnaean hierarchial classification based on families and genera. Accordingly in listing the various groups of viruses in Table 1 the family and genus names recommended by the International Committee for Taxonomy of Viruses have been indicated as well as the more widely used vernacular 'group' names. Plant virologists have not used the taxa 'family' and 'genus', and have designated only groups.

Table 1 is not designed to embrace all known viruses and indeed does not include all the internationally accepted families or groups. It is intended only to give some brief indication of the grouping and properties of the viruses whose structures will be discussed in the following sections.

VI. Virus Structure

A. Icosahedral Viruses

Figure 2 shows an electron micrograph of an adenovirus particle in negative contrast so that the background is dark and the particle light. The surface structure of the protein coat is clearly revealed by the penetration of negative stain between the subunit structures. The protein shell of a virus is termed the *capsid* and the morphological units comprising it are called *capsomeres.* It will be clearly seen that most of the capsomeres are surrounded by six nearest neighbours — such capsomeres are named *hexons.* However some capsomeres have only five nearest neighbours and these are termed *pentons.* The pentons consist of a *penton base* to which is attached a *fibre* and *knob* (Figure 2b). Figure 2(a) shows that the pentons are arranged on the vertices of equilateral triangles and the model of Figure 2(c) derived from the micrograph indicates that the arrangement can be derived by piling table tennis balls (representing hexons) in close packed array on to the faces of an icosahedron (Figure 1) and placing pentons on each of the 12 vertices of the icosahedron. The number of hexons may be easily computed. There are four hexons on each edge of the icosahedron between the vertex pentons. The icosahedron has 20 triangular faces with three edges. Since each edge is shared between two faces the total number of edges = $\frac{1}{2} \times 20 \times 3 = 30$. (The numbers of faces, edges, and vertices of a polyhedron are related by Euler's theorem Faces + Vertices = Edges + 2, i.e. here $20 + 12 = 30 + 2$), so we have $30 \times 4 = 120$ hexons on the edges. On each face we have $3 + 2 + 1 = 6$ hexons so there are $6 \times 20 = 120$ such hexons in the whole icosahedron.

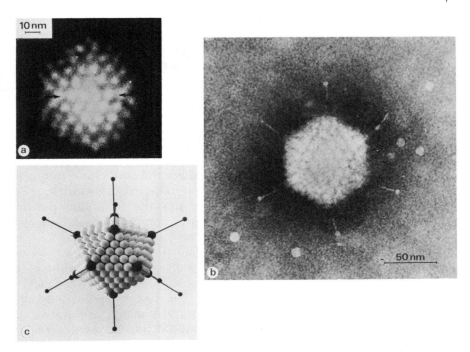

Figure 2. Negatively stained particle of adenovirus showing (a) regular capsomere array — pentons (surrounded by five nearest neighbours) arrowed, (b) particle showing fibres and knobs attached to penton base. A model of the structure is shown at (c). (a) Reproduced by kind permission of Professor R. W. Horne and Academic Press. (b), (c) reproduced by kind permission of the late Dr R. C. Valentine and Academic Press

The total number of hexons is thus 120 + 120 = 240 and together with the 12 pentons we see that there is a total of 252 capsomeres. Although, as expected from the earlier discussion, the capsomeres are arranged in icosahedral symmetry their number is plainly not a multiple of 60. However in the earlier discussion we placed the first unit (and by implication all the others) in a general position on the face, i.e. not on one of the axes of symmetry. Here the pentons are placed on the vertices, that is on the fivefold symmetry axes. Strict considerations of symmetry would then demand that the pentons are composed of five smaller subunits termed *structure units*. Since there are 12 pentons there are then 60 penton structural units, and the number of hexons (240) is already a multiple of 60, so the number of structure units must be a multiple of 60 irrespective of the number of structure units in a hexon.

We may compute the number of hexons in a more generalized virus model with n capsomeres ($n - 2$ hexons and 2 pentons) lying on the edge. Proceeding as we did with adenovirus there will be a total of 30 $(n - 2)$ hexons on edges. On each face there will be $1 + 2 + 3 + \ldots + (n - 3)$ hexons, i.e. $[20 \times (n - 3)(1 + n - 3)]/2$ (from the formula for summation of an arithmetic series). Remembering that there are 12 pentons the total capsomere number will be

$$12 + 30(n - 2) + \frac{20(n - 3)(n - 2)}{2} = 10(n - 1)^2 + 2.$$

As we shall see capsomere numbers of 12, 162, and possibly 92 have been observed.

All adenoviruses of man as well as of dogs and birds have this same structure although different serological subgroups may show differences in detail such as length of fibre. It is possible to obtain purified preparations of the different morphological components and there is accordingly a wealth of information on the biological properties of the different morphological units. Hexons can be crystallized and X-ray diffraction studies on these crystals show hexons to have a threefold but not a sixfold axis of symmetry as would be expected were they to consist of six identical structure units. The hexon has a molecular weight of 360,000 from ultracentrifuge measurements, and polyacrylamide gel electrophoresis of hexons disrupted by detergent reveals a polypeptide molecular weight of 120,000, suggesting that three such polypeptides make up the hexon. Hexons apparently carry the group-specific antigenic determinant common to human adenoviruses but also carry a type-specific determinant unique to each adenovirus type since antiserum to hexon will neutralize the infectivity of the homologous but not heterologous types.

Antisera to fibres also neutralize infectivity but antisera to penton bases do not. Purified fibre protein inhibits cellular synthesis of DNA, RNA, and protein.

Figure 3 (a and b) shows the two morphological types (enveloped and naked) of particle found in preparations of herpes simplex virus and of other herpesviruses, e.g. varicella (chickenpox). The naked particle possesses an icosahedral array of capsomeres whose centres are filled with negative stain and are therefore presumably hollow. It is not so easy to discern icosahedral faces on herpesviruses as on adenoviruses. The particles are much less angular and seem to be more truly spherical than icosahedral in outline. The structure can be imagined as arising from packing of capsomeres on faces of an icosahedron inscribed on a spherical surface — such a construction will be familar to *Match of the Day* enthusiasts (Figure 3d). Because of the curvature of the particle surface it is more difficult to pick out a triangular arrangement of pentons, but by analysis of a number of micrographs by empirical model-building procedures the structure is considered to be that shown in the model of Figure 3(c). It will be seen that here we have two

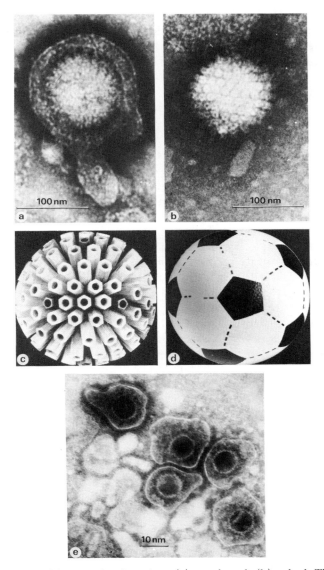

Figure 3. Particles of herpes simplex virus: (a) enveloped, (b) naked. The model at (c) shows the detailed capsomere structure deduced from a number of micrographs. The particle has icosahedral symmetry although not an icosahedral shape; it may be considered as being derived by inscribing an icosahedron on a spherical surface as shown in (d) where the icosahedral edges are represented by dotted lines and pentagons have been placed on the vertices. (e) Enveloped particles of herpes simplex virus grown in BHK-21 cells agglutinated by antiserum to BHK-21 cells. (a), (b), (e) reproduced by permission of Academic Press. (c) reproduced by kind permission of Professor P. Wildy and Academic Press.

pentons + three hexons per edge and so the capsomere number is given by the value of $10(n-1)^2 + 2$ for $n = 5$, i.e. 162.

In the enveloped particle the naked particle is surrounded by another layer. Sometimes this layer is impervious to negative stain but partially damaged particles permit penetration of the negative stain which thus delineates the internal capsid structure. The surface of the envelope has a fringe of fine surface projections — such projections from the surface of a virus envelope are called *peplomers*. Immunological techniques show that the envelope contains host cell antigen since antiserum to the cell in which the virus is grown will agglutinate enveloped but not naked particles (Figure 3e). However the envelope is not unaltered cell membrane; the particles may also be agglutinated by antiserum made against virus grown in a heterologous cell showing that the envelope contains virus specific material. Analysis of enveloped particles, labelled either before or after infection, by polyacrylamide gel electrophoresis shows that the envelope does not contain prelabelled (i.e. host cell) protein so the host cell component must be carbohydrate or lipid. The envelope does contain proteins labelled after infection which differ from any pre-existing host cell protein and are therefore virus specific. These proteins are glycosylated and we shall see that surface proteins of virus envelopes are invariably glycoproteins.

The naked particle is assembled in the cell nucleus and appears to acquire the envelope on passing through the nuclear membrane. The layer between the capsid and the envelope has been called the *tegument* and it has been suggested that this gives the capsid an affinity for the nuclear membrane necessary for envelopment. There may be other concentric shells inside the capsid. More certain is the existence of a nucleoprotein core seen as an electron dense region in thin-sectioned particles. This core has been described as toroid in form with the DNA wound round a central cylinder.

Figure 4(a,b) shows particles of a rotavirus. This virus, like all reoviridae, has two capsid shells one within the other. The outer capsid layer of reovirus probably consists of 92 capsomeres ($n = 4$) although counts of the number of peripheral capsomeres (about 20) are inconsistent with this model. Each capsomere has a large central hole and this, combined with the small intercapsomeric spaces, gives the illusion that the capsid consists of an array of 92 holes. The core or inner capsid of reovirus can be obtained by enzymic digestion of complete particles. Its capsomere number is uncertain although like the outer capsid it seems to have 20 peripheral capsomeres. The outer capsid contains three polypeptides σ_3, μ_2 and σ_1. On treatment with chymotrypsin σ_3 is lost first. Virus infectivity is unaffected by this removal but is markedly decreased in the next stage of digestion when μ_2 is lost.

Microviridae (the ϕX bacteriophage group) have an icosahedral structure with 12 capsomeres, i.e. pentons only ($n = 2$).

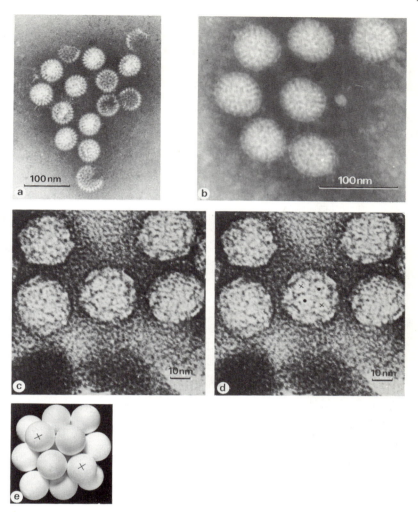

Figure 4. (a) Core particles of rotavirus; (b) double shelled particles of rotavirus. Note capsomeres with large central hole. (c) Eggplant mosaic virus, a tymovirus. (d) Same field as (c) with particle marked to indicate pentons (cross) and hexons (dot). (e) Model of 32 capsomere structure consistent with (c) and (d). The pentons are marked with a cross. (a), (b) reproduced by kind permission of Dr T. H. Flewett, (c), (d) by kind permission of Dr I. M. Roberts and Scottish Horticultural Research Institute.

When we pass to the structure of the plant viruses known as tymoviruses (Figure 4c,d) we encounter a structure which cannot be reconciled with an icosahedral structure of the $10(n - 1)^2 + 2$ series. Each six coordinated capsomere is surrounded by five coordinated capsomeres and *vice versa*. The structure seems to be consistent with the model of Figure 4(e) which shows one capsomere on each icosahedral face (sited on the threefold axis) and capsomeres on each vertex. Plainly there are 20 capsomeres on the faces and 12 on the vertices making 32 in all. This structure not only has a capsomere number not predicted by the formula $10(n-1)^2 + 2$, but is based on a completely different orientation of the hexagonal packing in relation to the icosahedral edges. In the $10(n-1)^2 + 2$ structure the icosahedral edges lay parallel to the close-packed direction of the hexagonal array. This meant that the line joining pentons always passed through a close-packed line of hexons. In the tymovirus structure the close packed direction seems to lie along the bisectors of the angles of the triangular face; correspondingly the line joining the pentons on one face

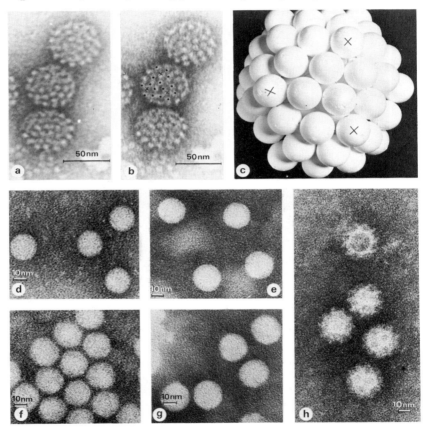

Figure 5. (a) Particles of human wart virus; (b) same field as (a) with hexons on one particle marked with dot and pentons with cross. (c) Photograph of 72 (dextro) capsomere structure corresponding to human wart virus. Pentons marked with cross. (d) Encephalomyocarditis virus; (e) foot-and-mouth virus; (f) swine vesicular disease virus; (g) human rhinovirus; (h) vesicular exanthema virus. (a), (b) reproduced by kind permission of Dr J. T. Finch and Academic Press. (d) to (h) reproduced by kind permission of Dr C. J. Smale and Cambridge University Press

does not pass through a close packed line of hexons. As we shall see later it is possible to build a series of models on similar principles having capsomere numbers represented by the formula $30(n-1)^2 + 2$.

The parvoviridae probably also have a 32 capsomere structure.

The papovaviridae introduce an even more complex structure. Here the close packed direction is asymetrically oriented with respect to the icosahedral edge and to traverse from one penton to another on the same face it is necessary to make a 'knight's move' — that is to move along a row of hexons and then move off at an angle to reach the second penton $-5-6-6\searrow5$ (Figure 5a,b). This structure unlike the two previous types is not identical with its mirror image and enantio-morphic forms are therefore possible. The form adopted by polyoma virus and human papilloma (wart) virus is the *dextro* form (Figure 5c) while rabbit papilloma is in the *laevo* orientation where the path between pentons may be represented $5-6-6\nearrow5$. Although all the papovaviridae have 72 capsomeres the polyomaviruses are smaller than the papillomaviruses (45 and 55 nm respectively) and this corresponds to a lower DNA molecular weight (3 and 5 x 10^6 respectively).

A large number of these so-called skew structures may be formulated but it is not clear how many of them actually occur. *Tipula* iridescent virus seems to have such a structure probably constructed from 1472 capsomeres.

Horne and Wildy pointed out that all possible capsomere numbers may be given by the formula $10x(n-1)^2 + 2$ where $x = 1$ for the close packed form, $x = 3$ for the other symmetrical structure, $x = 7$, 13 . . . for skew structures. We shall return to the theoretical basis for these numbers below.

The picornaviridae are all spherical but it has not been possible to resolve clearly their precise capsomere structure. Enteroviruses and rhinoviruses have a similar morphological appearance (Figure 5d—g). Caliciviruses such as vesicular exanthema virus (Figure 5h) have a different and very distinctive appearance; there are probably 32 capsomeres whose form is very similar to that of the reovirus capsomeres. They differ also from other picornaviruses in being unstable at low pH, in RNA content, and in polypeptide composition. Enteroviruses such as poliovirus are notable for their existence in two distinct antigenic forms which may be correlated with different physical structures. Heated poliovirus differs antigenically from native particles. The heated particles appear empty in the electron microscope — i.e. their cores were penetrated by the negative stain. Immune electron microscopy shows that an antiserum containing antibodies to both determinants agglutinates each kind of particle into separate clumps. The heated particles appear to have lost one of the four polypeptides (VP4) of the native particle and it would therefore appear that this polypeptide is involved in the reaction with neutralizing antibody, since antibody to heated particles does not neutralize infectivity.

While a number of other isometric viruses clearly have capsomeres, their exact number and arrangement have not been conclusively determined (e.g. ribophages).

B. Theoretical Treatment of Icosahedral Virus Structure

We have seen that there are two symmetrical, and a series of skew, capsomere arrangements possible in icosahedral viruses and have noted that the capsomere numbers may be empirically given by the formula $10x(n-1)^2 + 2$. The different forms may be considered as being derived by rotating a hexagonally close-packed array of capsomeres with respect to fixed icosahedral faces but it is easier to consider this in terms of rotating or expanding a triangle corresponding to the icosahedral face over a hexagonally close packed array as in Figure 6(a). Fixing one vertex at the origin we wish to find all possible capsomere structures. Recognizing that the other two vertices must lie on capsomeres if we wish to have pentons at each vertex, it may be seen in Figure 6(a) that we can empirically pick out the packing arrangements for 12, 92, 162 etc., for 32 and for 72 capsomeres (Figure 6b–d). In Figure 6(a) the other vertices for any particular capsomere structure are labelled with the appropriate number, e.g. the capsomere arrangement for 162 is shown by drawing the icosahedral triangular face with vertices at the origin and at the two 162 balls. Two possible 72 triangles can be drawn and these give the dextro and laevo packing arrangements. A vertex may be defined in terms of coordinates (a,b) referred to hexagonal rather than the usual rectangular coordinates. Thus the point $(2,1)$ is found by measuring 2 units along the a axis then moving 1 unit parallel to the b axis to arrive at the point $(2,1)$. We can define the capsomere number in terms of the coordinates (a,b) of the second vertex since the third vertex always lies at the corresponding point in the next sextant. The solution is in fact an adaptation to the problem of virus structure of a mathematical treatment by Goldberg, who was concerned with investigating the number of faces on polyhedra derived by regular truncation of the vertices of a regular dodecahedron. This procedure produces 12 pentagonal faces with a number of hexagonal faces. In fact the solutions obtained by Goldberg to this problem correspond to the possible numbers of capsomeres arranged in patterns of icosahedral symmetry. Figure 6 shows that the number of capsomeres will be related in some way to the area of the triangular face. Now the edge length of face with a vertex at (a,b) will be the length of the line from the origin to (a,b). This line forms the third side of a triangle whose other sides have lengths a and b, these other sides enclosing an angle of $120°$.

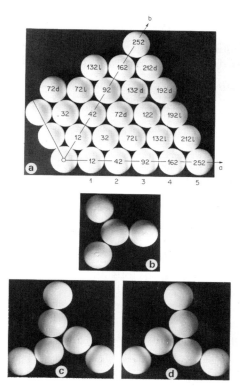

Figure 6. (a) Photograph of regular close packed array of spheres to illustrate principle of construction of virus particles with different capsomere numbers. The arrangement of capsomeres on the face of, for example, a 32 capsomere structure can be derived by placing the vertices of the face at the origin and on the spheres labelled 32 (as shown in b). (c) Arrangement of capsomeres on face of 72 *dextro* structure derived in analogous fashion; (d) arrangement for 72 *laevo* structure. The capsomere number may be calculated from the coordinates (a,b) on the hexagonal axes of the second vertex as $10(a^2 + ab + b^2) + 2$. The third vertex is seen to be located at the corresponding point in the next sextant

Accordingly, by trigonometry the length l of the line from the origin to (a,b) is given by $l^2 = a^2 + b^2 - 2ab\cos 120°$, i.e. $l^2 = a^2 + ab + b^2$.

Now the area of an equilateral triangle with edge l is $\frac{1}{2}l^2 \sin 60° = (l^2 \sqrt{3})/4 = (a^2 + ab + b^2)(\sqrt{3}/4)$. To determine the total capsomere number it will be sufficient to determine the number of hexons per face since the number of pentons is always 12. This number may be derived by dividing the area of the triangle available for packing of hexons by the area required for one hexon. Now this area = area of face − area occupied by pentons. We may estimate the latter area from the icosahedral face having $a = 1$, $b = 0$; from Figure 6(a) we see that this is the 12 capsomere structure which has no hexons. Area for pentons is therefore given by value of $(a^2 + ab + b^2)(\sqrt{3}/4)$ for $a = 1$, $b = 0$, i.e. $\sqrt{3}/4$.

Now the triangular face with $a = 1$, $b = 1$, has one hexon (Figure a,b) shows that this is the 32 structure) and this face has area $(3\sqrt{3})/4$.

$$\text{Net area on this face for one hexon} = \frac{3\sqrt{3}}{4} - \frac{\sqrt{3}}{4} = \frac{2\sqrt{3}}{4}$$

Therefore number of hexons on face with vertex at (a, b) is

$$\frac{(a^2 + ab + b^2)\dfrac{\sqrt{3}}{4} - \dfrac{\sqrt{3}}{4}}{2\dfrac{\sqrt{3}}{4}} = \frac{(a^2 + ab + b^2) - 1}{2}$$

There are 20 faces, therefore the total number of hexons will be $10(a^2 + ab + b^2) - 10$ and adding the 12 pentons the total capsomere number will be $10(a^2 + ab + b^2) + 2$. The similarity to our empirical formula $10x(n - 1)^2 + 2$ is obvious and we note that $x = 1$ corresponds to icosahedra with vertex coordinates $(a, 0)$; $x = 3$ corresponds to vertex coordinates where $a = b$ and other values of x corresponding to the skew structure are given by $a \neq b \neq 0$. It will be noted that $a = y$, $b = z$ gives a capsomere number the same as $a = z$, $b = y$, but the two forms are mirror images and represent the two enantiomorphic forms for skew structure.

It should be noted that the 'Goldberg solution' can be applied equally well to the situation where each ball is a cluster of structure units and indeed is also consistent with structures in which the array is built up with protein units placed in the spaces of Figure 6(a) with the balls of the model representing the spaces between them. Such arrangements would provide alternative explanations for the structures of, for example, reoviruses and caliciviruses, the large 'holes' we have noted in their structure being placed at ball positions on the 'Goldberg model' and protein units in the spaces between them, rather than envisaging them as being built of bulky capsomeres, with large holes, placed at positions of balls in the array of Figure 6(a).

So far then we have considered generation of icosahedral structures based on packing of capsomeres, while recognizing that these must consist of structure units. Caspar and Klug have considered the theory of formation of icosahedral shells from structure units. They recognized that it is impossible to place more than 60 identical units in an icosahedral structure in such a way that each is identically situated (just as in our capsomere model the pentons are not identically situated in terms of nearest neighbours to the hexons). They therefore adopted the criterion of quasi-equivalence where $60n$ units are arranged so that they are held together by the same bonding structure throughout, but this is deformed in slightly different ways so that the situation of each unit is only approximately equivalent. Their solution was inspired by the geodesic domes designed by Buckminster Fuller in which a spherical shell is triangulated, in the same way as the football in Figure 3(d), into 20 triangles meeting at fivefold vertices and then each triangular face is divided into smaller triangles arranged in hexagonal array. Stated in another way, these structures are derived by attempting to coat a sphere with a hexagonal array of triangles (Figure 7a). This can only be

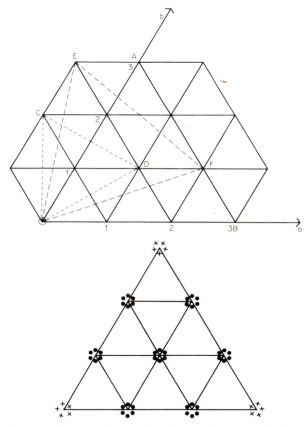

Figure 7. (a) Hexagonal array of triangles illustrating triangulation of icosahedral faces. The different arrangements of triangles on a face may be derived by placing one vertex at 0, the second at point (a,b) on hexagonal net (cf. Figure 6) with the third at corresponding point on next sextant. The triangulation number T (number of triangles per face) is given by $T = a^2 + ab + b^2$. The face OAB has $T = 9$ with triangles arranged with sides parallel to axes a,b. The face OCD has sides parallel to bisectors of the unit triangles of the hexagonal array and has $T = 3$ while face OEF represents a skew structure with $T = 7$ (*laevo*) and has edges of the face irregularly inclined to the triangular array. (b) $T = 9$ triangulation pattern with structure units placed in vertices of unit triangles producing hexamer clustering (dots) and pentamer clustering (crosses) and giving a 92 capsomere model

done if 3, 4 or 5 triangular faces, rather than 6, join at a polyhedron vertex. Caspar and Klug recognized that the departure from equivalence would be least when fivefold vertices were formed. The different possible structures can then be specified by defining the orientation of an icosahedral face to the hexagonal net of equilateral triangles. The different arrangements may be described in relation to the number of triangles per icosahedral facet, which is given by the quotient:

Area of triangular face ÷ area of 'unit triangle of the hexagonal net'

As before, the area of a triangular face is $(a^2 + ab + b^2)(\sqrt{3}/4)$ where a and b are the hexagonally based coordinates of the second vertex, relative to the origin at the first. The unit triangle has unit length (its second vertex is at point $(1,0)$) and therefore has area $\sqrt{3}/4$ (as above) and the number of triangles per face known as the triangulation number T is given by

$$T = \frac{(a^2 + ab + b^2)\dfrac{\sqrt{3}}{4}}{\dfrac{\sqrt{3}}{4}} = a^2 + ab + b^2.$$

Caspar and Klug then arrange structure units so that there are three structure units in each unit triangle, and consider that capsomeres result from clustering of structure units. Thus if the structure units are placed just inside the vertices of unit triangles, we see that clusters of six units (i.e. hexons) form at each intersection of the hexagonal net, while pentons will be formed at the vertices of the icosahedral face. In Figure 7(b) we see from the diagram that $T = 9$. The coordinates of the second vertex are $(3,0)$ so the formula gives us $T = 3^2 + (3 \times 0) + 0^2 = 9$, in agreement with our observation. Further the clustering of structure units into hexamers and pentamers has given two pentons and two hexons on each edge of the icosahedral face so $n = 4$ and the capsomere number is 92 from $10(n - 1)^2 + 2$. In the general case we see that if there are T triangles per face then there are $20T$ triangles per icosahedron and a total of $60T$ structure units. Sixty of these form the 12 pentons and there are $60T - 60$ structure units left to form $10T - 10$ hexons, so the capsomere number is $10T - 10 + 12 = 10T + 2 = 10(a^2 + ab + b^2) + 2$ as we found from our capsomere packing model.

Caspar and Klug considered that other clustering arrangements would be possible. Hence, if the structure units clustered round the centroid of each unit triangle, trimer capsomeres would result and in this structure the particle would consist of 'holes' at the conventional penton and hexon locations. We have seen that some workers have considered that the reovirus and calicivirus structures are built in this way with the large holes we have described (Figure 4b) representing spaces between capsomeres formed by trimer clustering rather than the large holes in 'conventional' pentons and hexons. It was also suggested that capsomeres could be formed at the mid points of edges of the unit triangle by dimer clustering of structure units placed on either side of the mid points of each edge.

Tomato bushy stunt virus is believed to consist of 180 structure units arranged in a $T = 3$ lattice into 90 capsomeres formed by dimer clustering.

It will be seen from Figure 7 that the unit triangles may be arranged with their faces parallel to the edges of the triangular faces of the icosahedron (as in Figure 7b) and these always have the second vertex on the a axis, i.e. $b = 0$. Another symmetrical pattern results from triangles packed with their edges parallel to the bisectors of the angles of the triangular face (e.g. triangle OCD in Figure 7a): these always have $a = b$ (e.g. $a = 1, b = 1$, as in Figure 7a gives $T = 3$ or 32 capsomeres for hexamer, pentamer clustering). Finally the unit triangles may lie in more general directions (e.g. for structure with triangular face OEF in Figure 7a) and these have $a \neq b \neq 0$ and generate skew structures. Thus the triangular face OEF in Figure 7a has vertex at point (2,1) and gives $T = 2^2 + 2 \times 1 + 1^2 = 7$. Hexamer–pentamer clustering gives 72 capsomeres for such a structure arranged in the *laevo* configuration, while the structure with vertex at the point (1,2) also gives $T = 7$ but 72 capsomeres arranged in the *dextro* configuration.

We have derived the formula for triangulation numbers in terms of the coordinates of the second vertex but Caspar and Klug in fact expressed T as Pf^2 where $f = 0, 1, 2 \ldots$ and $P = h^2 + hk + k^2$ where h,k may take any values >0 with the restriction that they must not have common factors >1 so that the possible values of P are 1, 3, 7, 13, 19, 21 \ldots P then corresponds to the x of our empirical capsomere formula $10x(n - 1)^2 + 2$.

We may note that the structure unit/triangulation model produces capsomeres only as the result of the clustering pattern of structure units and would perhaps be inconsistent with our knowledge of, for example, adenovirus where capsomeres can be found as real structures in the infected cell. More importantly the model assumes that adenovirus hexons consist of six structure units while all the experimental evidence suggests that there are only three.

C. Helical Viruses

Tobacco mosaic virus is the typical helical virus. This virus may be crystallized and X-ray diffraction analysis of the crystals showed that the virus consists of a helical arrangement of protein subunits round the RNA which is 'embedded' within the protein. The centre of the

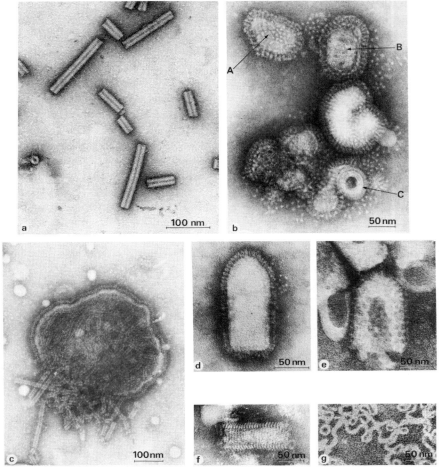

Figure 8. (a) Tobacco rattle virus showing long and short particles with helical symmetry. (b) Influenza virus particles — note spikes projecting from envelopes, triangular end view of haemagglutinin spike at A, side view of coiled ribonucleoprotein at B and end view of this at C. The background shows isolated haemagglutinin and neuraminidase spikes. (c) Measles virus particle — note threads of internal helical ribonucleoprotein and that the surface 'spikes' are less prominent than those of the influenza particle in (b). (d) Vesicular stomatitis particle; (e) rabies virus particle; (f) vesicular stomatitis virus particle after phospholipase treatment to reveal internal helical array of ribonucleoprotein; (g) internal ribonucleoprotein of vesicular stomatitis virus released by treatment of particle with sodium deoxycholate. (a) Reproduced by kind permission of Dr I. M. Roberts and Scottish Horticultural Research Institute, (b) reproduced by kind permission of Dr N. Wrigley; (c) reproduced by kind permission of Dr J. D. Almeida; (d) and (e) reproduced by kind permission of Dr C. J. Smale; (f) and (g) reproduced by kind permission of Dr C. J. Smale and Cambridge University Press

cylindrical rod is hollow. Figure 8(a) exhibits another helical virus, tobacco rattle virus which is remarkable for the fact that it has two characteristic particles, long and short. It is typical of 'divided genome' viruses. Both are necessary for successful infection, the RNA of the short particle apparently coding for the coat protein of both long and short particles. The RNA of the long particle is needed for manufacture of other necessary components in the growth cycle. The long particles are infective by themselves in that they can replicate their RNA, but such an infection does not produce particles since the short particle RNA is needed to make coat protein. Some other helical viruses (e.g. potato virus Y) have a more flexible structure and are seen as 'wavy' rather than straight rods.

A number of animal viruses have structural elements with helical symmetry enclosed within an outer envelope. Both orthomyxoviridae such as influenza and paramyxoviridae such as mumps and measles (Figure 8b,c) are in this category. The pleomorphic outer envelope with projecting spikes contains coils of ribonucleoprotein. The diameter of this is a distinguishing feature between the two families: the paramyxoviridae ribonucleoprotein helix is approximately 18 nm wide, that of myxoviridae only 9 nm. The spikes of the orthomyxoviridae are also more pronounced. In high resolution micrographs these spikes of orthomyxoviridae are seen to be of two morphological types: triangular prisms which correspond to the virus haemagglutinin and mushroom-shaped structures with a knob mounted on a thin stalk, corresponding to the neuraminidase. Freeze-etch techniques suggest that these spike peplomers may be arranged more regularly than is often suggested by negative staining. The spikes are of vital importance in the epidemiology of influenza because it is changes in these structures, both glycoproteins, which account for the antigenic variation of influenza virus either in relation to the major antigenic shifts (as for example resulted in the emergence of Asian strains in 1957 or of Hong Kong strains in 1969) as well as to the more gradual year-to-year antigenic drift.

The protein of the ribonucleoprotein determines the type of influenza — A, B, or C. It seems likely that the ribonucleoprotein exists in the virus as separate pieces. By contrast the ribonucleoprotein of paramyxoviruses may be extracted as a single coil 1 μm long. The envelopes of myxoviruses contain a host antigenic component which has been identified as carbohydrate forming part of the haemagglutinin and neuraminidase.

Rhabdoviruses such as rabies and vesicular stomatitis virus (Figure 8d,e) also contain an internal ribonucleoprotein component wrapped in an outer layer of characteristic bullet shape which gives the virus family its name (the name is derived from the Latin for a rod or index finger, the Romans not having needed a word for bullet). The envelope once again has projecting spikes of glycoprotein. The virus particle contains a RNA transcriptase probably associated with the internal ribonucleoprotein complex but not corresponding to its principal polypeptide. The internal arrangement of the ribonucleo-protein can be seen in particles treated with phospholipase (Figure 8f). The ribonucleoprotein can be released from the particle by treatment with deoxycholate (Figure 8g).

D. Other Structural Forms

The T even bacteriophages (myoviridae) have the very characteristic form depicted in Figure 9(a,b). The DNA is packed in the bipyramidal hexagonal head to which is attached the tail. This bears a sheath and wrapped round it are fibres which come into play when the virus approaches a susceptible cell. They attach to the cell and serve to draw the tail proper towards the cell wall. The pins at the end of the tail are brought into contact with the cell wall and by enzymic action penetrate it, allowing the DNA to be injected directly into the bacterium. Injection is accompanied by a characteristic retraction of the tail sheath.

The assembly of this intricate structure has been studied in some detail using a combination of electron microscopy with genetic methods. Certain mutants produce parts of particles, e.g. heads, tails, etc., and by mixing lysates from cells infected with appropriate mutants it is possible to assemble *in vitro* heads produced by one mutant with tails produced by another. Proceeding in this way infective particles can be produced by mixing appropriate components. However it is clear that some stages of assembly cannot be done in this way and require expression of certain gene functions not necessarily associated with gene products destined for incorporation in complete virus particles.

Poxviruses such as smallpox and vaccinia have a brick-shaped form (Figure 9c), quite clearly distinguishable from the herpes type particles of chickenpox with which smallpox could be confused clinically in its early stages. This led to electron microscopy being used as a diagnostic aid in differential diagnosis of chicken pox and smallpox before the virtual eradication of smallpox by the WHO vaccination campaign. The surface of the particles is covered with randomly arranged 'threads' or 'tubules'. Studies by freeze etching have suggested that each tubule is composed of a double row of spherical subunits. The viruses of the paravaccinia genus have a more regular arrangement of the external tubules, which appear to be arranged as a single thread wound round a cigar-shaped particle to produce a characteristic wickerwork appearance.

The internal structure of the poxviruses has been studied by

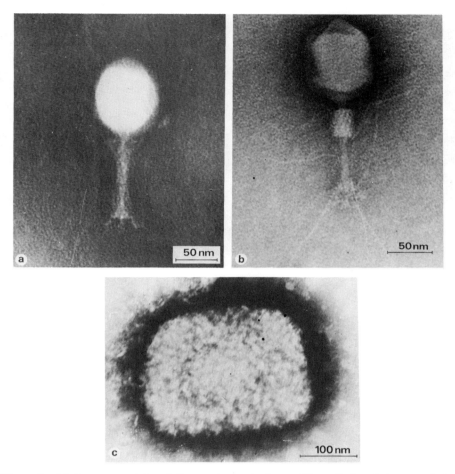

Figure 9. (a) Intact particle of T-even bacteriophage. The tail sheath is extended and the head filled with nucleic acid. The sheath shows regular striations and terminates in the baseplate. (b) Particle after treatment with peroxide which induces tail sheath contraction analagous to that occurring on attachment to host cell. The contracted sheath allows the internal tail core to be seen. Long fibres extend from the base plate. (c) Smallpox virus particle showing brick-shaped form and surface tubular structure. The core is faintly delineated within the particle by partial penetration of stain. (a), (b) reproduced by kind permission of Professor R. W. Horne and Academic Press

thin-sectioning methods. There is a well defined core with which are associated two oval shaped lateral bodies. The core is known to contain a DNA-dependent RNA polymerase which allows messenger RNA from early genes to be transcribed inside poxvirus cores.

The agent of hepatitis B (or serum hepatitis) has been associated with a double shelled particle (Figure 10a). Confusion existed at one time

4/30

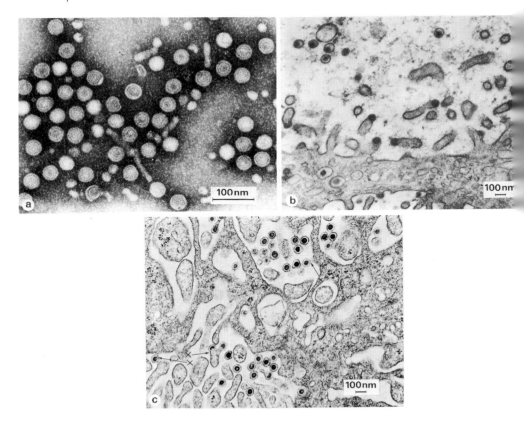

Figure 10. (a) Preparation of hepatitis B virus showing double-shelled particles thought to represent the actual virus particle, together with filaments and small spherical particles which represent aggregates of coat protein. (b) Thin section of type B particles of mouse mammary tumour virus particles in a chronically infected producer mouse mammary cell line. Note eccentrically situated nucleoid. (c) Type C particles of HL 223V-1 obtained by infection of rat cells with material from patient with acute myeloid leukaemia. Note particles 'budding' from cell membrane at X, centrally placed nucleoids. (a) reproduced by kind permission of Dr J. D. Almeida; (b) reproduced by kind permission of Dr C. Dickson; (c) reproduced by kind permission of Dr N. Teich and Messrs Macmillan Ltd.

over the different kinds of particles visible in the blood of affected patients. As well as the double shelled particles, filamentous particles and small spherical particles were observed. Immunological methods showed that all three could be agglutinated by specific antisera and that heterogeneous clumps of all three kinds could be seen. This showed all three to possess the same antigenic specificity (known as HBs, the s indicating surface). In fact the tubular and small spherical particles are now thought to be aggregates of excess coat protein. The double-shelled particles contain another internal or core antigen designated HBc and are thought to contain DNA.

Finally we may mention particles associated with oncoviruses. From observations on sectioned particles, Bernhard classified these morphologically into B- and C-type; B-type oncovirus particles (Figure 10b) such as mouse mammary tumour virus are formed by envelopment of spherical A particles found intracytoplasmically. The enveloped A particles degenerate to give B particles in which the densely stained nucleoid containing the virus RNA is eccentrically placed within the envelope.

C-type particles (Figure 10c) such as the murine leukaemia and avian leukosis viruses differ from B-type particles in having no intracellular precursor like the A particle — they apparently arise only in the final stage of assembly as the particle buds from the cytoplasmic membrane. Like B particles they contain a nucleoid within an outer envelope but it is not eccentrically situated. These particles are of considerable interest because candidate human leukaemia viruses (Figure 10c) have this morphology although it should be emphasized that neither the origin of these particles nor their exact role in human leukaemia is clear.

This chapter has only attempted to survey the structural principles of icosahedral and helical viruses and to give a few examples of the many viruses whose structure lies outside those classes. More extensive coverage of this topic will be found in the listed references.

VII. Acknowledgements

The author is greatly indebted to: Dr J. D. Almeida, Dr C. Dickson, Dr J. T. Finch, Dr T. H. Flewett, Professor R. W. Horne, Dr I. M. Roberts, Dr C. J. Smale, Dr N. Teich, the late Dr R. C. Valentine, and Professor P. Wildy, who provided illustrations, and to Dr F. Brown and Dr B. D. Harrison for arranging provision of illustrations. He would like to thank Miss Barbara E. Duncan for building the models, Dr R. W. Honess for drawing the figure and his long suffering secretary, Mrs M. Kidd, for miraculously converting his illegible hieroglyphics into respectable typed copy.

VIII. Further Reading

The Nature and Classification of Animal Viruses

Fenner, F., McAuslan, B. R., Mims, C. A., Sambrook, J., and White, D. O. (1974). *The Biology of Animal Viruses*, 2nd ed., Chapter 2. London: Academic Press.
Fenner, F. (1976) *Classification and nomenclature of viruses* (2nd Report of the International Committee on Taxonomy of Viruses). Also printed as Intervirology, Vol. 7, Nos. 1–2. Basle: Karger.

Virus Assays, One-hit Kinetics, etc.

Fenner, F., McAuslan, B. R., Mims, C. A., Sambrook, J., and White, D. O. (1974). *The Biology of Animal Viruses*, 2nd ed., Chapter 2. London: Academic Press.

Isaacs, A. (1957). Particle counts and infectivity titrations for animal viruses. *Advances in Virus Research*, 4, 112–158.

Cooper, P. D. (1961). The plaque assay of animal viruses. *Advances in Virus Research*, 8, 319–378.

Virus Structure: Theoretical Aspects

Crick, F. H. C. and Watson, J. D. (1957). Virus structure: General principles, in Ciba Foundation Symposium *The Nature of Viruses*, eds. G. E. W. Wolstenholm and E. C. P. Millar. Edinburgh: Churchill.

Caspar, D. L. D. and Klug, A. (1962). Physical principles in the construction of regular viruses. *Cold Spring Harbor Symposia on Quantitative Biology*, 27, 1–24.

Horne, R. W. and Wildy, P. (1963). Virus structure revealed by negative staining. *Advances in Virus Research*, 10, 102–170.

Virus Structure — well illustrated comprehensive accounts of structure of virus particles

Fenner, F., McAuslan, B. R., Mims, C. A., Sambrook, J., and White, D. O. (1974). *The Biology of Animal Viruses*, 2nd ed., Chapter 3. London: Academic Press. Press.

Horne, R. W. and Wildy, P. (1963). Virus structure revealed by negative staining. *Advances in Virus Research*, 10 102–170.

Horne, R. W. (1974). Virus Structure. London: Academic Press.

Dalton, A. J. and Haguenau, F. Eds. (1973). *Ultrastructure of Animal Viruses and Bacteriophages* (Vol. 5 of series Ultrastructure in Biological Systems). London: Academic Press.

Techniques for Electron Microscopy in Microbiology

Kay, D. (1977). Electron microscopy of small particles, macromolecular structures and nucleic acids. In *Methods in Microbiology*, Ed. J. R. Norris, 9, 177–215. London: Academic Press.

Structure of Viruses in Relation to Function

Wildy, P. and Watson, D. H. (1962). Electron microscopic studies on the architecture of animal viruses. *Cold Spring Harbor Symposia on Quantitative Biology*, 27, 25–47.

Watson, D. H. (1968). The Structure of Animal Viruses in relation to their biological functions. Symposia of the Society for General Microbiology. XVIII. *The Molecular Biology of Viruses*, pp. 207–228. Cambridge: Cambridge University Press.

Form and Function — IV. Protozoa

C. R. CURDS and C. G. OGDEN

British Museum (Natural History)

I. Introduction to the Phylum Protozoa

The name Protozoa means 'first-animals' and they are still defined in some traditional textbooks as simple, microscopic, single-celled or acellular animals. It will be shown in the current text that protozoa are not simple, not necessarily single celled nor acellular, and some are certainly not animals. The organisms referred to as protozoa (see Figure 1) include a diverse assemblage of nucleated microorganisms that, by a generally accepted compromise, have been fitted into a single phylum.

Protozoa possess the attributes of both cells and entire organisms. As cells they are eukaryotic, that is to say they have a nucleus with a nuclear envelope and are elaborately differentiated by a series of membrane systems, such as endoplasmic reticulum, Golgi apparatus, mitochondria, plastids, and locomotory organelles (cilia and flagella). Prokaryotic cells such as bacteria and blue-green algae are much simpler in their construction and do not contain these membrane systems. As entire organisms, protozoa must be capable of reproduction, feeding, movement, excretion, respiration, and so on. Most protozoa reproduce by some means of binary fission involving a mitotic division of the nucleus, but sexual processes are also known. In addition to the nucleus, certain cytoplasmic organelles are able to replicate either just before or at fission. The variety of nutrition and locomotion mechanisms reflects the diversity of organisms included in the phylum, but all have to perform these complex physiological processes within the confinements of a certain microscopic size. The majority of protozoa are much less than 200 μm in diameter, although foraminifera are often larger and *Stannophylum* attains several centimetres.

Protozoa are found in all moist habitats. Although many may survive arid conditions by the formation of a resistant cyst, none can feed in the absence of water, hence protozoa are common in the sea, soil, and freshwaters, whilst examples of parasitic protozoa may be found in most animal groups. Some protozoa have solved the problem of living in a changing environment by becoming adaptable in their morphology or nutrition, and parasitic protozoa commonly modify their morphology and physiology to cope with a change of host. For example, in the malarial parasite *Plasmodium* (Figure 1 F), the stimulus for the production of male gemetes is the drop in temperature that occurs on transfer from the warm-blooded mammalian host to that of the mosquito. Similar examples may be found in free-living protozoa: the soil amoeba *Naegleria* secretes a resistant cyst in dry weather, is a naked amoeba in moist soils, but produces flagella when flooded with water.

The distribution of cysts in the atmosphere and of trophic forms in the seas and freshwaters has resulted in the cosmopolitan spread of free-living species throughout the world. However, the geographical distribution of higher-animal hosts and some of Man's activities (such as land-drainage schemes and the use of insecticides to reduce the incidence of invertebrate hosts) means that many parasitic species have a restricted distribution pattern.

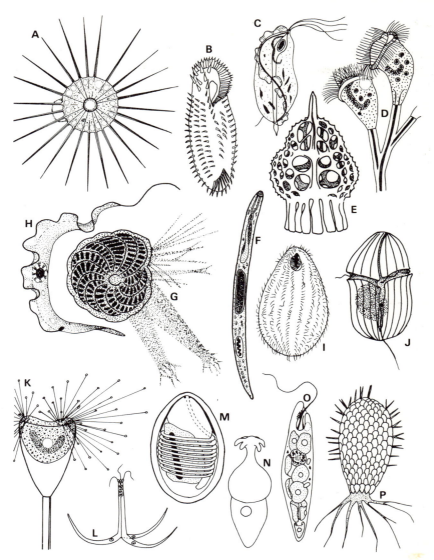

Figure 1. A diverse assemblage of protozoa. A, *Actinophrys* (diam. 50 μm), a heliozoon; B, *Paraurostyla* (250 μm long), a hypotrichous ciliate; C, *Trichomonas* (30 μm long), a polymastigote zooflagellate; D, *Carchesium* (zooids 100 μm long), a peritrichous ciliate; E, *Sepalospyris* (200 μm long), a radiolarian; F, *Plasmodium* (10 μm long), a sporozoite of the malarial parasite; G, *Elphidium* (diam. 450 μm), a foraminifer; H, *Trypanosoma* (20 μm long), a parasitic zooflagellate; I, *Tetrahymena* (50 μm long), a holotrichous ciliate; J, *Gymnodinium* (80 μm long), a dinoflagellate; K, *Acineta* (zooids 200 μm long), a suctorian ciliate; L, *Triactinomyxon* (diam. 150 μm), a cnidosporidian spore; M, a microsporidian spore (about 5 μm long); N, *Rhopalonia* (160 μm long), a gregarine; O, *Euglena* (120 μm long), a phytoflagellate; P, *Euglypha* (80 μm long), a testate amoeba. (L and M reproduced with permission from Kudo (1966). *Protozoology*. Springfield, Illinois, C. C. Thomas, Publisher.)

Table 1. Abbreviated classification of the phylum Protozoa

Subphylum 1   SARCOMASTIGOPHORA
Flagella and/or pseudopodia present, spores not produced

Superclass
1. Mastigophora.   Flagella
typically present, division by
longitudinal binary fission.

Class
1. Phytomastigophorea.   Plant-like flagellates typically with
chromatophores. Amoeboid forms in some orders. Mainly
free living, e.g. *Euglena*.

2. Zoomastigophorea.   Animal-like flagellates. Chromatophores absent, one or many flagella. Amoeboid forms with
or without flagella in some orders. Mainly parasitic, e.g.
Trypanosoma, Hypermastigida.

Superclass
2. Opalinata.   Binary fission takes place between rows of the cilia-like organelles which cover
the organism; two or many monomorphic nuclei, e.g. *Opalina*.

Superclass
3. Sarcodina.   Pseudopodia
typical, flagella restricted to
developmental stages when
present.

Class
1. Rhizopodea.   Locomotion by pseudopodia. Includes the
naked (*Amoeba*) and testate amoebae (*Arcella*), the
foraminifera (*Astrorhiza*) and the slime-mould amoeba
(*Dictyostelium*).

2. Piroplasmea.   Small, pyriform, round or amoeboid
parasites of vertebrate blood cells with ticks as vectors.
Locomotion by gliding or body flexion.

3. Actinopodea.   Spherical, typically floating forms.
Pseudopodia delicate with axopodia. Some naked, others
with tests of chitin, silica or strontium sulphate. Includes
Radiolaria, Acantharia, and *Heliozoia*.

Subphylum 2   CILIOPHORA
Cilia or compound ciliary organelles present in at least one stage of life cycle.
Two types of nuclei present and binary fission transverse.

Class
Ciliatea

Subclass
1. Holotrichia.   Body ciliature simple and typically uniform;
buccal ciliature inconspicuous, e.g. *Paramecium* and
Tetrahymena.

2. Peritrichia.   Body ciliature absent in adult; oral ciliature
conspicuous winding anticlockwise to cytostome. Sometimes colonial, often stalked, e.g. *Vorticella*.

3. Suctoria.   Adults without cilia but with tentacles;
ciliated larva produced by budding. Often sessile, e.g.
Acineta.

4. Spirotrichia.   Body ciliature sometimes sparse; cirri present
in one order. Buccal ciliature conspicuous winding clockwise to cytostome, e.g. *Euplotes, Stentor*.

Subphylum 3   SPOROZOA
                Simple spores without polar filaments but with one or many sporozoites. All
                parasitic.

Class   Subclass
1. Telosporea.   Spores present, 1. Gregarinia.   Trophozoites large and extracellular. Parasites
   flagellated microgametes.   of digestive tracts and body cavities of invertebrates.
   Pseudopodia usually absent,
   but if present are only for   2. Coccidia. Trophozoites small and typically intracellular.
   feeding.

2. Toxoplasmea.   Spores absent, pseudopodia and flagella absent at all stages.

Subphylum 4   CNIDOSPORA
                Spores with one or more polar filaments and one or more sporoplasms. All
                parasitic.

Class
1. Myxosporidea.   Spore of multicellular origin; one or more sporoplasms; two or three
   (rarely one) valves.

2. Microsporidea.   Spore of unicellular origin; single sporoplasm, single valve.

Certain protozoa are of great medical and veterinary importance. In man, *Plasmodium* is responsible for the disease malaria, while *Trypanosoma* (Figure 1 H) cause sleeping sickness in Africa and Chaga's disease in South America. In domestic animals *Trypanosoma* are the causative organisms of the disease nagana in cattle, and *Eimeria* cause coccidiosis in poultry. Free-living protozoa cannot be forgotten: the autotrophic forms are primary producers at the base of many food chains; the predatory activities of the holozoic protozoa provide the vital link in the food chain between the heterotrophic bacteria and many metazoan invertebrates and subsequently vertebrates.

It will be evident from what has been said above that it is not possible to make any all-embracing definition of the phylum Protozoa. In Table 1 we give an outline classification of the phylum adapted from that of Honigberg and coworkers (1964). All organisms are not dealt with at the same taxonomic level, but we have attempted to include all major groups. It should be pointed out to the microbiologist that unlike bacteria, protozoan classification is based almost exclusively upon their internal and external structures, hence the scheme may be used as a preliminary guide to the form of protozoa.

II. Skeletal and Protective Structures

All protozoa have skeletal or supporting structures of some form which may protect the organism from harmful external influences, sustain a definite shape, or act as the focal point for the exertion of cytoplasmic forces. The first two functions predominate and usually involve structural, external or internal, modifications to the surface layers.

A. External Coverings of Protozoa

The cytoplasm of a protozoan cell is enclosed by a unit membrane or plasmalemma; this controls the cell permeability and sensitivity to external factors, and the loss of substances from the cytoplasm. The membrane appears to have a degree of elasticity, allowing the cell to expand or contract, while contributing to the maintenance of cell shape. It is in its simplest form in amoebae, where the production of pseudopodia cause a continual change of body shape without affecting the volume of the cytoplasm. A glycocalyx of fringe-like processes is often present on the outer surface of amoebae, and this is thought to play an important role in pinocytosis or in adhesion of the cell to the substratum. In tectinous allogromiid foraminifera, a thicker modified glycocalyx, consisting of fibrillar material and often with an outer electron-dense layer, is present. Further specialization of this layer is seen in *Gromia oviformis*, where organically bound ferric iron is selectively retained and a complex system of honeycombed membranes is present.

The cell membranes of flagellates and ciliates are continuous with their locomotory organelles. The chrysomonad flagellates are frequently covered, even the flagella, with a layer of siliceous scales while the coccolithophorid flagellates have similar but calcareous scales. In the phytomonads the unit membrane is enclosed within a homogenous layer containing cellulose and pectin which is rigid. Flexibility of the cell is achieved in some species of *Euglena* by the pellicle being divided into ridges separated by grooves. A continuous skeleton of cellulose is present in some dinoflagellates (Figures 1 J and 2 J) which may take the form of elaborately embossed armour. The pellicle of *Tetrahymena* is typical of many ciliates. Below the surface membrane are two further membranes, an outer lying close to the surface membrane and an inner which is underlain by an electron-dense layer of cytoplasm called the epiplasm. Some ciliates secrete a pseudochitinous lorica; exoskeletons of this type are sometimes found in sessile peritrichs (Figure 2 I), planktonic tintinnids (Figure 2 H), and suctoria (Figure 1 K).

An exoskeleton in the form of a shell is constructed by many rhizopod protozoa. The shells may be composed of proteinaceous material alone, but more often siliceous or calcareous mineral deposits are incorporated in the organic matrix. Testate amoebae construct a single-chambered shell (Figure 1 P), usually with one aperture, which may be proteinaceous, agglutinate, siliceous or occasionally calcareous. In contrast, the foraminifera construct shells that are often multi-

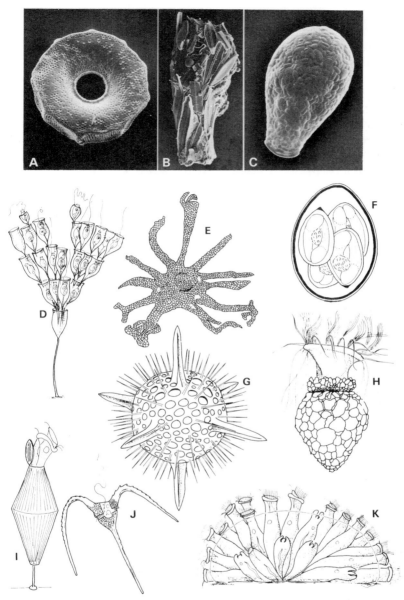

Figure 2. External coverings of protozoa. A, *Arcella* (50 μm long), a testate amoeba with proteinaceous shell; B, *Difflugia* (150 μm long), a testate amoeba with an agglutinate shell; C, *Nebela* (60 μm long), a testate amoeba with captured shell plates; D, *Stylobryon* (individuals 50 μm long), a colonial phytoflagellate with pseudochitinous lorica; E, *Astrorhiza* (diam. 15 mm), a foraminifer covered with sand grains; F, *Eimeria* (22 μm long), a sporulated oocyst; G, *Hexacontium* (diam. 210 μm), a radiolarian with radially arranged spicules; H, *Tintinnopsis* (50 μm long), with lorica composed of sand grains; I, *Pyxicola* (300 μm long), a ciliate with lorica having a retractable operculum; J, *Ceratium* (diam. 250 μm), an armoured dinoflagellate; K, *Ophrydium* (colony diam. 800 μm, individuals 300 μm long), a colonial ciliate living within a gelatinous matrix

chambered (Figure 1 G), with one or more apertures, which are mainly calcareous although there are some proteinaceous and agglutinate forms.

While the proteinaceous shells of the tectinous foraminifera are usually flexible, those of testate amoebae are rigid. The proteinaceous testacean shells may be divided into those which have a thin continuous envelope and those whose walls are constructed of numerous alveoli (Figure 2 A). Animals with agglutinate shells also fall into two categories, namely those that select their shell material and those that are non-selective. Examples of species that are selective include the foraminifer *Technitella thompsoni* which constructs a shell entirely of echinoderm plates, and the testate amoeban genus *Nebela* (Figure 2 C) whose species produce shells constructed from the siliceous shell plates of other testate amoebae which they ingest. Those that are non-selective include some species of the foraminifera genera *Schizammina* and *Astrorhiza* (Figure 2 E), and some of the testate genus *Difflugia* (Figure 2 B). Perhaps the greatest diversity of shells is displayed by the calcareous foraminifera which usually have many chambers and are formed as an organic layer reinforced with calcite. The shells grow by the addition of new chambers either over existing chambers or simply by the addition of a chamber to the previous chamber. The resultant shapes exhibit a range of forms including long tubular shells (*Rhabdammina*); oval sacs (*Saccammina*); funnel-shaped structures (*Jaculella*); regular spirals (*Ammodiscus*), and plaited braids (*Textularia*).

B. Internal Supporting Structures

While exoskeletons are commonly found in protozoa, rigid internal skeletons are more rare and, with few exceptions, limited to the heliozoa and radiolaria (Figures 1 E and 2 G). In these groups the skeleton is composed mainly of siliceous spicules which usually fuse to form elaborate and often beautifully symmetrical structures, but the spicules may sometimes remain separate, and freely distributed throughout the radiolarian cytoplasm. Although the shape of the skeleton is varied, they may be divided into two main groups, the asteroid type with radially arranged spicules (Figure 2 G) that more or less converge on the centre, and the spheroid type consisting of latticed spheres or tangential layers distributed around the cell's periphery (Figure 1 E).

Although hard internal skeletons are relatively rare, many protozoa contain soft cytoskeletal elements. These structures are known as pellicular microtubules and often lie in a zone of peripheral cytoplasm from which other organelles are absent.

The walls of the buccal cavities of ciliates and some flagellates are lined with microtubules where they may be involved with the packaging of food vacuoles. Furthermore, the axostyles of trichomonad flagellates (Figure 1 C), which usually traverse the length of the cell, are composed of numerous microtubules.

C. Cysts

Many protozoa form resistant cysts at certain times of their life cycle. These cysts may protect them from adverse environmental conditions such as desiccation, exhaustion of food supplies, and anaerobiosis. In parasitic species the developmental stages are often transmitted from host to host within a cyst. Indeed intestinal protozoa comprise about one-half of the parasitic species, and in most cases they enter the alimentary tract as a resting cyst, hatch in a suitable region, and then leave the host within a resting cyst. Most cysts are resistant and can survive for long periods outside the host. Asexual reproduction in some flagellates and ciliates is associated with the formation of a cyst, and sexual reproduction of sporozoa invariably takes place within a common cyst.

The cyst wall is secreted as a closely fitting extracellular coat; it may consist of several layers and have a special region differentiated for excystment. It is composed of a proteinaceous material which may be tanned, or of two different layers, as in the amoeba *Acanthamoeba*, which has an outer phosphoprotein layer and an inner cellulose one. The cytoplasm is commonly attached to the cyst wall at one or more points, is reduced in volume and remains dormant.

D. Other Protective Structures

Some protozoa are able to protect themselves by means other than the possession of solid external coverings; certain ciliates secret a mucilage from subpellicullar vesicles called mucocysts, and several protozoa defend themselves by the expulsion of harpoon-like trichocysts.

Toxicysts have a thread-like tubular structure, with an occlusion at the distal end which may contain the toxin. When the toxicyst is discharged the toxin is distributed along the surface of the thread. Haptocysts, which occur on the tentacles of suctoria are smaller than trichocysts, and are thought to be used to contact and immobilize prey.

III. Movement

All protozoa move in some way; this may simply involve the transportation of cellular contents around the body, a temporary change in shape, or result in the locomotion of the entire cell. While movement may be in response to a wide range of external physico-chemical stimuli, earlier claims of neural or neuroid systems being involved (e.g. fibrillar systems in ciliates) have not been substantiated. The presence of non-striated microfilaments in contractile organelles such as the myonemes of *Carchesium* (Figure 1 D), stalks and in the M-bands of *Stentor*, has led to the belief that these filaments are contractile in nature. Microfilaments appear to be constructed either of a single row of globular protein molecules or of several of these rows woven together. There is now an accumulation of evidence which suggests that many protozoan microfilaments are actinomyosin-like in nature and reaction. That is to say, they may contain F-actin similar to the actinomyosin protein system of muscle which can be made to contract using adenosine triphosphate (ATP) as the energy scource. The presence of actomyosin-like proteins is now well established in a wide range of protozoa and the cilia, flagella, and contractile organelles of widely different protozoa have been shown to use, and be stimulated or accelerated by, ATP.

A. Pseudopodia and Amoeboid Locomotion

The rhizopods and some flagellates possess locomotory and feeding structures in the form of flowing cytoplasmic extensions of the cell known as pseudopodia. Four major types, known as lobopodia, filipodia, reticulopodia, and axopodia, may be recognized by their morphology. Lobopods (Figure 3 A) are broad, blunt pseudopodia, and most students will be familiar with this type from their observations of *Amoeba*. Filipodia (Figures 1 P and 3 B) may be found in some of the smaller amoebae, where they form a small or extensive branching system of tapering pseudopodia which resemble reticulopods, but do not anastomose and are less granular than the latter. Reticulopodia (Figures 1 G and 3 C) are best known in the foraminifera where they characteristically form a fine, anastomosing, granular network of pseudopodial strands. Axopodia (Figures 1 A and 3 D) are the long, straight, and slender pseudopodial structures that radiate from heliozoa and contain a central fibrous axis composed of microtubules.

Amoebae move in many different ways and the term 'amoeboid movement' usually refers to that of *Amoeba proteus* for which dozens of different theories have been proposed. While it is certain that amoeboid movement involves the cyclical conversion of a semirigid, fibrous outer tube of cytoplasm, the plasmagel, into a less fibrous and more fluid internal stream of cytoplasm, the plasmasol, there are problems when attempting to define the origin of the motive force. Conversion of gel to sol takes place at the rear of the cell and the sol flows forward within the tube of gel to revert back to gel at the anterior. Thus while the plasmagel is stationary relative to the sub-

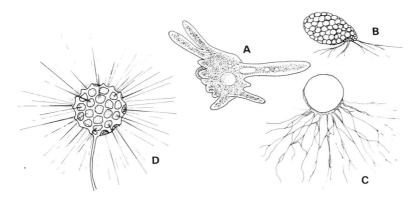

Figure 3. Pseudopodia in the rhizopods. A, Lobopods of *Amoeba*; B, filopods of testate amoeba *Trinema*; C, reticulopods of *Gromia*; D, axopods of the heliozoon, *Clathrulina*

stratum, the continual cyclical conversion of gel to sol to gel at extreme ends of the cell results in the movement of the entire amoeba. There are two major theories concerning the source of the motive force. The most popular is the contraction—hydraulic theory which maintains that the conversion to sol squeezes the fluid plasmosol forward. The alternative hypothesis – the fountain zone frontal eversion theory – is that the site of contraction of the plasmagel is at the anterior where the plasmagel tube reforms from the streaming fountain of sol, so that the animal is pulled forward.

B. Gliding

Gregarines and certain euglenoid flagellates may be observed to glide over a mucous track which is secreted by the cell. Although a realistic theory has not yet been evolved it would appear that the gregarine pellicle and adjacent layer of cytoplasm are involved.

C. Structure of Cilia and Flagella

Cilia and flagella are most commonly associated with locomotion but it should be remembered that they are just as important for the promotion of feeding currents. Both organelles have the same basic organization and it is remarkable how similar these structures are throughout eukaryotic cells.

Cilia and flagella (Figure 4) are cylindrical extensions of the cell that vary in length, although the latter tend to be longer than cilia. Each is composed of a bundle of microtubules, the axoneme, which runs straight along the longitudinal axis and is enclosed within a unit membrane that is continuous with the outer membrane of the cell. The axoneme originates below the cell surface from a basal body known either as the blepharoplast in flagellates, or the <u>kinetosome in ciliates.</u> Sections taken along cilia and flagella reveal that structural changes occur at intervals. Deep in the region of the basal granule there is a circle of nine peripheral groups of microtubules, each consisting of three fibrils a, b, and c (Figure 4) which share adjacent walls. The groups are not aligned along the radius of the organelle but are set at an angle to it. Three strands connect the microtubule groups to each other giving the overall appearance of a cartwheel. One of the microtubules (c) terminates higher up the basal granule leaving nine doublets (a and b) that extend up the shaft of the organelle. At the level of the cell surface there is a granule or transverse plate, and this may fill much of the area within the circle of nine doublets. It is from this plate that two central microtubules emerge and these extend the complete length of the organelle. Thus for the most of the length of the shaft there is a peripheral ring of nine doublets and a central separate pair of micro-tubules. At the tip of the flagellum a single peripheral microtubule (b) may terminate before the other (a), and at the extreme tip often only the central microtubules remain, resulting in narrowing of the tip.

It is probable that the central microtubules are not directly involved

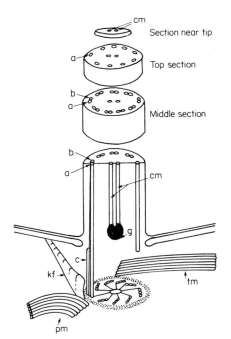

Figure 4. Diagrammatic representation of a cilium and its associated microtubular structures in *Tetrahymena*. Cilium is 0.3 μm in diameter and other structures are approximately to same scale (a, b and c — peripheral microtubules; cm — central microtubules; g — granule; kf — kineto-desmal fibril; pm — post-ciliary micro-tubules; tm — transverse microtubules)

in the bending of the flagellum, and it has been suggested that the mechanism underlying the actual movement of the organelle is the differential activity of the peripheral microtubules. There is evidence to suggest that while microtubules do not contract they might slide past one another, and this has led to the assumption that bending is based on a sliding-tubule mechanism comparable to that of muscle contraction.

In ciliated protozoa the body cilia are often arranged in longitudinal rows known as kineties. These may be stained using silver impregnation methods, and the kinetosomes and associated fibrillar structures are commonly referred to as the 'silver-line system'. The actual arrangement of the kinetosomes and fibrils varies considerably but that of *Tetrahymena* may be used as a fairly typical example (Figure 1 I). In this ciliate there are three ectoplasmic fibrils associated with the inner end of the kinetosomes (Figure 4). The largest, or kinetodesmal fibril, arises from the anterior right of the kinetosome, passing forward, right, and up towards the inner membrane of the pellicle. There it runs alongside the kinetodesmal fibril of the adjacent cilium before it terminates. Thus the kinetodesmos which is the longitudinal fibre travelling along the right hand side of the kinetosome row is not a continuous fibre as is apparent from the light microscope, but a series of overlapping fibrillar units.

D. Locomotion by Flagella

Flagella are usually long and few in number. While they are characteristic of the superclass Mastigophora (Figure 1 C, H, J, and O), they may also be found in some developmental stages of rhizopods such as swarmers of radiolaria, gametes of foraminifera, and the flagellate stage in some amoebae, and also appear in some sporozoa, e.g. microgametes of coccidia. Flagella are morphologically diverse and many have laterally projecting 'hairlets' which may be either thin, flimmer filaments, or thick and stiff, mastigonema.

When an organism has more than one flagellum, they are commonly arranged differently or may be of different lengths or diameters. Flagella move in an undulating manner and the waves may be in a single plane (planar) or be out of plane to various degrees (helical). The undulations may travel in either direction although it is more common to find the wave initiated at the base rather than the tip. Movements passing along the flagellum result in a force upon the surrounding water which acts along the flagellar axis, the force being in the direction of the undulation when the flagellum is smooth but it is reversed when coated with mastigonemes.

The locomotory flagellum of *Trypanosoma* (Figure 1 H) emerges from a pocket at the posterior end of the cell and is attached along its length to the organism except for an anterior free portion which extends beyond the body. When the flagellum beats, the pellicle is pulled up into a fold and this plus the flagellum is known as the undulating membrane which is believed to be an adaptation to moving in a viscous fluid, such as blood.

E. Locomotion by Cilia and Their Coordination

The stroke that produces the greatest force upon the surrounding water (the effective stroke) is the stiff forward movement of the cilium in one plane, bending taking place only at the base. In the recovery stroke, the cilia sometimes swing slightly to the left, and the bending zone moves progressively up the shaft towards the tip so that they unroll towards their starting point. When a cilium moves, it carries with it a surrounding layer of water, the extent of which is dependent upon the viscosity of the liquid and the velocity of the cilium. Hence the fast 'effective' stroke carries more water with it than the slower 'recovery' stroke. It is well known that cilia beat in a coordinated fashion but there is some controversy on how this is achieved. The cilia beat at about the same frequency but each cilium is slightly out of phase in the beating cycle with its immediate neighbours. This produces the effect known as metachronal rhythm and ensures that an approximately equal number of cilia at any phase in the beating cycle are spread over the organism's body. An equal spread produces a smooth propulsive force rather than the jerky erratic movement that would result if all cilia beat in unison.

There are two theories on how metachronal rhythm is induced. One suggests that ciliary coordination is achieved via nerve-like or neuroid impulses passing along the kinetodesmata. This theory has now lost favour and there is an increasing tendency to accept the mechanical theory, which states that when two cilia lie sufficiently close together so that their transported water layers overlap, the cilia become hydro-dynamically linked.

F. Compound Ciliary Organelles

In some ciliates the cilia are modified to form compound organelles that are specialized for locomotion, feeding, or both. These are called undulating membranes, membranelles or cirri according to their structure. An undulating membrane is simply a line or arc of cilia set close together, in a single row, so that they more or less permanently coalesce into a membrane. They are set on the right of the buccal cavity and although generally small, in some holotrich genera such as *Uronema* and particularly *Cyclidium*, the membrane is generally enlarged to act as a feeding net.

Conversely, membranelles are composed of two or three rows of cilia which adhere together to form triangular or trapezoidal-like flaps. They are typically arranged on the left of a buccal cavity and beat towards the cytostome or mouth. Membranelles in the subclass Spirotrichia (Figure 1 B) are greatly enlarged usually with a corresponding reduction in the body cilia. A collection of membranelles is commonly referred to as the adoral zone of membranelles (AZM).

The third type of ciliary organelle — the cirrus — is also found in the spirotrichs and is a dominant feature of the hypotrichs. Cirri are complexes of numerous long cilia loosely grouped together to form stout, tapering organelles that are rounded in cross-section. Although coordinated, cirri do not beat uniformly but are used for 'walking' over solid surfaces.

G. Organelles of Attachment

The ability of a protozoon to maintain its position in a favourable environment is useful when there is a flow of liquid, such as river water or gut contents, which might dislodge it, or when the organism is epizoic on a mobile host. Adhesion can be achieved by the use of specialized anchor-like organelles, stalks, threads, elaborate sucker-like structures or by cellular secretions of adhesive.

Some phytoflagellates possess adhesive threads of similar dimensions to flagella called haptonemes (Figure 5 B). These structures may rapidly contract and slowly uncoil but the former condition appears to be the relaxed position. Haptonemes originate between two flagella bodies as a hexagonal array of microtubules which is reduced progressively from nine to eight and then seven before the shaft emerges above the cell surface to be formed into a loose ring surrounded by three concentric membranes.

Stalk production is a common method of attachment, for example some flagellates form colonies (Figures 2 D and 5 D) on branched stalks after the settlement of an individual swarmer, and many ciliates, particularly the suctoria and peritrichs (Figures 1 D and 2 I), secrete stalks. The stalk consists of a wall and a system of tubular fibres with a static or elastic function. In the latter, the fibres are arranged in bundles called spasmonemes (myonemes) and are responsible for the contractile nature of these stalks, as found in *Vorticella, Carchesium,* and *Zoothamnium*.

Mobile peritrichous ciliates such as *Trichodina* (Figure 5 E), are parasitic on fish and have a large adhesive or sucker-like basal disc, often with a complex arrangement of tooth-like denticles; the sucker

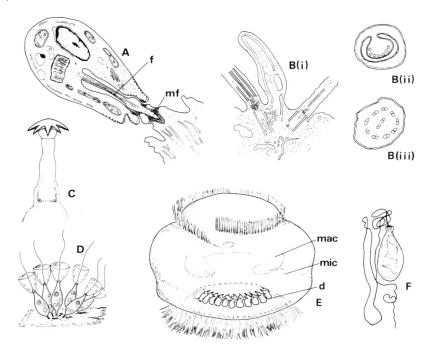

Figure 5. Organelles of attachment in protozoa. A, Section through *Crithidia* attached to gut of *Anopheles* showing concentrations of micro-filaments (mf) beneath adhering membrane of the flagellum (f); B, haptoneme (i) showing origin between two flagella, (ii) section through haptoneme (diam. 0.3 µm), (iii) section through flagellum (diam. 0.3 µm); C, epimerite (100 µm long) of the gregarine *Hoplorhynchus*; D, stalked choanoflagellates; *Monosiga* (18 µm long); E, *Trichodina* (diam. 80 µm), an epizoic ciliate parasite of fish; F, *Stempellia*, a microsporidian spore (12 µm long) with expelled polar filament

enables the parasite to change its position on the host's surface and gills. Internal parasites often have anchor-like organelles, for example the anterior region of a gregarine carries an organelle called an epimerite (Figures 1 N and 5 C) which serves to attach the young parasite to the host. Epimerites may be simple knobs, proboscis like, frilled or hooked. In each case the parasite retains a hold such that its body projects into the lumen of the host's alimentary tract. Cnidocysts are a feature of the spores of the Cnidosporida. They take the form of coiled threads within polar capsules (Figure 1 M) which, when ejected (Figure 5 F), act as temporary attachments to the host tissue.

Trypanosome flagellates adhere to the gut wall of their insect hosts, to debris, and to each other in rosette groups by means of their flagella. At the site of attachment (Figure 5 A), there are concentrations of microfilaments beneath the adhering membrane of the flagellum which form the adhesion plaques or desmosomes.

IV. Cell Physiology

A. Nutrition

Organisms may be divided into two groups: those which can use simple inorganic components to build larger molecules are known as autotrophs and those which can only utilize food materials manufactured by other living organisms are known as heterotrophs. Examples of both modes of nutrition are commonly found in the phylum Protozoa.

1. Autotrophy

Autotrophic organisms may be subdivided into those that require sunlight as an energy source (phototrophs) and those which rely upon chemical reactions (chemotrophs). Chemotrophy is not known in protozoa but phototrophy is the usual and common method of nutrition in the class Phytomastigophorea. The most obvious structural feature of autotrophic flagellates (Figure 1 O) is the presence of chlorophyll-bearing plastids called chloroplasts. The latter are limited by two unit membranes within which are enclosed numerous lamellae, and it is within these lamellae that the photosynthetic pigments are situated. The number and size of chloroplasts varies; phytomonads have a single cup-shaped chloroplast, whilst the other flagellates have several which are usually smaller and evenly distributed throughout the cytoplasm. In some chrysomonads and cryptomonads the chloroplasts are enclosed in another double membrane which is connected to the nuclear envelope. Chloroplasts contain 5–8 per cent chlorophyll, but carotenoids f and b_3, lipids, proteins, ribonucleic acid (RNA), and deoxyribonucleic acid (DNA) are also present. The presence of the two nucleic acids has led to the theory that chloroplasts were originally mutualistic autotrophic microorganisms which invaded protozoan cells early in evolutionary times. The chloroplasts of some flagellates contain dense regions of the matrix known as pyrenoids. These form the centre for the production of reserve carbohydrates and may be surrounded or capped by large grains of starch or other polysaccharide material. While the pyrenoid is usually contained within the chloroplast it can occasionally be a stalked projection from it.

Few flagellated protozoa are completely autotrophic in their nutrition and many that are photosynthetic are also known to be able to grow heterotrophically in the dark using organic carbon sources such as acetate.

The autotrophic mode of nutrition is also important to some other protozoa since several contain mutualistic photosynthetic algae. The ciliate *Paramecium bursaria* is a well known example that contains *Chlorella* cells. The ciliate obtains some photosynthate from the alga which apparently utilizes the nitrogenous wastes of the host.

2. *Heterotrophy*

Heterotrophs are more widespread in the phylum than are the autotrophs and they too may be subdivided into two groups. Those which absorb large organic molecules are said to be osmotrophic (sometimes called saprozoic). When the molecules are sufficiently small they may enter, by diffusion or by active transport, direct through tne cell wall while those too large are taken in by pinocytosis (Figure 6). The latter may be defined as the ingestion of fluids by a cell. In amoebae and other protozoa a long pinocytosis channel is formed during this process, which subsequently pinches off to form vesicles.

The other group of heterotrophic organisms are those which ingest particulate food materials and these are said to be phagotrophic (holozoic). Phagotrophy is perhaps the most common protozoan method of obtaining food and is an animal characteristic. Particulate foods are usually other organisms which vary in size from bacteria to small crustacea and flatworms. There is a wide variety of ways in which the particles enter the cell but in all cases a food vacuole or phagotrophic vesicle is formed (Figure 6). The prey organisms are killed and digested by the enzymic contents of lysosomes which surround the food vacuole. Lysosomes are thought to be formed in the region of the Golgi complex (Figure 6). They have a single unit membrane and contain a variety of hydrolytic enzymes with a common optimal acid pH capable of attacking the major classes of organic molecules as well as simpler ones such as amino acids and monosaccharides. During digestion, pinocytosis vesicles may be seen to form around the food vacuoles and the latter shrink slightly. Eventually undigested remains are discharged from the cell at its surface; in some organisms such as amoebae, discharge may take place via any part of the cell membrane while in others, such as ciliates, a permanent pore or cytopyge may be present.

3. *Food capture*

Protozoa capture their food in a variety of ways and many specialized structures have developed to carry out these activities. In all cases, however, the food enters the cell at the cytostome which may be

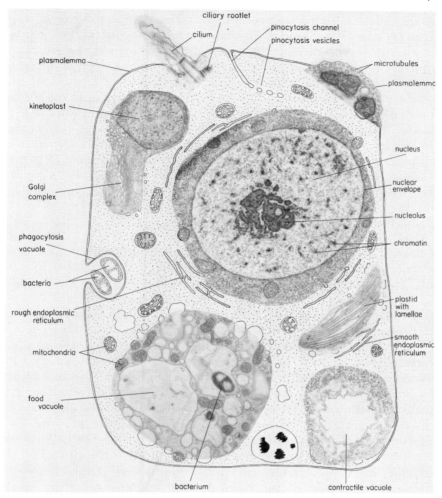

Figure 6. A generalized protozoan cell, a collage of electron micrographs and line drawings. It should be noted that all organelles have a double membrane whereas only the plasmalemma is depicted as such in the diagram

a permanent structure as in some ciliates, flagellates, and sporozoa, or it may be temporary, as in the amoebae. Some protozoa have a single ingestion site while others, such as suctoria and some sporozoa, may have several.

In rhizopods, particulate food is captured by the formation of a cup of cytoplasm around the prey, which completely engulfs it. The majority of small amoebae feed upon bacteria but some of the larger species, such as *Amoeba proteus* and *Pelomyxa carolinensis*, ingest larger prey, such as flagellates, ciliates, and even rotifers, but it is not known how these active prey are immobilized while being engulfed.

Intracellular parasites rely upon the products of the cytoplasm of the host as a nutrient. Telosporidians, piroplasms, and zoomastigotes possess a cytostome but the microsporidia and haplosporidia do not. The cytostomes or micropores of the telosporidia and piroplasms are usually permanent structures which take the form of circular concavities in the cell surface. In many haemosporidia the host-cell cytoplasm is ingested in small vacuoles which are pinched off from parts of the cytostomial cavity. Several of these vacuoles, containing ingested host tissue, then pass to a digestive vacuole where electron-dense spheres (probably lysosomes produced by a Golgi-like body) appear around the periphery.

Many protozoa produce feeding currents by the beat of their cilia or flagella. In the more advanced of the ciliate orders, certain cilia are specialized to become powerful organelles of feeding and locomotion. This is perhaps most obvious in the peritrichs such as *Vorticella*, and spirotrichs such as *Stentor* and *Euplotes*, where the cilia coalesce to form a large adoral zone of membranelles. In other ciliates, such as *Pleuronema* and *Cyclidium*, there is an undulating membrane enlarged to form a prominent sail or net-like structure around the buccal overture to trap food particles. Feeding nets are produced by some sarcodines, such as heliozoa, radiolaria, foraminifera (Figure 1 G) and dinoflagellates — but in these cases the net is formed by an array of pseudopodia that sometimes anastomose.

In the suctoria, the adult cells do not have cilia but have suctorial tentacles which extend out around the perimeter of the cell (Figure 1 K). When a suitable prey collides with a tentacle it is captured by the discharge of small bodies known as haptocysts. These organelles develop in the body of the suctorian and migrate to the tip of the tentacle where the cell membrane is protruded by the haptocyst. Discharge is caused by the collision of an appropriate object and the haptocyst penetrates the cell wall of the prey. The contents of the prey are then sucked out through the hollow tentacle leaving a husk of cell wall which is discarded. Suctorian tentacles are supported internally by a hollow cylinder of microtubules through which the prey passes. Some suctoria have tentacles designed for different functions; *Ephelota gemmipara*, for example, catches the prey with long tapering tentacles which bend to pass the food to blunt, short, contractile feeding tentacles. Electron microscopy of these two types of tentacles has revealed that the prehensile ones are supported by two or three continuous groups of microtubules separated by a septum, while the feeding tentacles only have a single group of longitudinal microtubules.

It is argued that the microtubules are responsible for the bending and contraction.

B. Energy Production

In protozoa energy may be derived either from sunlight or by the breakdown of organic molecules; it is stored in the chemical bonds of organic compounds within the cell. Energy is biologically most useful in the form of adenosine triphosphate (ATP) in which it is carried in energy-rich phosphate bonds. Some of the most elaborate membrane systems are concerned with energy production and ATP is built up largely in the mitochondria (Figure 6) of the heterotrophic protozoa and in the plastids of autotrophs.

Mitochondria are granular or filamentous structures of variable size that may be carried around the cell by cytoplasmic streaming although they tend to be concentrated in regions of high energy consumption such as near to the basal bodies of cilia and flagella. All mitochondria are bounded by two unit membranes of which the inner forms elaborate internal invaginations, called cristae, to increase the surface area, and the interior is filled with a ground matrix. Tubular cristae are found in most protozoa but discoid or plate-like cristae occur in some flagellates. Cell-free extracts of mitochondria have shown that they contain many enzymes, particularly those of the citric acid (Krebs) cycle and of the respiratory chain. Anaerobic protozoa, such as rumen ciliates, flagellates in termites, entamoebae, etc., may either have simple sacks that lack cristae or have no mitochondria. Mitochondria multiply by fission and some contain a self-replicating mitochondrial DNA.

The characteristic kinetoplast (Figure 6) of certain flagellates such as *Bodo* and *Trypanosoma*, may be regarded as a specialized region of what is usually a single mitochondrion. They are limited to one per cell except in *Costia*, which has several, are larger than most mitochondria, and are always located behind the basal body of the flagellum. These are self-replicating bodies that contain and synthesize DNA. Their fine structure is very similar to that of mitochondria but the interior contains a fibrillar material which represents the DNA content. The DNA content of the mitochondria of protozoa has led to the theory that mitochondria like chloroplasts may have been derived from a mutualistic association with another microorganism at a very early stage in the evolution of eukaryotic cells.

Other organelles involved in energy production are the plastids, described earlier in the section dealing with autotrophy, and the peroxysomes. The latter are small bodies, often called microbodies, containing an inner matrix of finely granular material limited by a single membrane.

C. Osmoregulation

Whilst contractile vacuoles (Figure 6) are most commonly found in free-living, fresh-water protozoa they also occur in some marine and parasitic species. Typically, contractile vacuoles are positioned close to the plasma membrane where they swell slowly (diastole) before suddenly collapsing (systole) to release their contents to the external environment. In many protozoa the vacuoles occur in specific places in the cell and there may be permanent pores leading outside and canal systems draining towards the vacuoles. It is difficult to decide if the vacuole forms in a specific position in many amoebae, but in *Thecamoeba verrucosa* which has a definite polarity, it always forms in the posterior of the cell. The formation of contractile vacuoles in some ameobae takes place by the fusion of numerous small vesicles; however, this is not simply an accumulation of liquid since in some they have been shown to be bounded by an elastic, semipermeable unit membrane. Hence the pulsating appearance and disappearance of the vacuoles is connected with the continuous formation and breakdown of membranes. Although the structural basis for contraction remains unclear, there is evidence that in some amoebae the cell membrane may supply the expulsion force while in others fibrillar structures have been observed bordering the vacuoles.

The contractile vacuoles of ciliates are more specialized; they are constant in number, have definite pores and are often served by long canals. In *Paramecium* there are two vacuoles which pulsate alternately; each consists of a central vacuole surrounded by several radiating canals which terminate distally in a wickerwork system of branched tubules. The canals fill and open into the central reservoir. The latter apparently fills and contracts by means of bundles of fibres running spirally out from the discharge pore over the reservoir to the canals.

While osmoregulation seems to be the primary function of contractile vacuoles, it is often assumed that they are also concerned with the excretion of nitrogenous wastes. Although urea has been identified in the vacuolar fluid of the ciliate *Spirostomum*, biochemical evidence suggests that most protozoa are ammonotelic, thus most nitrogenous waste is liberated as ammonia which rapidly diffuses away.

V. Growth and Reproduction

Although the majority of higher animals reproduce by sexual means, as a general rule single-celled organisms increase their kind by asexual reproduction usually dividing by binary fission after the cell has grown to a certain size and this is considered to be a primitive characteristic. This is not to say that sexual processes are absent in protozoa, indeed many are able to undergo both asexual and sexual processes, and in some parasitic forms there may be an asexual phase in one host and a sexual phase in another.

A. Organelles Associated with Growth and Reproduction

Several organelles in the protozoan cell are associated with growth and reproduction and the most obvious is the nucleus (Figure 6). All nuclei contain the genetic material DNA in the chromosomes of the nucleoplasm. The resting or interphase nucleus is composed of a granular matrix, chromosomes, one or more nucleoli, and is surrounded by a nuclear envelope. The nucleus has two major functions, the replication of the genetic material of the cell and the release of information to the biosynthetic machinery of the cell. Replication involves the synthesis of new DNA to duplicate the chromosomes and subsequently the separation of chromosomes into new daughter nuclei immediately before cell division.

Deeply staining areas of the nucleus represent the chromatin (DNA) bodies which in electron micrographs may appear as aggregations of an electron-dense material in the resting nucleus. During mitosis or meiosis the chromatin condenses into the rod or thread-shaped chromosomes which are the only nuclear structures capable of identical reduplication. The number of chromosomes is constant in each species and they may be helically coiled to varying degrees. The cycle of chromosomal condensation and decondensation during nuclear division depends upon the coiling and uncoiling of these threads. The size of chromosomes varies, they are small in cryptomonads and amoebae but are large structures in hypermastigine flagellates and some radiolaria. Some protozoa have a haploid (single) set of chromosomes, others a diploid (double) set, and some a polyploid (more than double the haploid number) set. For example, the polyploid *Amoeba proteus* has more than 500 chromosomes. Nuclear division normally takes place by mitosis so that each daughter nucleus receives the same number of chromosomes as the original parent nucleus, but in sexual reproduction a reduction division known as meiosis is necessary to produce haploid gametes which then fuse to form a diploid zygote.

Nuclei contain one or more bodies known as nucleoli (Figure 6) and these are often associated with one or more of the chromosomes. They are rich in RNA but lack DNA, and are composed mainly of small particles similar to ribosomes. Ribosomes are the sites of protein synthesis and it is now widely believed that ribosomal RNA originates in the nucleolus. Ribosomes may lie free in the cytoplasm but are also attached to the membranes of the nuclear envelope and endoplasmic reticulum.

The two membranes which surround the nucleus make up the nuclear envelope that is characteristic of eukaryotic cells. The membranes are punctured by regular pores which may connect the nucleoplasm with the cytoplasm although there is some doubt whether or not there is a free exchange between the two. The outer of the two membranes is continuous with the elements of rough endoplasmic reticulum which normally surround the nucleus and in some phyto-flagellates may be continuous with the outer membrane of the plastid. It is thought that the vesicles of the endoplasmic reticulum are used to manufacture and process substances and that the reticulum surfaces serve as sites upon which enzyme systems may be arranged. Both smooth and rough endoplasmic reticulum are found in protozoa. The smooth variety is often found around the contractile vacuoles, and in some ciliates such as *Paramecium* may connect with the tubular branches of the vacuole. Rough endoplasmic reticulum is studded with ribosomes and is therefore a site of protein synthesis activity. The rough variety may connect with the nuclear envelope which often also bears ribosomes on its outer membrane coat.

The Golgi complex (Figure 6) or dictyosome is another membranous structure associated with both nucleus and endoplasmic reticulum. While they are found in almost all protozoa they vary in number and in the degree to which they develop. They are highly developed and numerous in flagellated protozoa but may be absent or only weakly developed in ciliates. Dictyosomes are membranous collections of cisternae that are always free from ribosomes, the outer margins are usually inflated and are pinched off as vesicles which are replaced by the growth of membranes. They have a distinct polarity; the proximal cisternae are closely bunched but they become more widely separated towards the distal face where the vesicles are detached. Occasionally the cisternae of endoplasmic reticulum may be seen in close proximity to the Golgi complex and there may be some exchange of materials between them. Golgi complexes are usually found close to the nucleus and it would appear that they are under nuclear control since the experimental removal of the nucleus in *Amoeba proteus* results in the disappearance of the Golgi. However they may be regenerated following the transplantation of another nucleus. It is thought that materials synthesized by the ribosomes are transported to the face of the dictyosomes. Within the cisternae the materials are transformed into their final form before being transported away from the mature face. In protozoa the Golgi complexes take part in the formation of polysaccharide secretions which are transported to the cell surface, and also in the production or packaging of hydrolytic enzymes (lysosomes). The scales or platelets of those flagellates and amoebae which have external skeletons are formed within the Golgi complex. The scales are then transported to the cell surface where they are incorporated into the skeletal system. The parabasal bodies of the flagellate *Polymastigina* are similar to Golgi complexes. These bodies are rod or sausage shaped and contain a parabasal filament which is directed towards the nuclear envelope and is then connected to a Golgi-complex-like stack of membranous sacs.

B. Asexual Processes

1. Nuclear Fission

Mitosis is the usual method of nuclear division in protozoa. At the metaphase stage in the mitotic division of higher organisms, asters and a spindle are evident, the nuclear envelope is withdrawn into the endoplasmic reticulum, and the nucleolus disappears. However, in both protozoa and metazoa the purpose of mitosis remains the same, that is the replication of the genetic material of the nucleus and the distribution of the two identical genomes between the two daughter cells by means of the spindle.

The division of the nucleus is accompanied by the division of certain cytoplasmic organelles which are said to have genetic continuity since if lost they cannot be regenerated. There is increasing evidence that all of these self-replicating organelles contain DNA and examples include kinetoplasts, plastids, mitochondria, pyrenoids, centrioles, and the blepharoplast of flagellates and the kinetosome of ciliates which give rise to the flagellum and cilium respectively. Plastids and mitochondria are believed to replicate by fission while other cytoplasmic organelles undergo a generative replication. The latter term means that the new organelle develops beside, but not confluent with, the parent organelle. There is good evidence to suggest that some of the activities of these self-replicating organelles are gene controlled and are not completely independent of the nucleus.

2. Division of the Cytoplasm

a. *Binary Fission.* The simplest form of binary fission is found in the amoebae where the pseudopodia are withdrawn before the nucleus divides. After nuclear division the amoeba elongates and constricts in the centre across the division plane of the nucleus. Large pseudopodia are then formed at opposite poles of the dividing cell and these pull the daughters apart. Testate amoebae are rather more complex in their mode of fission and this is directly related to the type of shell that they possess. In those with soft shells, the division plane is longitudinal along the major body axis and the test constricts into two halves. In those

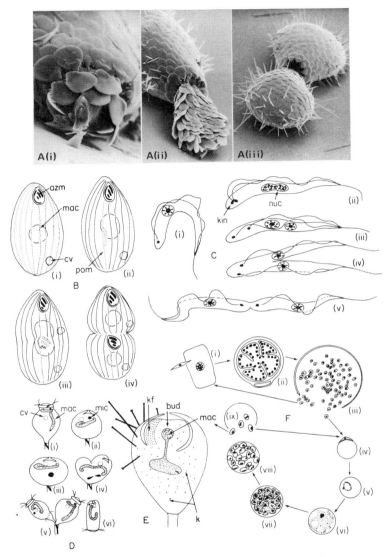

Figure 7. Asexual reproduction in protozoa. A(i)–(iii), Division stages in the testate amoeba *Euglypha*; B(i)–(iv), transverse binary fission in the ciliate *Tetrahymena pyriformis*, showing development of new oral structures of the opisthe from the postoral meridian of the parent, and the division of the macronucleus (azm – adoral zone of membranelles; cv – contractile vacuole; mac – macronucleus; pom – postoral meridian); C(i)–(v), longitudinal binary fission in the zooflagellate *Trypanosoma*, note the replication of the nucleus and kinetoplast (kin – kinetoplast; nuc – nucleus); D(i)–(vi), budding or teletroch formation in *Vorticella*, note the division of the two nuclei and that the telotrochs in v and vi bear a ring of aboral

where the shell is more rigid, part of the cytoplasm protrudes from the aperture to secrete a new shell over its surface. In *Euglypha*, where the shell is composed of many small platelets, reserve platelets are formed in the mother cell before the onset of nuclear division. These are then laid down in a precise order such that the apertures of the two cells face each other (Figure 7 A). It is not until the shell is formed that nuclear division proceeds and is completed by cytoplasmic fission.

The way in which protozoa divide is used as a taxonomic feature. In flagellates the plane of division is longitudinal along the major body axis (Figure 7 C) while it is transverse across the longitudinal axis in ciliated protozoa (Figure 7 B). This feature was used when deciding the taxonomic position of the opalinids which are covered with a continuous coat of cilia. Although other features were taken into account, the fact that the oblique fission takes place between the kineties rather than across them was the major factor which suggested that these organisms are flagellates rather than ciliates. There are several exceptions to this general rule of division plane, for example many dinoflagellates divide either obliquely or even transversly.

In ciliates (Figures 5 E and 7 D) there are two types of nucleus — the macronucleus, which is responsible for the vegetative processes, is derived from the micronucleus which is involved in sexual processes. During asexual fission the micronucleus divides, normally by mitosis, while the macronucleus undergoes DNA synthesis and constricts into two parts without the regular reduplication of the chromosomes. While this is sometimes referred to as 'amitosis' this tends to mask the fact that each daughter micronucleus contains the full complement of genes.

The presence of cilia and other complex organelles in ciliates has far reaching consequences at division. These structures are partly formed anew and are partly derived by the transformation of the existing structures in the mother cell. Renewal of the ciliature, for example, begins with an increase in the number of kinetosomes. In some this may

cilia (cv — contractile vacuole; mac — macronucleus; mic — micronucleus); E, endogenous budding in a suctorian ciliate showing the division of the macronucleus, the presence of erratic kinetosomes in the parent and the regularly arranged field of kinetosomes in the bud (bud — endogenous bud; k — erratic kinetosomes; kf — field of kinetosomes; mac — dividing macronucleus); F(i)–(ix), multiple fission in the malarial parasite *Plasmodium*; (i)–(iii), the pre-erythrocytic and exo-erythrocytic cycles in the human liver; (i), sporozoite penetrates liver and undergoes (ii), multiple fission to produce (iii), large numbers of merozoites that are released into blood stream; some reinvade liver cells; (iv)–(ix), erythrocytic cycle in red blood cells; (iv), merozoites invade red blood cells which (v)–(vi) grow and undergo (vii)–(viii) multiple fission to produce (ix), many merozoites; the latter may reinvade red blood cells or become gametocytes. (F Reproduced, with permission from Vickerman and Cox (1967). *The Protozoa.* London: John Murray) take place where the transverse division furrow is subsequently formed, and this zone gives rise to the ciliature of the posterior half of the anterior daughter (proter) as well as the anterior ciliature of the posterior daughter (opisthe). Furthermore it is a general rule that specific areas or fields of kinetosomes give rise to specific ciliary structures. In *Tetrahymena pyriformis* (Figure 7 B), the mouth of the opisthe always develops from the post-oral ciliary meridian. In others, such as *Euplotes*, scattered kinetosomes lying close to the mouth of the mother cell give rise to the oral ciliature of the opisthe.

b. Budding. It is tempting in a textbook of microbiology to use the term 'budding' in the same sense as mycologists who use it to describe the unequal fission of yeast cells. In protozoology, however, it is more usual to reserve its use to describe the varied processes by which sessile protozoa produce motile offspring. In several of these cases it is not obvious that the bud is any smaller than the mother cell which remains attached to the substratum. Some form of budding is found in all sessile ciliates and is used to disseminate the species while the mother cell remains *in situ*. In the solitary peritrichs, such as *Vorticella* (Figure 7 D), an unequal division produces a motile daughter known as a telotroch which develops a ring of aboral cilia. The telotroch breaks free from the mother cell and swims away to settle and secrete a stalk. In the suctorian ciliates budding is rather different; here the adults do not possess cilia but they do produce ciliated buds (Figure 7 E) which may be internal (endogenous) or external (exogenous). It is of interest to note that the cilia of the buds are derived from free kinetosomes that lie scattered within the cytoplasm of the adult which, on budding, congregate in regular rows to form a ciliary field.

c. Multiple Fission. In binary fission only two daughter cells are produced at division but in multiple fission large numbers of progeny may be formed. Usually the division of the cytoplasm is preceded by multiple fission of the nucleus, which is closely followed by the simultaneous cleavage of the cytoplasm into a corresponding number of offspring. Multiple fission is not as widespread as is binary fission but it often occurs in addition to the latter form of division. Specific examples may be found in the flagellates (*Noctiluca*), and in the parasitic amoebae (*Entamoeba histolytica*), while it is also common in the heliozoa, the radiolaria, and the foraminifera. Perhaps the best known examples of multiple fission are found in the sporozoa (Figure 7 F) where it is known as schizogony and serves to spread the parasites quickly within the host.

C. Sexual Processes

Sexual reproduction is a feature of animals so it is not surprising to find it widespread in the phylum Protozoa. While in every class there are at least some orders which contain representatives that are known to reproduce sexually, sexual processes have not yet been observed in certain groups of flagellates and rhizopods even though exhaustive studies have been made.

Sexual processes require that some form of meiosis should take place in the life cycle of the organism in order to produce haploid nuclei in the sexual cells which are known as gametes. The fusion of two gametes. is called fertilization and the resultant cell is a zygote. Gametes of the two sexes may be morphologically indistinguishable (isogamy) or may be distinguished (anisogamy) by differences in their size (micro- and macrogametes), form, structure or behaviour. Three basic forms of fertilization processes may be recognized in the phylum: gametogamy, gamontogamy, and autogamy.

1. Gametogamy

This is the type of fertilization where gametes are produced before mating and fusion takes place between the two free gametes. The phytomonad flagellates are perhaps the best known examples which exhibit this form of fertilization. Here in the simplest cases the gametes are released so that they fuse at a distance from the gamonts. Phytomonads are haploid so that fusion of the two gametes is followed by meiosis of the zygote to produce two haploid individuals. While the gamonts are not normally distinguishable by their morphology, their gametes sometimes are and a whole range of examples is furnished within the genus (*Chlamydomonas*). Most species of the genus have gametes that are morphologically similar to each other but in some the sexes may become apparent at fertilization. For example, one of the pair of flagella may cease to beat while those of the mate continue, in another species one partner sheds its cellulose wall after fusion and in *C. braunni* there is a definite anisogamy with micro- and macrogametes. Some phytomonads are monoecious, that is they produce gametes which can fuse with gametes produced by the same gamont or from gamonts of other cells within the same clone. Others are dioecious where sexually compatible gametes are only produced by different clones or gamonts. In the genus *Volvox*, some species are monoecious while others are dioecious and in that colonial genus only specialized cells (gonidia) are capable of gamete production. In the colourless zooflagellates that are found in the alimentary tracts of termites and cockroaches, the sexual processes are under the control of the insect moulting hormone, ecdysone.

The occurrence of free-swimming gametes has been established conclusively in only a few foraminifera which are the only representatives of all the rhizopods to do so. In the foraminifera which exhibit sexual reproduction there is an alternation of generations where a sexually reproducing generation (gamonts) alternates with an asexually reproducing (agamont) generation. The two generations may be morphologically similar or may differ to a greater or lesser extent. The sexual generation arises from multiple fissions of agamonts. After the gamonts have grown, they produce asexual agamonts by multiple fission where the entire cytoplasm divides and departs to leave an empty shell.

The production of free-swimming gametes in the sporozoa is virtually limited to the coccidia in which the gamonts arise either directly from the sporozoites or from the merozoites of a previous schizogony (an asexual multiple fission). The presumptive 'macro' gamont develops directly into an immobile macrogamete while motile flagellated microgametes are derived from the gamont after multiple fission.

2. Gamontogamy

This is the type of fertilization in which the gamonts mate before the gametes are formed. Gametes, in fact, are not always formed but where they are it is by multiple fission. In those where gametes are not formed, the gamonts fuse directly and gamete nuclei are formed.

Some foraminifera undergo gamontogamy; during the sexual generation the gamonts aggregate and their shells fuse to form a common internal chamber. In some species more than two gamonts may form the aggregate and an organic layer is secreted to fuse the shells together. Nuclear division takes place to produce several gametes which lie freely in the internal chamber. The gametes fuse in pairs to form a zygote which is then released in the form of an agamont or asexual generation of the life cycle.

Sporozoa may also exhibit gamontogamy. In the eugregarines, for example, the gamonts are derived by the growth of the sporozoites and mating can take place either before or after the growth phase is complete. The gamonts fuse together in syzygy, usually in a head-to-tail formation and often in chains, before dividing into pairs. Paired gamonts secrete a gametocyst within which multiple fission produces large numbers of gametes. Fusion of the gametes takes place within the cyst and each resultant zygote forms a spore from which sporozoites develop.

In ciliated protozoa, the process of fertilization is known as conjugation. In most cases the gamonts look alike but there are some in which

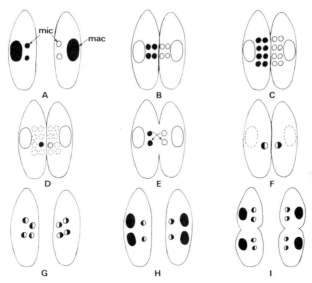

Figure 8. Conjugation in *Paramecium aurelia*, see text
for detail (mac — macronucleus; mic — micronucleus)

the gamonts are different either at the onset, as in the sessile peritrichs,
or change during the course of mating, as in the suctoria. The best
known example of conjugation in ciliates is in *Paramecium aurelia*.
Mating can occur only when two mating types of a variety are mixed
together, where a variety or syngen is now recognized as being equivalent to a species and mating type equivalent to a sex. All ciliates have
types of nucleus, a macronucleus which is vegetative in function and
deals with the day-to-day running of the cell, and a much smaller
micronucleus (Figure 8 A). The latter, which carries the genetic
information, is essential for sexual processes but not for growth or
asexual binary fission. During the first stages, the two cells or gamonts
join along their oral surfaces and remain as such until conjugation is
complete (Figure 8 B). The macronucleus begins to break down and
eventually disappears completely (Figure 8 F). In *P. aurelia* there are
two micronuclei and these both divide twice by meiosis (Figure 8 B and
C) to produce eight haploid micronuclei of which seven disintegrate
(Figure 8 D). The remaining micronucleus in each gamont divides by
mitosis to produce two gametic nuclei in each cell. One of each pair
(the migratory nucleus) migrates to the other cell (Figure 8 E) where it
fuses with the remaining stationary nucleus. After fusion (Figure 8 F),
the two cells separate and the zygotic nucleus divides twice mitotically
(Figure 8 G); two of the four products become macronuclei, the
other two remaining as diploid micronuclei (Figure 8 H). At the next
asexual binary fission, one macronucleus passes to each daughter cell,

while the micronuclei each divide mitotically so that the daughters have the original nuclear complement (Figure 8 I).

3. Autogamy

Autogamy is the type of fertilization process in which the gametes or gametic nuclei are derived from the same gamont and is therefore a form of self-fertilization. Although autogamy has been shown to take place in heliozoa, some foraminifera, and in some polymastigote flagellates, it is best known in the ciliates such as *Paramecium*. Here, although the macronucleus disintegrates and the micronuclei undergo the same divisions as in conjugation there is no pairing of cells. Thus autogamy in *Paramecium* is similar to cytogamy but without the pairing of gamonts.

VI. Further Reading

Aikawa, M., and Sterling, C. R. (1974). *Intracellular Parasitic Protozoa.* New York: Academic Press.

Baker, J. R. (1969) *Parasitic Protozoa.* London: Hutchinson University Library.

Buetow, D. E. (1968). *The Biology of* Euglena. New York: Academic Press.

Chen, T.-T. (1967—72) *Research in Protozoology.* Vol. 1, 1967; Vol. 2, 1967; Vol. 3, 1969; Vol. 4, 1972. Oxford: Pergamon Press.

Dogiel, V. A. (1965) *General Protozoology.* Oxford: Oxford University Press.

Elliott, A. M. (1973) *Biology of* Tetrahymena. Stroudsburg, Pennsylvania: Dowden, Hutchinson and Ross Inc.

Grell, K. G. (1973) *Protozoology.* Berlin: Springer Verlag.

Hedley, R. H. (1964). The Biology of *Foraminifera.* In *The International Review of General and Experimental Zoology.* Ed. W. J. L. Felts and R. J. Harrison. Vol. 1, 1—45. London: Academic Press.

Honigberg, B. M., Balamuth, W., Bovee, E. C., Corliss, J. O., Gojdics, M., Hall, R. P., Kudo, R. R., Levine, N. D., Loeblich, A. R., Weser, J., and Wenrich, D. H. (1964). A Revised Classification of the Phylum Protozoa. *J. Protozool.*, 11, 7—20.

Jeon, K. W. (1973). *The Biology of Amoeba.* New York: Academic Press.

Jones, A. R. (1974). *The Ciliates.* London: Hutchinson University Library.

Kudo, R. R. (1966). *Protozoology.* Springfield, Illinois: C. C. Thomas Publ.

Lumsden, W. H. R. and Evans, D. A. (1976). *Biology of the Kinetoplastida.* London: Academic Press.

Mackinnon, D. L. and Hawes, R. S. J. (1961). *An Introduction to the Study of Protozoa.* Oxford: Oxford University Press.

Pitelka, D. R. (1963). *Electron-Microscope Structure of Protozoa.* International Series of Monographs on Pure and Applied Biology, Vol. 13. Ed. G. A. Kerkut. Oxford: Pergamon Press.

Sleigh, M. A. (1962). *The Biology of Cilia and Flagella.* International Series of Monographs on Pure and Applied Biology, Vol. 12. Ed. G. A. Kerkut. Oxford: Pergamon Press.

Sleigh, M. A. (1973). *The Biology of Protozoa.* London: Edward Arnold.

Tartar, V. (1961). *The Biology of* Stentor. International Series of Monographs on Pure and Applied Biology, Vol. 5, Ed. G. A. Kerkut. Oxford: Pergamon Press.

Vickerman, K, and Cox, F. E. G. (1967). *The Protozoa.* London: John Murray.

The Chemistry and Composition of Microorganisms

PAULINE M. MEADOW

Department of Biochemistry, University College, London

I. Introduction

The chemistry of life is based on that of carbon and is similar throughout the living world. Why then should we consider the chemistry and composition of microorganisms separately? There are two main reasons. Firstly the variety in the shapes, sizes, structures, and metabolic pathways of microorganisms presents a unique challenge. Microbiologists interested in the ways in which living things develop, evolve, and survive, and the uses to which their activities may be put, will seek an explanation. Secondly some microorganisms cause disease in animals and society needs ways of preventing their growth without injury to the hosts. One obvious approach is to look for some component which is unique to microorganisms and is essential for viability. There are in fact several such compounds some of which have already been exploited in the design of antimicrobial drugs, as will be discussed later.

The word microorganism can be applied to any living thing that is reasonably small. It is usually limited to the Protista which, although they may be multicellular, are characterized by their lack of differentiation. They can be further subdivided into the eukaryotes (from the Greek, true nucleus) which include yeasts, fungi, protozoa, and most algae, and the prokaryotes, which are the bacteria and the blue-green algae. Viruses do not belong in either class since they are not independent organisms. In an essay of this length it is impossible to cover the whole range and I have been selective. I have chosen to concentrate on the prokaryotes whose composition and structure can be clearly differentiated both from that of the eukaryotes and of higher plants and animals. Their elementary and molecular components will be discussed and then the ways in which they are assembled first into macromolecules, and then into ordered structures, particularly those which are unique to bacteria. Many are extremely complicated and an understanding of their chemical composition has to some extent depended on the development of inhibitors which prevent their synthesis or function. Many of these inhibitors are antibiotics, substances that are themselves synthesized by microorganisms. They include penicillin and streptomycin, both of which owe their selective activity to the unusual composition of bacterial structures.

II. General Composition of Microorganisms

All living organisms consist largely of water. Man contains about 60 per cent, the jelly fish 96 per cent and microorganisms in the vegetative state from 70 to 90 per cent. Their water content depends not only on the species but on the growth conditions which in turn affect the proportions of different cellular constituents. Some microbes can also exist as relatively anhydrous heat-stable forms known as spores. Even here, although the core contains very little water, there may be as much as 70 per cent water in the spore cortex.

Of the 27 elements found in most living things, four — hydrogen, carbon, oxygen, and nitrogen — make up 99 per cent of their mass and only 12 others — sulphur, phosphorus, magnesium, calcium, sodium, potassium, chlorine, iron, manganese, copper, zinc, and molybdenum, occur in all species. Their role in microorganisms is similar to that in higher plants and animals. The first four elements together with sulphur and phosphorus make up the main macromolecules. Many of the others, including those present in only trace amounts, are required to stabilize or activate the macromolecules. The key to cellular composition must therefore be its constituent macromolecules. The metabolic state of the organism will naturally affect its composition but a few generalizations can be made.

Proteins are the most abundant molecules not only in amount (anything up to 50 per cent of the dry weight) but in the variety of molecular species present. In an organism like *Escherichia coli*, for example, there may be as many as two or three thousand different kinds of proteins in the cell at any one time. Quantitatively the next most important components are the nucleic acids with up to ten times more ribonucleic acid (RNA) than deoxyribonucleic acid (DNA). Carbohydrates and lipids comprise most of the rest of the organism with smaller amounts of low molecular weight metabolites and inorganic ions. The four main macromolecular types exist in combination with others as for example lipoproteins, lipopolysaccharides, glycopeptides, and nucleoproteins, but all are derived from a few relatively simple molecules which act as the precursors of the complex polymers.

III. Main Classes of Polymers

A. Proteins and Peptides

Proteins consist of L-amino acids joined by peptide bonds linking the amino group of one amino acid to the carboxyl group of the adjacent amino acid. This results in long polypeptide chains most of which contain between 100 and 1000 amino acids. A single protein may contain one or more polypeptide chains each of which is folded into a particular conformation and is held in that shape by both covalent and non-covalent bonds.

The amino acids found in microbial proteins are usually no different from those of higher animals and plants. They can be divided into classes by the number and type of side chains attached to the carbon atom carrying the amino and carboxyl groups (see Table 1). Differences in protein structure arise from the order in which the amino acids are joined together (primary structure) and this results in differences in secondary and tertiary structure.

Many microorganisms contain, in addition to the amino acids in

Table 1. Types of amino acids found in proteins (all are L-isomers; standard abbreviations in brackets)

Aliphatic	Aromatic	Acidic
Alanine (Ala)	Phenylalanine (Phe)	Aspartate (Asp)
Glycine (Gly)	Tryptophan (Trp)	Glutamate (Glu)
Isoleucine (Ile)	Tyrosine (Tyr)	
Leucine (Leu)		
Serine (Ser)		
Threonine (Thr)		
Valine (Val)		
Basic	Sulphur-containing	Imino Acid
Arginine (Arg)	Cysteine (Cys)	Proline (Pro)
Histidine (His)	Methionine (Met)	
Lysine (Lys)		

proteins, other amino acids which are sometimes unique to them and do not necessarily belong to the L-series. Some are components of bacterial walls, others are found in antibiotics. None of these amino acids occurs in proteins even though they may be joined together by peptide bonds. They usually form parts of small peptides or glyco-peptides and it seems likely that their assembly into the polymer is quite different from that of proteins.

Although not strictly an amino acid and certainly not a component of proteins, the amino acid precursor dipicolinic acid (Figure 1) should perhaps be included in this section. It is a precursor of lysine in some microorganisms, but also occurs commonly in bacterial spores where it may constitute up to 5 to 10 per cent of the dry weight. It is found as calcium dipicolinate and it may play some part in the heat resistance of bacterial spores. Whether it is directly responsible for heat resistance or whether it merely helps to stabilize established heat resistance remains to be seen. It is rapidly lost on germination and it is possible that its major function might be to maintain dormancy or to initiate germination.

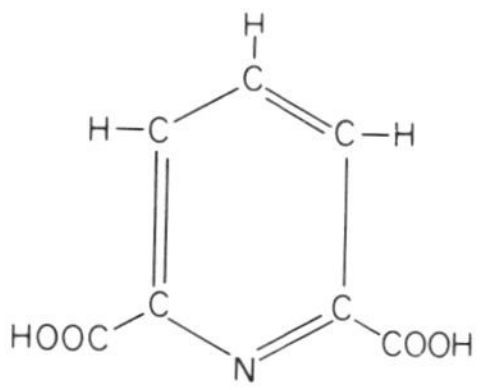

Figure 1. Dipicolinic acid

B. Nucleic Acids

The genetic information defining each species is stored in the nucleic acids, usually DNA, although in some viruses this function is assumed by RNA instead. Viruses differ from all other microrganisms in that they contain only one kind of nucleic acid. They have no independent metabolism and can replicate only with the aid of the host's synthetic machinery. All other microorganisms contain in addition to their DNA genome several species of RNA including the messenger, transfer and ribosomal RNA molecules needed for protein synthesis.

Both DNA and RNA are made up of long chains of five carbon sugars (pentoses) linked together by phosphodiester bonds. They differ in the component pentose and in the bases which are attached to them (Table 2). They may also differ in three dimensional structure. DNA usually occurs as a double-stranded molecule in which pairs of purine and pyrimidine bases are stabilized opposite each other by hydrogen bonds but a few bacteriophages (bacterial viruses) contain single-stranded DNA. In those viruses and bacteriophages where RNA is the genetic material it may be either double- or single-stranded according to the species. Replication always involves the formation of a double-stranded intermediate. Other forms of RNA (t-RNA, m-RNA, and ribosomal RNA) are usually single-stranded but there may be complex folding of the strand and even some base-pairing which can hold the structure into fixed loops. Although the most common bases are shown in Table 2, many others occur in smaller amounts both in microbial nucleic acids and those of higher animals and plants. They include hydroxymethyl cytosine which is found in place of cytosine in some bacteriophages. The hydroxymethylation of the cytosine in the DNA enables the bacteriophage to resist attack by enzymes (deoxy-

Table 2. Major components of nucleic acids

Deoxyribonucleic Acid (DNA)	Ribonucleic Acid (RNA)
Sugar backbone	
β-D-Deoxyribose	β-D-Ribose
Phosphate	Phosphate
Purine bases	
Adenine	Adenine
Guanine	Guanine
Pyrimidine bases	
Cytosine	Cytosine
Thymine	Uracil

ribonucleases) which degrade bacterial (cytosine-containing) DNA. Other modified bases are 6-aminopurine (6-methyl adenine) which occurs widely in microbial DNA, and the methylated bases which occur in most RNA molecules and which may play some role in stabilizing their secondary structure or in providing recognition sites for enzymes.

The hydrogen bonds between adenine and thymine and between guanine and cytosine in the double-stranded DNA molecule means that all such DNA's will contain approximately equal numbers of adenine and thymine residues, and the number of guanine molecules will be equal to the number of cytosine molecules. This was discovered by Chargaff as long ago as 1950 but it was not until Watson and Crick showed how this could result from base-pairing that we had any understanding how it occurred. Despite the overall similarity between the DNA's from all species, these molecules carry the genetic information necessary to distinguish between them. In higher animals the total base composition of DNA's from different species is fairly constant and differences depend almost entirely on the sequence in which the four bases are arranges in the chains. In the microbial world there is much greater diversity in terms of total base composition and this could be a reflection of their wider range of metabolic capabilities. The DNA composition of different species of microorganisms can be conveniently compared by considering the percentage of the bases which are guanine—cytosine (GC) pairs. In different species this can vary from about 25 to 80 per cent GC and is one of the criteria used to classify them. The same GC% does not necessarily mean that the DNA's are identical but simply that they contain the same proportions of the bases. The DNA molecules would only be identical if the sequence of the bases was also the same and this can be estimated by making hybrids between the single strands of two different DNA molecules. The amount of the strand which forms a stable double molecule with its opposite number is a measure of the degree of homology between them. Such experiments have been used to show evolutionary relationships between different families. Complete homology means identical DNA structure both in total base composition and in base sequence and must mean identity of species.

C. Polysaccharides

Most of the naturally occurring polysaccharides are composed of five-, six- or seven-carbon sugar molecules. Despite the numbers of possible isomers resulting from the asymmetric carbon atoms, it is striking that almost all the microbial six-carbon sugars are derived from only four of them: the D-isomers of glucose, galactose and mannose, and the L-isomer of galactose. The only common ketose is fructose. D-ribose and its derivatives are the most important of the pentoses. In microorganisms too the central metabolic pathways involve a limited number of hexoses and pentoses, but many more can be and are metabolized so that they can be fed into the central pathways. Moreover many microorganisms contain derivatives of the common hexoses which do not normally occur in higher animals. In mammals the D-isomers of the sugars predominate, but although quantitatively these are more important, microorganisms also contain sugars of the L-series. Most of these, together with the more unusual derivatives such as dideoxyl and diamino-sugars, occur as parts of the outer layer of microorganisms where their resistance to the normal metabolic enzymes no doubt provides added protection.

The hexoses and their derivatives almost all exist as six-membered rings (pyranose) in combination with other molecules whereas the pentoses and fructose form five-membered (furanose) rings. This ring structure produces another asymmetric carbon atom and hence two more stereoisomers defined by the sterochemistry around the reducing carbon. These are known as α and β isomers. In free solution the α and β isomers are freely interconvertible through the open chain form of the sugar but they become fixed in combination with other molecules and lead to very different properties in the resulting structure. For example starch consists of α 1—4 linked glucose units, that is glucose units joined together from carbon 1 of one molecule to carbon 4 of the adjacent one both being in the α configuration. It is a soluble polymer, readily hydrolysed by most species, and a common storage polymer in both plants and microorganisms. Cellulose is an identical polymer except that it consists of β 1—4 linked glucose units but it has quite different properties and functions. It is an insoluble rigid molecule which forms the structural polymer of many plants and is not metabolized by the amylases which degrade starch. This means that mammals, other than ruminants whose microbial flora hydrolyse cellulose, cannot utilize cellulose as a source of energy. This is unfortunate in view of the world food shortage. On the other hand the affluent society has been able to exploit this and celluloses are often added to slimming foods to provide bulk without calories.

D. Lipids

Long-chain fatty acids are the basic building blocks of most neutral lipids and phospholipids. In the neutral lipids they are esterified to the hydroxyl groups of glycerol which in the phospholipids is usually replaced by glycerol-3-phosphate or a derivative thereof. In general the microbial fatty acids are simpler and fewer than those occurring in higher plants and animals. The bacteria, in particular, seem to be rather limited in their fatty acid components, the commonest being the fully saturated straight-chain acids with 12 to 18 carbon atoms, and the mono-unsaturated palmitoleic and oleic acids. Fatty acids containing more than one double bond do not occur in bacteria but have been reported in yeasts and fungi. Many bacteria produce cyclopropane fatty acids by adding a methyl group across the double bond. This seems to be a method whereby the unsaturated acids are protected against oxidation, the behaviour of cyclopropane fatty acids in membrane structures being very similar to that of the unsaturated acids themselves. Another unusual group of fatty acids is produced by *Corynebacterium, Nocardia,* and *Mycobacterium* spp. They are very long chain fatty acids containing up to 88 carbon atoms, often with methyl side groups and at least one double bond. Their chemistry is complex and though the structures of many have still to be elucidated, they have recently attracted more attention from microbiologists because they provide a specific site of attack for antimicrobial compounds. The mycolic acids (from the mycobacteria) which are some of the longest of these compounds appear to be synthesised from acetate like other fatty acids and their synthesis is inhibited by the antituberculous drug isonicotinic acid hydrazide. This compound specifically inhibits the growth of mycobacteria which are resistant to most other antibiotics. It must therefore inhibit a function essential to growth but the role of the mycolic acids in this is still not known.

The distribution of fatty acids between the neutral fats and phospholipids is different in the pro- and eukaryotes. The yeasts and fungi may contain as much as 50 per cent of their dry weight as mono-, di- or triglyceride storage polymers depending on their metabolic state. Most bacterial fatty acids are found in the phospholipids and form part of the wall and membrane structure as I shall discuss later. One unusual feature of bacterial lipid composition is the complete lack of sterols. These complex lipids are essential components of all other living organisms including the eukaryotic microorganisms, yet although bacterial membranes seem very similar to those of higher organisms they do not require sterols. The only comparable structure in bacterial membranes is a long chain isoprene derivative of 55 carbon atoms which is used as a carrier in the biosynthesis of the outer layers of the organism, but there are similar molecules of different chain lengths which perform the same function in the eukaryotes. Whereas it is easy to see the chemical relationship between the bacterial isoprene derivatives (which include carotenoids and coenzyme Q as well as the

C_{55} carrier lipid) and the cholesterol found in many other membranes, their functional relationships are much harder to define. These must await further information.

Progress in determining the structures and function of lipid-containing macromolecules has been much slower than that in other branches of biochemistry. This probably arises because most techniques were developed for studying water-soluble substances and all the early metabolic and biosynthetic work was done using soluble enzymes in an aqueous environment. The concept of proteins which can carry out their enzymic reactions in a completely anhydrous lipid environment has been forced upon us by studies of the biosynthesis of insoluble complex polymers, particularly in microorganisms. It is now clear that they exist and that some may even contain lipids as an integral part of their structure. Others require lipids and phospholipids for activity. The number and variety of lipid-containing macromolecules in various microbial species is growing rapidly as biochemical techniques improve and we are now able to separate and to study some of the complex glycolipids and lipoproteins.

IV. Chemical Composition of Microbial Cell Structures

The variety of chemical compounds found in and made by micro-organisms is overwhelming. Much of their interest lies in their utility to the organism and this discussion is limited to the chemical composition of cell structures. We must first define the structures to be discussed and these are shown diagrammatically in Figure 2. Not all bacteria have

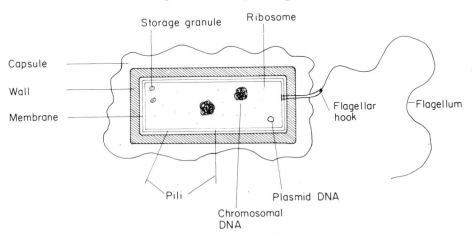

Figure 2. Diagrammatic representation of a bacterial cell

all these features and the presence and composition of a particular structure may be one of the ways of distinguishing different species.

A. Flagella

Bacterial flagella are quite different from the locomotive and contractile organelles of eukaryotes which consist of a system of microtubules arranged so that nine pairs surround two central tubules. The entire flagellar structure arises from within the cytoplasm and is covered with a membrane which appears to be an extension of the cytoplasmic membrane. Bacterial flagella consist of three main sections. These are the main shaft which has a characteristic wave form, the hook region, and the basal body. The latter anchors the flagellum in the wall and cytoplasmic membrane, and is the side of the motor activity which allows the rigid flagellum to rotate and to propel the bacterium. There is no evidence for a surrounding membrane to the flagellum itself. The main component of all bacterial flagella so far studied is a single species of protein known as flagellin. The flagellins of several different bacteria have been isolated and analysed. Although they differ in the detailed amino-acid sequence, they share a number of features and provide an interesting example of the relationship between protein structure and a specific function. All the flagellins have molecular weights between 30,000 and 60,000. They lack the sulphur-containing amino acid cysteine and have only a few copies of most of the aromatic and other cyclic amino acid . In contrast the acidic amino acids together with their amides (asparagine and glutamine) make up almost 20 per cent of the total amino acids of the protein. One of the flagellins has been completely sequenced but similar fragments have been found in flagellins from other species. The amino-terminal end of the protein is generally basic, most of the acid residues are in the centre of the molecule and the carboxy-terminal end is weakly acidic. These clusters of charged groups in the protein sequence suggest that interactions between adjacent molecules in the complete flagellum may depend largely on electrostatic forces. This view is supported by the ease with which the flagella disaggregate into flagellin monomers below pH 3, above pH 11, and at low ionic strength. Under suitable conditions the flagellin monomers in solution will reaggregate into helical structures of the correct waveform without the addition of any enzymes, ATP or any other cofactors. Aggregation is thus a characteristic of the protein itself and this is clearly shown in flagellins from bacteria grown with p-fluorphenylalanine. This amino acid can be incorporated into proteins in place of tyrosine and thus alter the way in which the macromolecule folds. Flagella synthesized in its presence have half the wavelength of the normal flagella and the flagellin proteins themselves will reaggregate from solution to the same short wave form. There are only one or two tyrosine molecules among the 300 amino acid residues in each flagellin protein and yet substituting p-fluorphenylalanine in this position has a dramatic effect both on the gross morphology of the flagellum and on its efficiency as an organ of locomotion.

Bacterial mutants with straight or curly flagella have been isolated and in every case the mutation has been shown to be in the structural gene coding for the flagellin protein. The flagella therefore seem to require their unusual amino acid composition in order to be assembled outside the metabolic activities of the cell. The high proportion of charged amino acids and the small numbers of bulky (ring-containing) amino acids may help the newly synthesized polypeptide chain to fold in a way which generates recognition sites for the subsequent binding of other flagellin molecules. The completed structure is further protected from changes in the redox potential of the surrounding media by the complete absence of cysteine.

We know very much less about the other parts of the bacterial flagella. Like the main shaft, the hook region appears under electron microscopy to consist of helically arranged subunits and the hook proteins of a few species have been isolated. They are chemically and antigenically distinct from the flagellins and contain more cyclic amino acids, but the significance of this in their mode of assembly is not known. The basal bodies of the flagella have not been analysed chemically. They appear by electron microscopy to be structures related to that of the wall and membrane of the bacteria from which they are derived and several genes related to their function have been identified. They must be complex organelles and no doubt their structure will become clearer as their function is more precisely defined.

Despite the amount of work devoted to the relationship between structure and function in bacterial flagellar movement, the chemistry of the process is still only partly understood. It is clear that motility is linked with chemotaxis, that is movement of bacteria towards attractant chemical molecules and away from repellant ones, but the movement is not necessarily related to metabolism of the compound concerned. Recently evidence has been accumulating which supports the view that the helical waveform of the flagellum is fixed and that movement results from the rotation of this helix clockwise or anticlockwise. The basal granule must therefore provide some kind of universal joint allowing complete rotation of the hook and shaft of the flagellum. What causes the rotation is by no means understood, but there is some evidence that membrane proteins, presumably in or around the basal granule, are modified to change their conformation in the process. The whole process is quite different from that in other motile cells where movement occurs by contractions travelling in waveform down the length of the filament and involves ATPase activity.

B. Pili

The other filaments projecting through the outer layers of some bacteria, particularly the enteric and other Gram-negative bacteria, are the pili. They are thinner, shorter, and straighter than flagella but resemble them in that they consist of helically arranged protein molecules. Like flagella they are disaggregated by extremes of pH into their constituent protein pilin. They can also be reformed from solutions of pilin under suitable conditions of temperature, pH, and ionic strength. Presumably the pilins will, like the flagellins, turn out to have very carefully designed amino-acid sequences for their purpose. One of the difficulties in studying them chemically is to obtain them in large enough amounts and to distinguish between the different kinds which may be present at the same time in a particular bacterium. Some pili can be recognized by the property they confer on their owner and others are distinguished by their ability to bind particular bacteriophages. Furthermore, it is interesting that certain bacteriophages bind only at the tip of the pilus whereas others bind only along their length.

The different types of pili are genetically determined and so can be lost by mutation. The ability to produce them can be transferred by genetic exchange. However, until more pili are isolated and their amino acids sequenced, it is difficult to know how much their different properties reflect differences in primary sequence and whether there is any homology between the various types.

C. Capsules

In several species the cell is surrounded by a loose aggregate of material which may not be strictly part of the organism but is certainly associated with it. In others this material may be rather more ordered, resulting in a clearly recognizable capsule. The amount of the capsule, or slime layer, is related to growth conditions, its production often being favoured by highly aerobic conditions. Whether there is a capsule at all and what chemical type is formed depends on the species. None requires a capsule for viability under favourable growth conditions, but the capsules may provide a protective layer enabling the organism to survive in otherwise unfavourable environments. One of the best known examples is the capsular layer of the pneumococcus which provides protection from phagocytosis and hence plays a part in virulence. Non-virulent variants which lack capsules have been isolated and these can be transformed into virulent encapsulated strains by the introduction of purified DNA from the latter. Such experiments provided the first direct proof that the genotype of an organism could be changed by material from heat-killed bacteria and finally, that purified DNA was the active principle.

The capsule in most species probably plays some part in the survival of the bacteria that carry them. The capsular structures are extremely varied but species-specific and often complex. They require the synthesis and participation of special enzymes which play no other part in the life of the organism and it is hard to escape the conclusion that the ability to produce capsules would have been lost during the natural processes of evolution if they had no useful function. One of the more unusual functions of these extracellular aggregates may be to provide a raft in aerobic bacteria. *Acetobacter xylinum*, which converts ethyl alcohol into acetic acid and is one of those used to produce wine vinegar, requires highly aerobic conditions. It synthesizes enough cellulose (β 1—4 linked glucose units) to form a loose pellicle which enmeshes the actively growing bacteria and allows them to remain floating at or near the surface instead of settling at the bottom of the vessel.

Many different chemical structures occur in capsules. In the yeasts and fungi they are often polysaccharides containing a wide variety of hexoses, pentoses, and their derivatives. Polysaccharides are probably also the most common type of bacterial capsule but, as shown in Table 3, other varieties exist. Many microbial capsules have proved useful to man. Since they are on the outside of the organism, they can often be detected immunologically and because they are species specific, they can be used to distinguish between different closely related groups of organisms. The dextrans produced by *Leuconostoc*

Table 3. Capsules and slime

Type	Constituents	Organism
Polysaccharide	Glucose (β 1—4 linked)	*Acetobacter xylinum*
	Hexoses, uronic acids, amino sugars	Pneumococci
	Mannose	Yeasts
	Glucose	Fungi
Protein	L-Amino acids	Streptococci
Polypeptide	D-Glutamic acid	*Bacillus anthracis*

mesenteroides are useful as plasma expanders in hospitals and in the laboratory as gels for filtration. The use of extracellular glycans and mannans produced by certain yeasts and fungi, as fillers and lubricants is increasing since these organisms can often be easily and cheaply grown on materials which would otherwise be waste products.

D. Walls

None of the structures discussed so far is absolutely necessary for growth, and some species lack them all. The wall, however, is essential for growth in all circumstances except in media of high osmotic pressure. It provides a rigid mechanical barrier to protect the osmotically fragile cytoplasmic membrane from changes in pressure. It is a particularly characteristic part of free-living microorganisms which must have developed this structure to protect them against extremes of the environment.

Apart from the Mycoplasmas, which completely lack walls, and the Halobacteria whose wall structure is unusual, the walls appear to be responsible for the shape of the organism from which they were derived. If the wall is removed, by enzymes for example, the cell is converted into an osmotically fragile round form known as a protoplast. In most species shape appears to be governed, at least in part, by a rigid layer whose chemical nature varies between species. Some of the more striking differences between the walls of the yeasts and filamentous fungi on the one hand and the bacteria on the other are listed in Table 4. The bacteria have been divided into the two main classes easily distinguished by their staining reactions, Gram-positive and Gram-negative. This division has been known ever since the last

Table 4. Comparison of wall composition

			Bacteria	
	Yeast	Filamentous fungi	Gram-positive	Gram-negative
Appearance	Microfibrillar	Microfibrillar	Lumpy	Lumpy
Fine structure	None	None	None	Layered
Rigid layer	Glucan	Chitin	Peptidoglycan	Peptidoglycan
Other components	Mannans Proteins	Glycans Proteins	Teichoic acids Teichuronic acids	Lipopoly- saccharide Proteins Lipoproteins Phospholipids

century and has proved useful not only in identifying bacteria, but in predicting something about their response to inhibitors and growth substrates. There was no indication of the reasons for the differential staining reaction for at least 50 years after its discovery. However, with the isolation of bacterial walls and the improvement in techniques for studying both their structure and chemical composition, it became clear that there were fundamental differences between the walls of Gram-positive and Gram-negative bacteria, and that these might explain the observed differences between them. Further investigation by electron microscopy showed that whereas the walls of Gram-positive bacteria were about five times thicker than those of the Gram-negative organisms, they showed little or no fine structure. Gram-negative bacterial walls, however, appeared to have a double-track outer membrane fraction very similar in appearance to the inner cytoplasmic membrane common to both. It is now clear that the only type of molecular structure found in the walls of *both* Gram-positive and Gram-negative bacteria is that mainly responsible for their rigidity.

1. Rigid layers

In each of the organisms listed in Table 4 a single component is mainly responsible for the shape and rigidity of the wall, but its nature is different.

The rigid component in yeasts consists mainly of glucose units linked together by β 1—3 links with occasional β 1—6 branch points. This provides a linear fibrillar molecule with much the same mechanical properties as cellulose in which the glucose units are β 1—4 linked and which is, of course, one of the main plant structural polymers. The yeast wall is therefore a tough structure with the added advantage that it is not readily degraded by enzymes since there are few which hydrolyse β-linked glucose polymers and fewer still which hydrolyse β 1—3 and β 1—6 linked glucose units. Nevertheless such enzymes do exist and one of the better known ones occurs in the edible snail. Treatment of yeast cells with this enzyme releases an osmotically fragile protoplast which lacks the glycan polymer.

The fungal cell wall contains another variation on the β-linked glucose theme. Here the rigid polymer is chitin which is a β 1—4 linked polysaccharide like cellulose, but made up of the *N*-acetyl derivative of glucosamine (2-amino-D-glucose) in place of glucose itself. The net result is an even more rigid crystalline type of structure. The long helical chains of glucosamine units are linked together in piles by hydrogen bonds. Within each pile the chains run in the same direction but adjacent piles have chains running in opposite directions, thus reinforcing the rigidity of the entire structure. The only other species to use chitin as a structural component are Invertebrates where the exoskeleton is almost entirely chitin. It is perhaps surprising that the fungi should have developed this type of structure as well. It is synthesized by a single enzyme chitin synthase which polymerizes units of N-acetylglucosamine from its UDP-derivative. This process is inhibited in a number of plant pathogenic fungi by an antifungal compound called polyoxin, which is a nucleoside antibiotic and shows many structural similarities to UDP N-acetylglucosamine. It appears to act by competing with the latter for the active site of the chitin synthase enzyme and so prevents chitin synthesis.

The main structural polymer of bacterial walls, known variously as peptidoglycan, mucopeptide or murein, can be considered as a complex chitin derivative. Like the other rigid polymers it is a β-linked polysaccharide and like chitin is made up of β 1—4 linked N-acetyl amino sugars which form long chains extending over the entire length of the cell. However, it differs from chitin in that alternate amino sugar units are not N-acetylglucosamine itself but its lactyl ether (Figure 3). This compound occurs only in bacterial walls, or materials derived from them, such as the peptide released by germinating spores. Indeed this is where it was first discovered by R. Strange and G. Dark who showed it was an amino sugar but its detailed structure took longer to determine. For some years this compound rejoiced in the distinctly sinister title 'unknown amino sugar (Strange and Dark)'! It is now called muramic acid (from the Latin, *murus* = wall). The carboxyl group of the muramic acid is substituted by short peptide side chains which may themselves be cross-linked. The bacterial peptidoglycan is therefore even more rigid than the comparable structures in yeasts, fungi, and plants. In addition to long parallel chains of β-linked amino sugars, there are short peptide chains which link them covalently to form a rigid three-dimensional network.

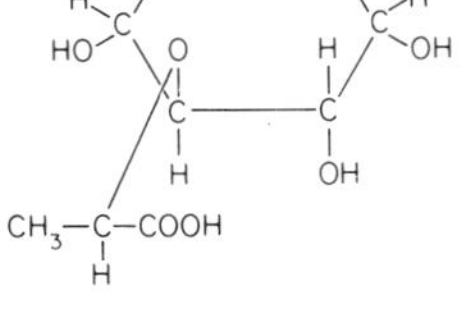

Figure 3. Muramic acid

Escherichia

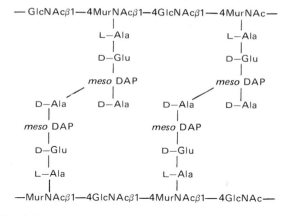

GlcNAc = *N*-acetylglucosamine; MurNAc = *N*-acetylmuramic
acid; *meso* DAP = *meso* diaminopimelic acid; Gln = glutamine
Other abbreviations as in Table 1.

Figure 4. Peptidoglycan structure

The exact sequence of amino acids in the peptide chains varies
between species, the most common being that found in all Gram-
negative bacteria and shown diagrammatically in Figure 4. It consists of
alternating L-and D-amino acids and includes an amino acid found
exclusively in bacterial walls, diaminopimelic acid (DAP). In the Gram-
positive bacteria there are several species-specific modifications of the
basic pattern, and that of *Staphylococcus aureus* is shown in Figure 4.
Here again the peptide chains consist of alternating L- and D-amino

acids, but D-glutamic acid is replaced by D-glutamine and lysine itself is the diamino acid allowing cross-linking between adjacent peptide chains. There is a further difference in that the short chains are cross-linked by a different short peptide, this time consisting of five glycine molecules. Other bacterial species may contain different diamino acids such as ornithine, and there are many possible combinations of amino acids in the cross-linking peptides.

All peptidoglycans seem to be synthesized in much the same way, starting with the formation of a nucleotide-linked muramyl pentapeptide of alternating L- and D-amino acids ending with a D-alanyl dipeptide. It should be recalled that few of the peptidoglycan amino acids are protein constituents. Proteins contain only the L-series of amino acids and do not contain diaminopimelic acid, ornithine or homoserine, all of which occur in peptidoglycans. It is therefore not surprising that the synthesis of short peptide chains from these unusual amino acids is quite different from that of proteins. Peptidoglycan synthesis involves the stepwise addition of free amino acids to the UDP muramyl precursor. The sequence is governed by the specificity of the enzyme catalysing each consecutive reaction both for its amino acid substrate and its UDP muramyl peptide acceptor. The peptide sequence is still genetically determined in that the sequence of amino acids in the enzyme catalysing the reaction is responsible for the specificity of that protein, and the amino acid sequence is itself determined by the sequence of nucleotides in the DNA of the gene coding for the enzyme. However, unlike the reactions of protein synthesis, peptidoglycan synthesis does not require activation of the amino acids by transfer-RNA, their assembly on a messenger RNA molecule or the participation of the ribosomes. The complete peptidoglycan molecule requires the assembly of long chains of alternating units of glucosamine and muramyl pentapeptide. This process is much more akin to the biosynthesis of carbohydrate or amino sugar polymers which require the nucleotide precursors. A disaccharide of N-acetylglucosamine-N-acetylmuramylpentapeptide is first formed and this is added on to the growing chain. In the completed peptidoglycan however the peptide chain linked to muramic acid consists of four, and not five, amino acids. The final stage in the formation of this complex macromolecule is the cross-linking of adjacent peptide chains with the loss of one molecule of D-alanine for each link formed.

This unique series of reactions has attracted much attention in recent years because it provides a possible site of attack for antibiotics. Just as the polyoxins provide possible specific inhibitors of fungal wall synthesis, there are several antibiotics which appear to inhibit bacterial growth by interfering with the biosynthesis of different stages in peptidoglycan synthesis. Cycloserine, for example is a structural anologue of D-alanine and prevents its incorporation into the UDP muramyl pentapeptide precursor. The penicillins and cephalosporins owe their antibacterial activity to their inhibition of the final stage in the cross-linking reaction. This causes the walls to weaken and finally the bacteria to burst. Bacitracin too prevents the assembly of the completed peptidoglycan polymer in one of the later stages of the synthesis. These antibiotics owe their usefulness to the fact that no similar reactions occur in man, and they are hence relatively non-toxic. Like all antibiotics, however, they suffer from the disadvantage that the bacteria may become resistant to their action either by destroying them or by preventing their reaching the target enzyme in some other way.

The biosynthesis of insoluble complex polymers outside the limits of the metabolic capabilities of the cell must pose problems for the yeast, fungi, and bacteria as well as for the microbiologist. We are just beginning to see the ways in which they have been solved. The final cross-linking reaction in peptidoglycan synthesis involves the loss of an extra D-alanine not required in the polymer, and this presumably allows the formation of a new peptide bond without requiring inaccessible cofactors such as ATP. Another problem is that of transferring the precursor from the inside of the cell where it is assembled, through the lipophilic cytoplasmic membrane, to the existing bacterial wall. This seems to have been resolved by attaching the precursor to carrier lipid molecules in the membrane so that the precursors are able to swivel through the membrane lipid mosaic.

2. Teichoic Acids

The other major components of Gram-positive bacterial walls are teichoic and teichuronic acids (from the Greek: *teichos* = wall). They may occupy as much as 50 per cent of the dry weight of the wall, or about 10 per cent of the dry weight of the cell. There are several different structural types of teichoic acids but they all contain either glycerol or ribitol phosphate. The simplest consist of long chains of either ribitol or glycerol units joined by phosphodiester bonds and substituted further with sugars, amino sugars or D-alanine. In many species, however, the polyphosphate backbone contains other substituents such as sugars or amino sugars as an integral part of the backbone. The basic structure of teichoic acid from *Bacillus licheniformis* is shown overleaf.

$$(-\text{galactose}-\text{glycerol}-\text{phosphate}-\text{galactose}-\text{glycerol}-\text{phosphate}-)_n$$

In addition to the polyphosphate polymers some bacteria contain other polyanions in which the charge of the molecule is provided by the carboxyl groups of a uronic acid rather than by a phosphate group. They are called teichuronic acids and that of B. *licheniformis* is shown below.

$$(-N\text{-acetylgalactosamine}-\text{glucuronic acid}-)_n$$

Not only do many bacteria contain several different teichoic acids and teichuronic acids, but the relative amounts of the polymers may vary with growth conditions. For example, in a medium in which phosphate limits growth all the teichoic acids of B. *licheniformis* can be replaced by teichuronic acids, suggesting that the two polymers serve the same function. They appear to be covalently attached to the other major wall component, the peptidoglycan. A short glycerol linkage unit joins them to the muramic acid, and it is likely that synthesis is closely coupled to that of the peptidoglycan.

The exact functions of the teichoic and teichuronic acids are still not completely defined. For the epidemiologist they provide a useful means of identification since they are antigenic and different types can be identified serologically. They also play some part in bacteriophage sensitivity, particularly in the bacilli and staphylococci, where the type and linkage of the sugar residues in the backbone affects susceptibility, however, neither of these properties confers an obvious advantage on the bacterium. Yet the structure of the teichoic acid is a genetically determined feature, and its biosynthesis is a complex and finely regulated process. It is therefore likely that it plays an important role in the life of the organism. The teichoic acid must help protect the organism against changes in external environment and there is some evidence that bacteria defective in the synthesis of teichoic acids are more sensitive to some antibacterial compounds. However, the changes in amounts of teichoic acids present in the wall under different growth conditions suggest that their major function may be as cation-binding compounds. Essential metal ions such as magnesium and calcium might be held near to the cytoplasmic membrane by the teichoic and teichuronic acids before being transported into the cytoplasm. A second possible role may be in the regulation of autolytic enzymes.

3. Lipopolysaccharides

In the Gram-negative bacterial walls which contain no teichoic acids their role may be taken over by the lipopolysaccharides. These complex macromolecules form part of the outer membrane of Gram-negative bacteria and, except in organisms with capsules, provide the first line of defence against antibacterial compounds, phagocytosis, and other host defence mechanisms. Like the teichoic acids they can be extracted from isolated walls or intact bacteria by trichloracetic acid or by 45 per cent phenol in which they are soluble. They are the major heat-stable antigenic determinants of Gram-negative bacteria and are generally known as O-antigens (from the German, *ohne Hauch*, without whip) since the other major antigens are the flagellar H-antigens. The lipopolysaccharides also provide the chemical groupings of the receptor sites for many bacteriophages. They have therefore been widely studied as a means of identifying strains for epidemiological purposes particularly in the *Enterobacteriaceae* whose members are responsible for many of the outbreaks of food poisoning. As their name implies they consist of both polysaccharide and lipid and the two parts can usually be separated after hydrolysis with dilute acetic acid.

Lipopolysaccharide $\longrightarrow$ Polysaccharide + Lipid A
(heat stable O-antigenic   (O-antigenic)   (toxic)
endotoxin)

The polysaccharide part is responsible for the antigenicity while the lipid part is toxic. The lipopolysaccharides themselves are described as endotoxins because they are part of the bacterial cell as opposed to exotoxins, such as anthrax or tetanus toxins, which are excreted by the bacterium into the surrounding medium. Although different Gram-negative bacteria show great variations in toxicity their Lipid A's appear to be very similar. Most consist of a β 1—4 or β 1—6 linked glucosamine disaccharide carrying long chain fatty acids attached by amide and ester links, that is, attached both to the amino and hydroxyl groups. The fatty acids are mainly straight chain saturated acids but always include one with a hydroxyl group in the 3 position. The most common is 3-OH myristic acid, a 14-carbon acid, but others occur in different species. These 3-OH acids exist only in the lipopolysaccharides and not in other parts of the bacterial cell, but whether they are responsible for toxicity is not known. A general pattern for the overall structure of the lipopolysaccharides is now emerging. Most consist of five sections:

Side Chains — Core — Backbone — Link — Lipid A

and Table 5 shows some of their components. They include many of the unusual L-, deoxy-, dideoxy-, and diamino sugars mentioned earlier in this essay as well as heptoses (seven-carbon sugars) and an eight-carbon sugar, 2-keto-3-deoxy-octulosonic acid (KDO). This provides the link between the lipid and polysaccharide part of the

Table 5. General structure of bacterial lipopolysaccharide

Fraction	Components	
Side chains	Repeating oligosaccharides	O-antigen Strain specific Often contain unusual sugars
Core	A short specific sequence of sugars and amino sugars	Often common to groups of bacteria
Backbone	Heptoses, phosphate, ethanolamine	L and/or D heptoses may be present
Link	2-Keto-3-deoxy-octulosonate (KDO)	In most species
Lipid A	Glucosamine, fatty acids, hydroxy fatty acids	3-OH fatty acids are most common

molecule and is the reason why the two parts can be separated by dilute acid hydrolysis.

Some indication of the role played by the different parts of this complex structure has been deduced from the properties of mutants with defective lipopolysaccharides. The best known are the so-called 'rough' mutants of the salmonella. Wild-type strains of *Salm. typhimurium*, for instance, produce smooth shiny colonies on most media, but spontaneous mutants arise which have a rough colonial appearance. Many of these have lost their ability to survive in an animal host, that is they are non-pathogenic. Chemical analysis of the isolated lipopolysaccharides showed that the rough strains had lost their O-antigenic side-chains and in some cases parts of the lipopolysaccharide core as well. The most defective mutants (deep rough mutants) had even lost some of the backbone heptose and phosphate units. However, no mutants completely lacking all parts of the lipopolysaccharide have ever been isolated, suggesting that at least the lipid part of the molecule is essential for viability. It seems possible that it plays an essential part in the stability of the outer membrane layer of the Gram-negative bacterial wall. Deep rough mutants are often more sensitive than their smooth parents to antibiotics, detergents and other inhibitors suggesting that the O-antigenic side-chains may act as a partial barrier preventing access of the inhibitors to their sites of action.

The lipopolysaccharide structure of the *Enterobacteriaceae* and particularly of the salmonella has been assumed to be the pattern for all lipopolysaccharides, but there now appear to be both minor and major variations in different species. For example, some lipopolysaccharides contain amino acids, others lack KDO or heptose, and some have different types of Lipid A. All seem to play a part in the antigenicity of the organism, to contain both hydrophilic and hydro-phobic residues, and to provide some protection for the organism.

4. Lipoproteins

Attached to the peptidoglycan of the Gram-negative bacterial wall is a lipoprotein which was first discovered when purified preparations of *E. coli* peptidoglycan were found to contain amino acids such as lysine and arginine, as well as the peptidoglycan amino acids. Trypsin treatment yielded a protein of unusual composition in that it contained only 13 of the amino acids normally found in proteins. It was linked to the diaminopimelic acid of the peptidoglycan by a sequence of basic amino acids ending with lysine. At its other end was a single cysteine molecule carrying a fatty acid on its amino group and a diglyceride covalently linked to its sulphur-containing side chain. Braun and his collaborators have now worked out the complete sequence of this protein and shown that it consists of a series of repeating short peptide chains alternately 15 and 7 amino acids long. This forms a sequence, in which every three and a half amino acids (that is alternately every third and fourth), is hydrophobic such as valine, leucine or isoleucine. This and other physical evidence strongly suggests that this lipoprotein is predominantly in the form of an α-helix with the hydrophobic groups arranged around the outside of the molecule. The unusual structure of this protein has attracted attention since it appears to be a specially designed molecule and the problem is for what purpose has it evolved. Its structure obviously allows it to fit into a hydrophobic lipid environment but its function is less easy to determine. One role may be to anchor the peptidoglycan to the outer membrane, thus providing one of the few covalent linkages joining the various parts of the Gram-negative wall. Alternatively its major function may be to provide a pore through which water-soluble metabolites may pass through the outer membrane.

The lipoprotein is not the only protein found in the Gram-negative outer membrane. Most species contain at least four major and several minor outer membrane proteins. They are all an integral part of the structure and are only released by strongly denaturing conditions but they may not be covalently linked to it. Probably like the protein components of the cytoplasmic membrane they are floating in a layer of phospholipid (see under cytoplasmic membranes). Many of the proteins act as bacteriophage receptors and so can be fairly readily detected, but their role in the bacterium is largely unknown since mutants lacking each, and indeed all four of the main outer membrane proteins, have now been isolated and appear to suffer no disadvantage.

It is possible that their main function is structural and that they can be replaced by other proteins of similar properties.

E. Cytoplasmic Membranes

The cytoplasmic membrane is very similar in all living cells. Electron microscopy shows a double track structure and the suggestion that it might consist of a bimolecular leaflet of phospholipids arranged in rows with their hydrophilic phospholipid heads on the outsides and their hydrophobic fatty acid tails inside, has readily gained support. This led to the acceptance of the membrane as a relatively stable structure in which the phospholipids and proteins were fixed in position both with respect to each other, and with respect to the inside and outside of the cell. It is now clear that this was an oversimplified picture, and the Singer and Nicholson model of a fluid mosaic of phospholipids, in which the other membrane proteins float, is much more likely. Such a model would easily allow changes in organization and conformation of membrane proteins. This is necessary to allow solutes to pass through the membrane and enter the cell and also to allow waste metabolites and precursors of walls to be transported to the outside. The phospholipids which make up 30 to 40 per cent of the cytoplasmic membrane thus have a dual function. They form a permeability barrier holding internal solutes inside the cell and they also act as a lipid solvent for the globular proteins and lipoproteins which float in the sea of phospholipid. The proteins are probably free to diffuse laterally through the matrix and even their orientation with respect to the two sides of the membrane may not be fixed, though some may span the entire membrane. Most of the phospholipids probably diffuse rapidly and laterally, but a few may be associated with specific membrane proteins and thus are a little more fixed.

Although the basic pattern of membrane structure in microorganisms is the same as that in higher animals and plants, there are some features which deserve comment. The prokaryotic membranes are all very similar, differing only in the nature of the phospholipid components and the number and types of proteins. The eukaryotic microorganisms on the other hand contain sterols, in addition to phospholipids and proteins, and in this respect they resemble higher plants and animals. This difference explains the different responses of prokaryotes and eukaryotes to polyene antibiotics such as nystatin and amphotericin. These compounds bind to the sterols in fungal and yeast membranes causing them to become leaky and so killing the cell. Bacterial membranes contain no sterols, do not bind these polyene antibiotics and bacteria are therefore unaffected. The exception to this rule is provided by the Mycoplasmas which are unusual in that they lack walls and when grown in the presence of sterols can incorporate them into the cytoplasmic membrane. When grown without sterols, mycoplasmal membranes do not contain these compounds and the organisms are resistant to the polyene antibiotics. When grown in the presence of sterols and in the few sterol-requiring species of Mycoplasma, these organisms are sensitive.

The lipids of all cytoplasmic membranes are predominantly phospholipids and the fluidity of the membrane is governed by the chain length and degree of unsaturation of their fatty acids. This is particularly striking in microorganisms which can often grow in a wide range of temperatures. Their membrane fatty acids are markedly changed by growth conditions and especially by temperature. The availability of metabolites may also affect membrane composition. In *Pseudomonas diminuta* grown under phosphate limitation, the membrane phospholipids are almost entirely replaced by acidic and neutral glycolipids, with sugar molecules in place of the phosphate groups. This is the same kind of phenomenon as that discussed earlier in terms of the replacement of teichoic acids by teichuronic acids in phosphate-limited *B. subtilis*. Presumably in the pseudomonad the glycolipids can perform the same functions as the phospholipids both as a hydrophobic barrier and as a solvent for the integral proteins.

In the prokaryotes the cytoplasmic membrane is the only cellular membrane. It must not only act as a permeability barrier and the site of active transport but must also carry out the functions undertaken by the mitochondrial and nuclear membranes, the endoplasmic reticulum, and the Golgi apparatus of the eukaryotes. For this reason the membrane proteins of bacterial membranes are more numerous and more varied in activities than those of higher animals. Furthermore many of these proteins are under strict regulatory control, being synthesized only when required. This is particularly noticeable in the proteins involved in transport and binding of metabolites most of which have to be transported into the cell with the aid of specific proteins integrated into the cytoplasmic membrane. One of the first to be studied in detail was the M protein of *E. coli*. This membrane protein is under the same regulatory control as the enzyme β-galactosidase which metabolizes the lactose once it has been bound to the M protein and transported into the cell. Many of the other membrane proteins are enzymes of various kinds and include ATPases, the enzymes involved in phospholipid synthesis, and those concerned with the synthesis of the walls and capsules. In bacteria they include succinic dehydrogenase and the cytochromes.

One of the obvious differences between the bacterial and fungal membranes lies in their carbohydrate components. The cytoplasmic membranes of some yeasts and fungi contain polymers of mannose and other sugars. Their function is not known since some strains have no carbohydrates at all. All Gram-positive bacterial membranes have at least one carbohydrate polymer in common, a lipoteichoic acid. These are polymers of glycerophosphate linked to glycolipids or phosphoglycolipids. They were originally called membrane teichoic acids and appear to be an integral part of the membrane. Although their widespread distribution was soon recognized, no function could be assigned until recently when they were found to act as lipid carriers for teichoic acid synthesis. The nucleotide precursors of the teichoic acids are water-soluble and synthesized within the cytoplasm, whereas the final polymer is insoluble and synthesized within the cytoplasmic membrane linked to the peptidoglycan. The precursor has therefore to be transported through the membrane which behaves as a hydrophobic barrier. The lipoteichoic acids seem to provide a lipid anchor within the membrane for the polymerization of the precursors. Once completed they are transferred to the linkage unit also attached through the peptidoglycan precursor to another lipid carrier in the membrane, the C_{55} isoprene compound. The complexities of this process are still being unravelled but it is clear that there exist in the cytoplasmic membranes lipid carriers whose function is to transport precursors through the hydrophobic barrier and to assemble them into the complex macromolecules found in the outer layers. The enzyme proteins which catalyse the process are also closely associated with both the carrier lipids and the membrane phospholipids and in isolation may show a strict requirement for them.

F. Cytoplasmic Components

The cytoplasmic membrane is the limit of the metabolic activity of all cells and its disruption causes lysis and subsequent death. The cytoplasmic components thus include all those materials necessary for survival, metabolism, and replication and, with a few obvious exceptions, are similar in all species. The soluble components include a wide variety of metabolic enzymes whose properties and functions are similar throughout the living world. The more we know of metabolic pathways the more we see that the main ones are common to all forms despite the diversity of substrates used and end-products released. Most of these result from the existence of additional enzymes designed to channel substrates into the central metabolic pathways or to modify end-products. They do not reflect fundamentally different types of metabolism. Carbohydrates are generally oxidized anaerobically by glycolysis and the pentose phosphate pathways, often with minor modifications and aerobically by the tricarboxylic acid cycle. The synthesis and breakdown of fats, nucleic acids, and proteins all occur through the same types of reactions in all living organisms. Their enzyme activities will therefore be similar although the actual sequence of amino acids in the proteins concerned may be very different in different species. In addition there will be enzymes which synthesize the unusual components discussed earlier and metabolic enzymes carrying out particular reactions some of which are useful in species identification. The proteins responsible for these differences are too numerous and varied to be included here but they should not be forgotten. Current methods for studying proteins in the cell do not usually allow us to distinguish the positions of different soluble proteins within it. Those incorporated into insoluble structures can sometimes be detected but our main means of distinguishing soluble proteins by their enzyme activity or antigenicity cannot easily be applied to the whole cell, even with labelled substrates and antibodies. Only those present in large amounts can be detected in this way and we are still far from being able to locate the majority of individual proteins. Nevertheless it is possible, if not probable, that each protein has both a specific role and a specific site within the cell.

It is the insoluble cytoplasmic components which most clearly separate the prokaryotic and eukaryotic microorganisms. The mitochondria, endoplasmic reticulum, Golgi apparatus, and special organelles such as chloroplasts, are confined to the latter. They are similar to those of higher animals and plants and will not be discussed further. Both groups contain nuclear material, ribosomes, and storage granules which show interesting differences.

1. Nuclear Material

The nucleus of the eukaryotic microorganisms is enclosed in a nuclear membrane composed of phospholipid and protein appearing in the electron microscope very similar to the cytoplasmic membrane. Like the latter it is semipermeable but rather more porous. The DNA itself is arranged in linear chromosomes and consists of double-stranded DNA closely linked to basic proteins called histones. Each cell has only one nucleus. The genetic material of the prokaryotes behaves as a single molecule of double-stranded DNA. There may be more than one copy in a single cell and each molecule is tightly coiled. There is no nuclear membrane and no histone protein.

In addition to the chromosomal DNA some bacteria may contain extrachromosomal DNA which is replicated independently and can be lost completely without loss of viability. It exists as small covalently closed circles of double-stranded DNA which are called plasmids. Plasmids may code for such useful properties as resistance to antibiotics or the metabolism of unusual substrates such as camphor or octane. They can also code for the ability to transfer themselves to other cells and it is this property which has had such a profound affect on the spread of antibiotic resistance among bacterial populations.

2. Ribosomes

The granular appearance of many cells is largely due to the presence of ribosomes which are the organelles responsible for protein synthesis. Their overall structure, composition, and functions are similar in all species, but there are variations which are of practical importance and which distinguish bacterial, mitochondrial, and chloroplast ribosomes from most eukaryotic ribosomes. The best studied ribosomes are those of E. coli which have a molecular weight of about 2.5×10^6 and a sedimentation value of about 70S. They consist of about one-third protein and two-thirds nucleic acid and form a complex structure about 20 nm in diameter. They can be dissociated into two parts both of which contain protein and RNA. The smaller of the two parts has a sedimentation value of 30S and contains a single molecule of RNA with 21 different proteins arranged in a specific way to form the complete subunit. The larger (50S) fragment has two different RNA molecules together with about 35 proteins. This organization of the ribosome into two subunits distinguishable by size, composition, and function appears common to all species. Eukaryotic ribosomes are larger than those of bacteria. The complete structure has a sedimentation value of 80S and the two subunits are 60S and 40S. The smaller unit like the bacterial 30S subunit has a single RNA/molecule with many different proteins while the 60S subunit has three RNA molecules combined with its proteins. The ribosomes of mitochondria and chloroplasts are more closely related to the bacterial (70S) ribosome than to the 80S

Table 6. Inhibitors of ribosome function

Prokaryotes (70S Ribosomes)	Eukaryotes (80S Ribosomes)
Chloramphenicol	Cyclohexamide
Erythromycin	Diphtheria toxin
Fusidic acid	Fusidic acid
Puromycin	Puromycin
Streptomycin	
Tetracycline	

ribosome found elsewhere in the eukaryotes. These differences in size and sedimentation value are a gross reflection of many differences in the detailed molecular architecture of the ribosome. They have been exploited to inhibit bacterial protein synthesis without affecting the same process in the eukaryotic host. Similarly, although of no practical importance in the treatment of disease, specific inhibitors of protein synthesis in fungi and higher animals have proved useful in elucidating the sequence of reactions leading to the complete protein molecule. Some of the antibiotics and other inhibitors which interfere with protein synthesis at the ribosomal stage are shown in Table 6. Most of them owe their inhibitory activity to their ability to bind specifically to one of the ribosomal proteins many of which play a catalytic role in protein synthesis.

3. Storage Granules

The only other materials common to most microbial cells are storage polymers. They are all osmotically inert compact polymers which provide a convenient means of storing high concentrations of readily metabolizable substrates without disrupting the cell. Often they can be detected as granular aggregates within the cytoplasm and they may be surrounded by a membrane. Although their primary function is in protein synthesis, ribosomes can also be considered as storage particles. When protein synthesis is prevented by nutrient limitation, the number of ribosomes in a cell may fall from thousands to a few hundred. Existing ribosomes are degraded and their constituents used for energy and maintenance. Since they can occupy as much as 40 per cent of the dry weight of the cell, they must make a significant contribution to energy storage, but despite this many species contain granules whose sole function is to act as reserve material.

The most common storage materials in microorganisms are, as in plants and animals, lipids and carbohydrates. The yeasts and fungi use neutral fats as their reserve materials more often than do the bacteria. Here the most common lipid storage compound is poly-β-hydroxybutyrate. In some species such as *Bacillus megaterium*, this compound can occupy as much as 60 per cent of the dry weight, particularly after growth on acetate or butyrate which are sources of acetyl Coenzyme A. It is formed by the condensation of acetyl Coenzyme A forming acetoacetyl Coenzyme A which is then reduced to β-hydroxybutyryl Coenzyme A before polymerization. It is thus a rather unusual polymer with a reduction level between carbohydrates and fats.

Both eukaryotes and prokaryotes can use various carbohydrate polymers as storage materials. Starch granules containing α 1–4 linked glucose units with occasional α 1–6 branch points occur in many yeasts, fungi, protozoa, and algae, and also in certain bacteria such as clostridia and the coliforms. Other glycans including some β 1–3 linked polymers are found in some fungi. Although a wide variety of types exist, each species is usually limited to one storage carbohydrate.

In addition to the lipid and carbohydrate storage materials common to other species, some microorganisms can store phosphate as linear chains of inorganic polyphosphate. Granules of this material can be seen under the light microscope after staining with a basic dye such as methylene blue. They are known as volutin granules after *Spirillum volutans* in which they are particularly common. They also frequently occur in the corynebacteria especially *C. diphtheriae*, and in the mycobacteria, where they provide a useful identification character. They accumulate when nucleic acid synthesis is prevented by starvation and the phosphate is incorporated into nucleic acid when synthesis starts again.

In a few species, storage granules composed of sulphur may accumulate transiently when excess sulphide is available in the medium. They result from two different kinds of metabolic reactions. In the photosynthetic purple sulphur bacteria, hydrogen sulphide is used as an electron donor thus producing sulphur. Other groups use the oxidation of sulphide as their main source of energy. Both groups oxidize the elemental sulphur to sulphate when suitable oxidizing conditions arise. Although relatively uncommon now, this production of elemental sulphur by microorganisms may have been more important in the geological past under the reducing conditions then existing. Most of the world's sulphur deposits are the result of intense microbial activity during an era of warmth and sunshine. Even now sulphur lakes are formed where warm water rich in calcium sulphate is available from hot springs and is converted into sulphur by a combination of two groups of sulphur bacteria.

V. Endpiece

In this short essay I have barely touched on yeasts, fungi, and algae and have almost completely excluded viruses and protozoa. Nor have I considered a very wide range of microbial metabolites many of which are commercially useful. I hope I have shown that the chemical composition of the microorganisms I have described can be expected to contribute to studies of many different groups. We still do not know how far the basic pattern extends to other species and how far it changes in response to environmental pressure. In terms of pure chemistry it is hard to explain the properties of macromolecules even in the special structures I have discussed and it may be many years before we know what makes a coccus round or what puts the curves into a spirillum. Nevertheless techniques of biochemical and genetic analysis are rapidly improving and we can confidently expect that the microbiologists of the future, including perhaps some reading this, will solve these and other equally fascinating problems in Microbiology.

VI. Further Reading

I have assumed that most readers will be familiar with the structure and metabolism of the simple monomers discussed. If not they can be found in most basic Biochemistry and Microbiology texts. There are many such books and choice will depend on personal preference. Three with particularly good diagrams are *Biochemistry* by Stryer (1975), published by Freeman & Company; *Outlines of Biochemistry* by Conn & Stumpf (1976), published by John Wiley; *Short Course in Biochemistry* by Lehninger (1973), published by Worth.

More detailed accounts of the topics discussed in this essay are also given in a number of standard Microbiology texts and special articles. I have selected a few of the more specialized ones here.

Chemical Composition

Organic Chemistry of Biological Compounds by Barker (1971), published by Prentice-Hall, covers most of the compounds mentioned in this essay as well as many more. There is a comprehensive reference collection of Tables of microbial components in the *Handbook of Microbiology*, Volume II: *Microbial Composition* (1973) edited by Laskin and Lechevalier and published by the CRC Press.

Cell Structure

Cell Structure and Function by Loewy and Siekevitz (1970), published by Holt Rinehart, Winston, provides a general picture of many cell types. *The Bacterial Cell: Major Structures* by Reynolds (1973), in *Bacterial Growth*, edited by Mandelstam and McQuillen and published by Blackwells, is limited to bacteria, but discusses some aspects of biosynthesis as well.

Flagella

There are two excellent *Scientific American* articles: *How Bacteria Swim* by Berg (August 1975), and *The Sensing of Chemicals by Bacteria*, by Adler (April 1976). A rather more comprehensive and advanced approach is given in *Bacterial Motility and Chemotaxis* by Smith (1978), in *Companion to Microbiology* edited by Bull and Meadow, published by Longmans.

Walls and Membranes

There are several interesting reviews in *Bacterial Membranes and Walls* edited by Leive (1973), and published by Marcel Dekker. The fluid mosaic model of cell membrane structure is clearly described by Singer and Nicholson in *Science, 175, 720*. Two more detailed accounts of membrane structure and function are given by Salton (1971) in *CRC Critical Reviews of Microbiology*, p. 720, and by Ellar (1978) in *Companion to Microbiology* edited by Bull and Meadow, published by Longmans. The structure and biosynthesis of bacterial walls is described by Meadow (1974) in *Companion to Biochemistry* edited by Bull, Lagnado, Thomas, and Tipton, published by Longmans.

Antibiotic Action

The Biochemistry of Antimicrobial Action by Franklin and Snow (1975) provides a concise and clear account.

Dynamics of Microbial Growth

D. W. TEMPEST

Laboratorium voor Microbiologie, Universiteit van Amsterdam,
Plantage Muidergracht 14, Amsterdam-C, The Netherlands

I. Introduction

'Dynamics' may be defined as a study of forces acting upon a body and, in this sense, the 'dynamics of microbial growth' implies a study of those (environmental) forces that act either to promote or to impede the growth of microbes. Now the term 'microbe' also requires to be defined since this, and the term 'bacterium', are frequently used synonymously, but whereas all bacteria are microbes, not all microbes are bacteria. Indeed, any organism (prokaryote or eukaryote) that individually cannot be perceived by the naked eye is entitled to be called a microorganism. However, it must be recognized that studies of microbial growth kinetics have concentrated largely on bacterial species, and it follows, therefore, that in this chapter attention will be directed mainly to aspects of bacterial growth.

The capacity to grow, and ultimately to multiply, is one of the most fundamental characteristics of living cells. In fact, with microorganisms it is generally the sole criterion that is used to assess whether or not such creatures are alive. Hence, by definition, a non-viable microbe is one which is incapable of increasing in size and number when incubated in an equitable, growth-supporting environment for a prolonged period of time. However, despite the fact that growth is such a basic aspect of microbial behaviour, little attention was paid to the principles which underlie it until the advent of continuous culture techniques. Indeed, prior to the 1940's it was generally considered sufficient to record growth as being either evident $(+)$ or not $(-)$, with the occasional recourse to semiquantitative flights of fancy such as $+++$, $++$, and $\pm$! And whereas these essentially qualitative assessments often proved adequate, there can be little doubt that the more quantitative approach that has been adopted over the last 30 years has yielded a vastly more penetrating insight into the nature of microbial life.

Unfortunately, the manipulation of quantitative data, and their correlation with growth-associated processes, frequently demands the construction of mathematical equations — nowadays popularly called 'models'. These range from the very simple to the exceedingly complex, and often serve only to confuse those biologists whom they are intended to enlighten. Hence, in this chapter, every effort has been made to keep the mathematics to a minimum and to explain fully the steps involved in deriving each equation. Thus it is hoped that the small amount of mathematics necessarily involved in the treatment of this subject will not create an insuperable barrier to the understanding of what is, after all, a vitally important aspect of Microbiology.

II. Microbial Growth in a Closed Environment

Microbial growth is a process that requires the coordinated synthesis of a range of complex macromolecules, the energy for this process (and the necessary intermediary metabolites) being derived from the uptake and chemical transformation of a relatively small number of compounds and elements. Hence growth in any environment is only possible when there is present a complete mixture of those compounds and elements that are essential for cell synthesis and functioning. In this connection, all living cells seemingly contain carbon, hydrogen, oxygen, nitrogen, sulphur, phosphorus, magnesium, potassium, and a number of so-called 'trace' elements (Mn^{2+}, Cu^{2+}, Fe^{3+}, Zn^{2+}, etc.). Consequently, all these elements must be present in a utilizable form, along with water, and, if other environmental conditions are propitious, then those microbes that may be present will grow and multiply.

Bacteria reproduce by a process of *binary fission* in which, so far as is known, the contents of the mother cell are partitioned equally between the two daughter cells. One obvious consequence of binary fission that may be overlooked is that the mother cell is consumed as the daughter cells are created. Thus, one must exercise caution in specifying the 'age' of a bacterial cell since, logically, this can only be made by reference to the time that has elapsed since its birth. This may seem to be a trivial point, but there are to be found in the literature frequent references to 'old cells' and 'young cells', meaning bacteria taken from cultures that have been incubated for long and short periods of time, respectively. The confusion that this loose terminology causes has far-reaching consequences, particularly in the interpretation of physiological data.

Binary fission ought to lead to the birth of identical twins with identical physiological properties and growth potentials. Therefore one might expect these twins to grow at identical rates and to divide synchronously. This rarely happens, and in any culture derived from a single bacterial cell there is generally to be found a wide spread of individual generation times from a well defined minimum to a less well defined maximum. Hence, the culture as a whole increases in population density with a doubling time that is not precisely the same as the mean generation time. This is a point of detail that can be ignored for the moment; but it must not be forgotten.

A. Kinetic Aspects

When growing in a relatively constant and equitable environment, microbial populations multiply at rates that are, overall, constant. Therefore, even though there may be a spread of individual generation times, a number of organisms (N) will give rise to $2N$ progeny, and these to $2^2 N$ progeny, and so on, with an overall doubling time (t_d) that is constant. This can be represented as follows:

$$N \rightarrow 2N \rightarrow 4N \rightarrow 8N \rightarrow 16N \rightarrow 32N \ldots$$

or,

$$2^0 N \rightarrow 2^1 N \rightarrow 2^2 N \rightarrow 2^3 N \rightarrow 2^4 N \rightarrow 2^5 N \rightarrow 2^n N.$$

Here, n represents the number of doublings that have occurred after some time interval t. Thus,

$$n = t/t_d.$$

It follows, therefore, that the number of organisms present in a culture after t hours of incubation will be related to the initial population by the equation:

$$N_t = N_0 2^n = N_0 2^{t/t_d}.$$

Similarly,

$$N_t/N_0 = 2^{t/t_d}$$

and taking logarithms,

$$\ln(N_t/N_0) = (\ln 2)t/t_d,$$

or

$$(\ln N_t - \ln N_0)/t = 0.693/t_d.$$

Therefore, plotting the natural logarithm of the number of organisms against the time of incubation should yield a straight line whose slope will be numerically equal to $0.693/t_d$. This is found to be the case with many real bacterial cultures: an example is given in Figure 1. It should be noted, however, that the 'exponential' growth period is preceded by

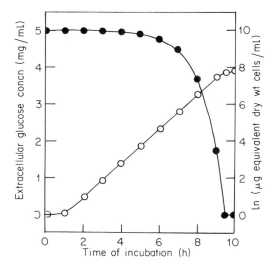

Figure 1. Changes in the concentration of organisms (○) and of glucose (●) with time of incubation of a culture of *Klebsiella aerogenes* growing in a simple salts medium at 37 °C (pH 6.8). Notice that cell concentration is plotted as its logarithm, but glucose concentration as its actual value.

a short 'lag' period (a period of metabolic adjustment) and that the exponential increase in the population density (here represented as biomass) ceases after a relatively small number of culture doublings.

We shall return to this so-called 'growth cycle' later, but first it is necessary to say a little more about exponential growth. As already mentioned, the slope of the line $\ln N$ versus t is numerically equal to $0.693/t_d$. Thus,

$$d(\ln N)/dt = 0.693/t_d.$$

But, $d(\ln N)/dt = d(\ln N)/dN \times dN/dt$, and since the differential co-efficient of $\ln N$ is $1/N$, then it follows that:

$$\frac{1}{N} \cdot \frac{dN}{dt} = \frac{0.693}{t_d}.$$

$1/N \cdot dN/dt$ is called the *specific growth rate* constant and is given the symbol μ. It is the rate of increase in cell numbers per unit of cell numbers.

Now the growth of each individual organism in a population proceeds through various stages of the cell cycle in which the mass of the organism increases up to some maximum value prior to cell division. Hence, in any population of organisms growing asynchronously, there will be a distribution of cell sizes. However, just as the population density as a whole increases exponentially with time, so too does the culture biomass concentration. Moreover, since biomass generally can be measured more accurately than can numbers of organisms, the basic microbial growth equations are generally expressed in terms of mass rather than number of cells. Thus, if x is the culture biomass concentration:

$$\mu = 1/x \cdot dx/dt = (\ln 2)/t_d = 0.693/t_d.$$

The use of natural logarithms in developing this particular growth equation is helpful in subsequently analysing the kinetics of microbial growth in continuous cultures. It is not, however, essential to use them, though if other logarithmic functions are employed then the specific growth rate constant μ has a different numerical value and ought to be ascribed a different symbol. For example, if logarithms to the base 10 were used, then:

$$\mu' = 0.301/t_d;$$

and if logarithms to the base 2 were used, then:

$$\mu'' = 1/t_d.$$

One final point of clarification: the units of time may, of course, be minutes, hours or days, but they *must* be specified. With bacterial cultures it is usual to express the doubling time in hours, and μ in reciprocal hours (h^{-1}). Cultures of algae, on the other hand, often grow at much slower rates and it is not unusual for their specific growth rate to be expressed in reciprocal days (d^{-1}).

As mentioned previously, the specific growth rate equation, derived above, adequately predicts the behaviour of many real microbial cultures, but only within certain limits. Indeed, unless these limits are clearly recognized, conclusions may be drawn that are palpably absurd. A simple example serves to make the point.

Escherichia coli (a common gut organism) is capable of growing at a rate such that the culture doubles in biomass every 20 minutes. One might ask the question, therefore, 'if 100 ml of nutrient medium is inoculated with about 1 μg of organisms (that is, about 5×10^6 cells) and subsequently incubated for 24 h, what quantity of organisms might one expect to be produced?'

$$\mu = 0.693/t_d = 0.693/0.333 = 2.08 \ h^{-1}.$$

$$\mu = (\ln X - \ln X_0)/t$$

therefore,

$$\ln(X/X_0) = 2.08 \times t = 49.92.$$

Hence,

$$X/X_0 = e^{49.92} = 4.8 \times 10^{21}$$

and

$$X = 4.8 \times 10^{21} \times X_0 = 4.8 \times 10^{21} \ \mu g.$$

Thus, starting with an initial inoculum of 1 μg organisms, and allowing the culture to grow for 24 h, should result in the production of an amount of organisms weighing 4.8×10^9 *metric tons*!

.Clearly the answer is ridiculous for, as shown in Figure 1, real cultures seldom attain a biomass concentration in excess of a few mg per ml. However, this kind of mathematical exercise does serve a useful purpose in that it concentrates attention not only on the enormous growth potential of many microorganisms, but also on the fact that there exist, both in Nature and in the laboratory culture, circumstances that limit the extent to which this growth potential is expressed. Such considerations lie at the heart of our understanding of the dynamics of microbial growth and will be briefly reviewed in the next section.

B. Cell–Environment Interaction

It follows from the fact that laboratory cultures of microorganisms seldom attain a biomass concentration greater than a few mg equivalent dry weight per ml that circumstances intervene to limit further cell multiplication. Two explanations seem plausible: (1) that organisms interact directly with one another to inhibit growth: such interaction is known to occur in cultures of some animal cells and is termed 'contact inhibition', and (2) that organisms interact with their environment in such a way as to render it no longer conducive to further growth.

The idea of contact inhibition limiting the size of bacterial populations has failed to receive convincing experimental support. On the contrary, cultures generally grow to much greater biomass concentrations if dialysed against a continuous flow of fresh medium. Moreover, if contact inhibition occurred, one would not expect organisms to produce large colonies when spread on a nutrient agar plate. It therefore seems most likely that the event(s) provoking cessation of growth in a batch culture reside in changes induced in the environment by the growth of the organisms. Again, one can imagine two possibilities: either growth leads to the depletion of some essential nutrient substance from the environment, or else growth leads to the accumulation of products that ultimately reach toxic concentrations. In this latter connection, although the formation of auto-inhibitory compounds during the growth of microbial cultures has been demonstrated in a number of cases, the most frequently occurring toxic products are hydrogen ions and hydroxyl ions which cause the culture pH value respectively to decrease or increase as growth proceeds. For example, when organisms are grown on an organic acid substrate (say acetate, succinate or lactate) with the pH value adjusted initially to neutrality, then growth of the organisms leads to a marked increase in the culture pH (unless this is deliberately controlled) due to the excessive uptake and metabolism of anions over cations.

Automatic adjustment of the culture pH value generally leads to a culture attaining a higher final population density, but the increase may be relatively small. With aerobic organisms, growth at high population densities becomes impeded by the decrease in available oxygen. It is insufficiently appreciated that it is only the oxygen that is dissolved in the culture fluid that is accessible to the organisms. Since the solubility of oxygen in water is low (5–7 μg/ml, at 20 °C), a high rate of oxygen consumption (as with rapidly-growing dense populations) means that the rate at which oxygen can dissolve in the culture fluids becomes the factor limiting the subsequent growth rate of the organisms.

Assuming the culture pH value to be automatically controlled and that growth is not restricted by the accumulation of auto-inhibitory substances or by the availability of oxygen, then ultimately growth will cease due to the depletion of some essential nutrient substance. This follows from the fact that, as growth proceeds exponentially, so there is an exponential rate of change in the concentrations of all essential nutrients. Indeed, the rates at which these essential nutrient substances are taken up from the medium have been found to be directly proportional to the rate of cell synthesis. That is:

$$1/x \cdot dx/dt \propto -1/x \cdot ds/dt$$

or,

$$\mu = Yq,$$

where q is the specific rate of substrate assimilation and Y is the proportionality factor or *yield value*. The molar yield value for glucose, found with cultures of *E. coli*, growing in a simple-salts medium, generally is in the region of 90 (g organisms synthesized per mol of glucose consumed). Hence, when growing at a rate (μ) of $1.0 \, h^{-1}$ (i.e. with a doubling time of 0.693 h, or 42 min), the specific rate of glucose consumption will be approximately 11.1 mmol per g-equivalent dry weight of organisms per hour. It follows, therefore, that if all other nutrients (including oxygen) were present in excess of the growth requirement, and glucose was added to a final concentration of 5 mg/ml (i.e. 27.8 mM), then the length of time that organisms would spend in the exponential growth phase would be critically dependent on the inoculum size. Hence, if the inoculum provided 0.001 mg biomass/ml (say, about 5×10^6 organisms/ml), then, assuming no lag period, the time taken to reach a state of glucose exhaustion can be calculated as follows:

(i) From the molar yield value (90), the concentration of organisms present when all the glucose has been consumed will be $90 \times 27.8/1000 = 2.5$ mg/ml.

(ii) Since $\ln(X/X_0) = \mu t$, and $\mu = 1.0 \, h^{-1}$,

$$\ln(2.5/0.001) = t$$

$$\ln(2500) \quad = t = 7.8 \, h.$$

It is clear, therefore, that if such a culture was incubated overnight (16 h) then, even assuming a lag period of 1–1.5 h, it would have reached a state of glucose exhaustion some 6–7 h prior to harvesting. Indeed, in order to ensure that such an overnight culture was still in the exponential growth phase after 16 h incubation, it would be necessary to decrease the inoculum size to 1×10^{-7} mg organisms/ml, or less (that is, to 5×10^2 organisms/ml, or less).

The inclusion of this small example serves to make one important point, and that is that microorganisms interact continuously with the environment and that, as growth proceeds, so the environment changes at an accelerating rate (Figure 1).

Thus, organisms in the exponential growth phase (particularly the late-exponential phase) are experiencing dramatic and precipitous changes in their environment, changes which lead inevitably to it being no longer able to support further growth. Hence growth ceases and the culture enters the so-called 'stationary phase'. In this connection, it should be realized (Figure 1) that the final doubling of biomass prior to the onset of the stationary phase lowers the extracellular carbon-substrate concentration from 50% of its starting value to zero. This precipitous drop in substrate concentration can occur within a period of 42 min, or less.

Further consideration of Figure 1 leads to two additional conclusions: (1) that the organisms can accommodate to marked changes in their environment and yet continue to express their full growth-rate potential right up to the moment when the limiting substrate is almost totally consumed, and (2) that the sequence of changes occurring in a batch culture, collectively referred to as the 'Growth Cycle' are not an expression of some inherent property of the organisms, but simply the inevitable consequence of the interaction of organisms with their environment in a closed system. And since most natural ecosystems are essentially open systems, it follows that the batch-culture growth cycle is essentially a laboratory artefact.

The above mentioned conclusions have far reaching consequences in our analysis of microbial growth kinetics and therefore will be considered in more detail.

C. Microbial Adaptability

Although biochemical research has, over the past three decades, tended to emphasize more and more the unity of life processes, nevertheless there are important differences between free-living micro-organisms and the cells of higher animals and plants. To appreciate this point, one must consider these different cells in relation to their environment. At one extreme one finds the cells of, say, the human body, and here we recognize the fact that the human animal possesses an elaborate array of organs whose functions, collectively, are to maintain a constant internal environment; as Claude Bernard realized, over 100 years ago, 'the constancy of the internal environment is a condition of life'. Therefore, if this environment is caused to vary beyond certain narrow limits (of temperature, pH, salt balance, nutrient concentration, etc.) the cells cannot accommodate — they cease to function and the animal dies.

At the other extreme one finds the free-living microbial cells which occupy environments that they cannot control directly and which may vary markedly and, often, rapidly. To cope with this sort of situation, microorganisms (we now know) have acquired in the course of evolution a whole armoury of sophisticated control mechanisms which allow them to change *themselves*, structurally and functionally, in order to accommodate to environmental changes and thereby to maintain their growth potential. Indeed, to such an extent can these creatures vary *phenotypically* (as it is said) — that is, both structurally and functionally — that it is virtually meaningless to talk of a 'normal' microbial cell. The state of 'normalcy', which has profound meaning so far as the cells of higher animals and plants are concerned, has little or no meaning with regard to the microbial cell. In short, then, it is impossible to define physiologically any microbial cell without reference to the environment in which it is growing. This is a most important (indeed fundamental) concept that is not widely appreciated and yet which fully accounts for the fact that microbes generally can be cultivated in the laboratory under conditions which may be far removed from those which they experience in natural ecosystems.

A detailed description of the structural and functional changes that have been found to occur in microbial cells, particularly when exposed to low-nutrient environments, is beyond the scope of this chapter. It is sufficient to state here that almost all components of the microbial cell vary quantitatively with changes in the growth conditions, and many components may also vary qualitatively.

It is now appropriate to return to the events occurring in a laboratory batch culture, and to reexamine these in the context of cell—environment interaction.

III. Microbial Growth in an Open Environment

It follows from the various considerations outlined above, that if one is interested in any one of a large number of aspects of microbial physiology (that is, cell structure and functioning), one cannot ignore the fact that the properties of organisms growing in a batch culture vary continuously throughout the growth cycle by virtue of the fact that the environment is varying continuously. In this connection, the tacit assumption that 'mid-exponential phase' or 'early stationary phase' meaningfully circumscribe the physiological state of the cells is clearly untenable since, as mentioned previously, these are periods during which the environment, and the properties of the cells, are changing dramatically.

How then can one get around this *impasse*? The answer is extemely simple once one takes into account the fact that the rate of change in each and every environmental parameter is linked directly to the rate of change of microbial biomass concentration (that is, to the growth rate of the organisms). Therefore, all one needs to do is to add fresh medium to the culture at a rate sufficient to maintain the culture population density at some prescribed submaximal value, thereby replenishing nutrients that are being consumed and, simultaneously, diluting out end-products that are accumulating at rates exactly sufficient to ensure that the environment no longer varies with time. Of course, if one continued to employ a closed system one would need to add fresh medium at an exponentially increasing rate, thus causing the culture volume to increase exponentially and soon reach unmanageable proportions. To overcome this difficultly, it is only necessary to insert an overflow tube (or weir) into the culture so that excess culture can flow from the growth vessel at the same rate that fresh medium is being added, thereby maintaining the culture volume constant. This is the operating principle of one type of continuous culture device the *Turbidostat* — and its essential features are illustrated in Figure 2.

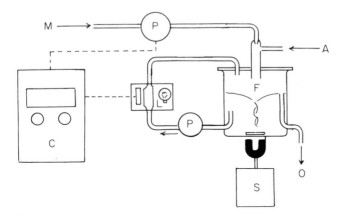

Figure 2.   Essential features of a turbidostat. The culture is contained in a flask (F) to which medium (M) is pumped at a rate determined by the photocell (L) and the controller (C). The culture is vigorously stirred by means of a magnetic stirrer (S) and aerated by means of injecting air (A) along with the medium. Excess culture is removed by means of an overflow device (O). Two pumps (P) are required, one to deliver medium to the culture, and the second to circulate culture through the photometer cell.

A. Kinetics of Growth in a Turbidostat

It is now important to look at the dynamics of microbial growth in such an 'open' system. With such a culture, the total biomass concentration is maintained constant by first monitoring the culture population density (usually the culture optical density at some convenient wavelength of light) and, with the aid of a controller, activating a pumping device that delivers medium at a rate sufficient to maintain that population density at some prescribed value. Since biomass will be constant, the specific rates of utilization of all nutrients will be constant. Therefore, fresh medium need no longer be added at an exponentially changing rate, but at a constant rate that will be specified by the total culture biomass concentration (that is, by $x \cdot V$, where x is the concentration of organisms per ml and V is the culture volume in ml). The specific rate of substrate consumption is, as mentioned previously, defined mathematically by $-1/x \cdot ds/dt$ and, for a turbidostat culture, can be determined experimentally by determining the concentrations of substrate in the inflowing medium and effluent supernatant. Thus:

$$\text{Rate of change} = \text{Input} - \text{Output} - \text{Consumption}$$
$$ds/dt = fS_R \quad - fs \quad - qxV$$
$$= f(S_R - s) \quad - qxV$$

where S_R is the substrate concentration in the reservoir, s the substrate concentration in the culture extracellular fluid, f the medium flow rate (ml/h), x the microbial biomass concentration (mg equivalent dry weight organisms/ml), V the culture volume, and q the specific rate of substrate consumption $(-1/x \cdot ds/dt)$. However, once the culture is in a steady state, the rate of change in substrate concentration will be zero. Hence:

$$qxV = f(S_R - s)$$
$$q = D(S_R - s)/x$$

where D is the ratio of the flow rate to the culture volume (f/V) and is called the *dilution rate*. Now the ratio $(S_R - s)/x$ is the reciprocal of the yield value Y; therefore, at steady state:

$$q = D/Y. \tag{1}$$

A further point of clarification is now necessary. With turbidostat cultures (as, indeed, with chemostat cultures; see below) fresh medium is added at a constant linear rate (f ml/h) and yet the culture is clearly growing exponentially. At first sight this seems anomolous, but then it must be remembered that the culture is, in effect, being diluted at an exponential rate by virtue of the fact that the overflow device removes excess culture at the rate of one drop per drop of fresh medium entering the growth vessel. The situation can best be appreciated by considering how the population density would change with time if the culture was suddenly sterilized by adding, say, a few drops of formalin, and yet the supply of fresh medium was maintained at the same rate as that used by the growing culture. The first drop of medium entering the culture vessel would dilute the culture by an infinitesimally small fraction, yet it would displace a drop of almost undiluted culture. The next drop of medium would dilute the culture by a second small increment, and again displace a drop of almost undiluted culture. Thus, by the time that the equivalent of one culture volume of fresh medium had been added, the culture would contain not half the initial concentration of organisms, but considerably less. In fact, the actual concentration of organisms present after the passage of one culture volume of medium through the vessel can be calculated as follows:

If the initial concentration of organisms was x_0 (mg-equivalent dry weight organisms/ml) and the culture volume V (ml), then the addition of one drop of medium would decrease the organism concentration by a factor $V/(V + dV)$. After the removal of one drop of culture and the addition of a second drop of medium, the dilution factor would be $(V/(V + dV))^2$, and so on up to $(V/(V + dV))^n$. Therefore, to find the extent to which the culture had been diluted after the addition of a number of drops collectively equalling one culture volume (that is, $n = V/dV$), one must determine the numerical equivalent of the expression: $(V/(V + dV))^{V/dV}$. This can be solved as follows:

$$\left(\frac{V}{V + dV}\right)^{V/dV} = \left(\frac{V/dV}{V/dV + 1}\right)^{V/dV} = \left(\frac{n}{n+1}\right)^n = \left(\frac{1}{1 + 1/n}\right)^n$$

or

$$(1 + 1/n)^{-n}.$$

And since, by definition, $\lim_{n \to \infty}(1 + 1/n)^n = e$ (or 2.71828 by calculation, after expansion by means of the Binomial Theorem), then it must follow that $\lim_{n \to \infty}(1 + 1/n)^{-n} = e^{-1} = 0.36788$. Hence, after the passage of one culture volume of medium through the vessel, the concentration of the (non-growing) population of organisms will have decreased from x_0 to $0.368x_0$. In general, then, and since the number of culture volumes pumped to the growth vessel over a specific period of time t will be tf/V (or Dt), the concentration of organisms present at any time t will be related to that initially present by the equation:

$$x = x_0\, e^{-Dt}. \tag{2}$$

Of course, this applies only to a non-growing culture. Where organisms are growing at a rate such that their steady state concentration is not changing with time (as in a turbidostat), then it is clear that the growth rate must be numerically equal to the wash-out rate. It should now be clear why it is preferable to express the specific growth rate μ in terms of its natural logarithm (i.e. $(\ln 2)/t_d$).

B. Kinetics of Growth in a Chemostat

It is now appropriate to consider what would happen with a turbidostat culture if the optical density control system was disconnected and the pump set to deliver medium to the culture at some rate *less* than that required to maintain the culture population density constant. Clearly, since the growth rate of the organisms would be greater than the dilution rate, the concentration of the organisms in the culture would increase. But it could not go on increasing indefinitely, for the same reason that it does not do so in a batch culture — that is, because of the growth-associated changes in the environment. On the other hand, growth would not ultimately cease, as it does in a batch culture, because the continuous addition of fresh medium to the culture would continuously replenish nitrients that were being used up, albeit at a suboptimal rate, and similarly, continuously dilute out end-products of metabolism that were accumulating. Clearly, a situation ultimately must be established in which environmental changes cause the growth rate to decrease to a point where once more it is equal to the dilution rate and steady state conditions prevail. This is the basic operating principle of a second type of continuous culture device — that is, the *Chemostat.*

Now, if the organisms were growing in a complex nutrient medium, such as tryptic digest of meat or casein hydrolysate, one would not know precisely the nature of the environmental changes that ultimately caused the growth rate of the organisms to decrease from their maximum value. However, if one constructed a more defined medium (say, a simple-salts medium) and arranged for one nutrient to be present at a concentration such that, in a batch culture, it would become depleted before any other nutrient was fully used up, then it would be the insufficiency of this nutrient that would cause the growth rate of organisms in a chemostat culture ultimately to decrease to a value equal to the dilution rate. In other words, it would be the rate of supply of this nutrient, added along with the other essential nutrients in the fresh medium, that would prescribe the rate of growth of organisms in the culture. If the limiting factor was the medium glucose concentration, one would term the chemostat culture 'glucose limited'. Alternatively it could be limited by the supply of utilizable nitrogen source or source of sulphur, or by the availability of phosphate, potassium or magnesium. With organisms having some auxotrophic requirement, growth may be limited by the medium content of, say, some purine, pyrimidine, vitamin or amino acid.

It is obvious that with such a nutrient-limited culture it will be the concentration of nutrient in the culture *extracellular fluids* that is actually limiting the rate of cell synthesis since the rate at which such a substrate is taken into the cell will be a function of its concentration. If the uptake process involves some enzyme-catalysed reaction, then one might expect that the relationship between uptake rate (and growth rate) and concentration of limiting substrate would be of the form of a Michaelis—Menten equation. That is,

$$V = V_{max}\left(\frac{s}{K_s + s}\right) \text{ or } \mu = \mu_{max}\left(\frac{s}{K_s + s}\right), \tag{3}$$

where V is the rate of penetration of substrate into the cell, V_{max} the potentially maximum rate of substrate uptake, s the actual concentration of substrate and K_s a saturation constant that is numerically equal to the substrate concentration that allows the uptake process to proceed at one-half its maximum rate. Similarly, μ is the specific growth rate expressed in the nutrient-limited chemostat culture, μ_{max} the growth rate that would be expressed if the substrate was present in a non-limiting concentration, and K_s a saturation constant that is numerically equal to the growth-limiting substrate concentration that would allow growth to proceed at one-half its potentially maximum rate. The above relationship was found, by Monod, to hold for many real cultures — at least approximately so, and though it may not be valid under all circumstances, it forms the cornerstone of much of the theory of microbial growth in chemostat cultures.

We have already shown by a process of reasoning that, with the dilution rate set below some critical value (D_c) at which the organisms express their maximum growth rate, steady state conditions ultimately must be established in which the specific growth rate (μ) and dilution rate (D) are equal. Further, we have suggested (equation 3) that, under such conditions, growth rate is actually limited by the concentration of some essential nutrient present in the culture extracellular fluid which causes a 'master' reaction to proceed at a submaximal rate. It is now necessary to restate these conclusions in a more formal way — a way which shows how a chemostat culture will respond to perturbations of the steady state.

As with a turbidostat culture, the change in concentration of organisms with time will depend on the balance between the growth rate μ and dilution rate D such that:

$$\begin{aligned} \text{Change} &= \text{Growth} - \text{Washout} \\ dx/dt &= \mu x - Dx \\ dx/dt &= x(\mu - D). \end{aligned} \tag{4}$$

Therefore, whenever $\mu > D$, the concentration of organisms in the culture will increase with time and whenever $\mu < D$ it will decrease with time. However at any fixed value of D, changes in the population density will be dependent on μ which itself is critically dependent on s (the growth-limiting substrate concentration in the culture extracellular fluids), since:

$$\mu = \mu_{max}\left(\frac{s}{K_s + s}\right)$$

Hence,

$$dx/dt = x\left[\mu_{max}\left(\frac{s}{K_s + s}\right) - D\right] \tag{5}$$

Similarly, changes in the concentration of the growth-limiting nutrient will depend upon the balance between the input rate, the rate of consumption, and the rate of loss in the overflow culture. Thus:

$$\begin{aligned} \text{Change} &= \text{Input} - \text{Output} - \text{Consumption} \\ ds/dt &= fS_R \quad - fs \quad - qxV \end{aligned}$$

where S_R is the concentration of growth-limiting nutrient in the reservoir, s its concentration in the culture extracellular fluids, q the specific rate of substrate consumption ($-1/x \cdot ds/dt$), f is the medium flow rate, and V is the culture volume. Thus, per unit volume of culture:

$$-ds/dt = D(S_R - s) - qx$$

and since

$$\begin{aligned} q &= \mu/Y, \\ -ds/dt &= D(S_R - s) - \frac{\mu x}{Y} \end{aligned} \tag{6}$$

It follows from equations (3), (4), and (5), that any sudden increase in the concentration of growth-limiting nutrient *in the culture* will cause an increase in the specific growth rate and therefore an increase in the culture population density. But this increase in growth rate and population density will effect an increase in the rate of substrate utilization such as to cause the concentration of the growth-limiting nutrient to fall back to its initial steady state value (equation 6). As the growth-limiting substrate concentration progressively diminishes, so too does the specific growth rate and specific rate of substrate consumption so that ultimately steady-state conditions are reestablished. In this connection, any overshoot that may occur in the rate of substrate consumption (due, for example, to there being an increased con-centration of organisms in the culture) will cause the specific growth rate to fall to a value less than the dilution rate. Hence wash-out will occur to a point where the concentration of organisms reaches such a value that the rate of consumption of growth-limiting substrate ($q \cdot x \cdot V$) again balances the rate of supply of that nutrient (fS_R) minus its rate of loss in the overflow culture (fs).

It can be seen, therefore, that perturbations of the steady-state conditions set up reactionary forces that ultimately (and often rapidly) reestablish the steady state. Once the culture is in a steady state, unique values exist for both the organism concentration (x) and the growth-limiting substrate concentration (s), which are given the symbols $\bar{x}$ and $\bar{s}$, respectively. Thus:

$$\bar{x} = Y(S_R - \bar{s}) \tag{7}$$

and,

$$\bar{s} = K_s \left(\frac{D}{\mu_{max} - D} \right) . \tag{8}$$

Substituting for $\bar{s}$ in equation (7), it follows that:

$$\bar{x} = Y \left[S_R - K_s \left(\frac{D}{\mu_{max} - D} \right) \right] \tag{9}$$

Therefore, since μ_{max}, K_s, and Y are constants, the principal effect of varying the dilution rate will be to change the concentration of growth-limiting substrate (equation 8) thereby effecting a change in the specific growth rate (equation 3) and doubling time of the culture. Moreover, assuming the yield value Y to be independent of dilution rate, and K_s to be small relative to S_R, it follows from equations (7), (8), and (9) that varying the dilution rate should produce a pattern of changes in the steady-state microbial concentration and concentration of growth-limiting substrate as depicted in Figure 3(A). On the other hand, if K_s is relatively large compared with S_R, then the pattern of changes of $\bar{x}$ and $\bar{s}$ with D should be more like that represented in Figure 3(B).

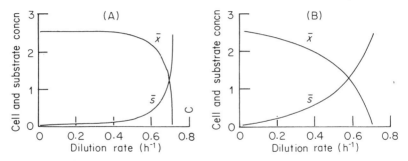

Figure 3. Theoretical changes in cell concentration $\bar{x}$) and growth-limiting substrate concentration $(\bar{s})$ with dilution rate in two cases: (A) when the K_s value is small relative to S_R, and (B) when K_s is large relative to S_R. Reproduced by permission of Academic Press Inc. (London) Ltd.

C. Other Quantitative Relationships: Yield Values

It was stated earlier in this chapter that growth involved the uptake and chemical transformation of a number of essential compounds and elements, and that the *rates* of uptake of these metabolites were proportional to the growth rate, thus:

$$\mu \propto q$$

$$\mu = Yq.$$

The proportionality factor (Y) is termed the *yield value*; it is the ratio of two rate processes (μ and q), but itself has no time component (e.g. g cells formed per mol substrate consumed). It is now necessary to examine the relationship between this yield value and growth-associated properties of the microbial culture.

Now equations (6) and (9), which define the behaviour of populations in chemostat culture, imply that the yield value is constant and independent of the growth rate. Hence, at all dilution rates, μ and q should bear the same relationship to one another, and to the dilution rate D. That is:

$$D = \mu = Yq.$$

Hence, as $D \to 0$, so too must μ and q such that Y (which is proportional to $\bar{x}$, equation 7) remains constant. In practice, this is rarely found to be the case, and with substrates as disparate as sugars (and other carbon-containing substrates), phosphate, ammonia, potassium, magnesium, and oxygen, the yield value (as indicated by the steady-state bacterial concentration) varies markedly with the growth rate (Figure 4A).

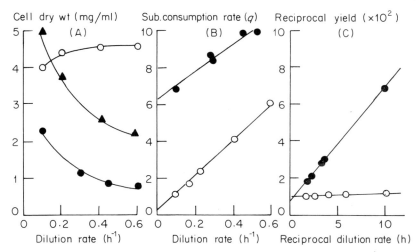

Figure 4. (A) Influence of dilution rate on the concentration of *Klebsiella aerogenes* organisms in (○) a glucose(50 mM)-limited culture; (●) a phosphate(0.75 mM)-limited culture, and (▲) a magnesium (0.25 mM)-limited culture. (B) The influence of dilution rate on the specific rate of glucose uptake by (○) a glucose-limited and (●) a phosphate-limited culture of *K. aerogenes*. (C) The relationship between the reciprocal of the yield value for glucose and the reciprocal of the dilution rate for cultures of *K. aerogenes* that were (○) glucose limited and (●) phosphate limited.

With carbon substrates, and with oxygen, it is generally found that their specific rates of consumption q are indeed linear functions of D, but that, on extrapolation to $D = 0$, the lines do not pass through the origin but intersect the ordinate at finite values of q (Figure 4B). And since $Y = \mu/q$ (or, at steady state, D/q) it is clear that as D approaches zero, so too must Y. In order to explain this variation in yield value with growth rate, it was proposed that a portion of the carbon substrate (and of oxygen) was required to deliver up energy that was needed for growth-independent 'maintenance' functions (for example, for the maintenance of solute gradients across the plasma membrane, and for turnover of macromolecular components of the cell). Hence, the extrapolated substrate uptake rate at zero growth rate could be taken as a direct measure of this maintenance energy requirement and could be subtracted from the actual rate of substrate consumption to derive an evaluation of the 'true' growth-associated substrate requirement. Thus:

$$q_{actual} = q_{growth} + q_{maintenance}.$$

Dividing by μ (=D at steady state),

$$q_a/\mu = q_g/\mu + q_m/\mu$$

and since q/μ is the reciprocal of the yield value Y, then:

$$1/Y = 1/Y_g + (q_m \times 1/\mu) \qquad (10)$$

where Y_g is the 'true' growth yield constant (that is, the yield value corrected for maintenance losses). Therefore, plotting $1/Y$ against $1/\mu$ (or $1/D$) should give a straight line with a slope equal to q_m that intersects the ordinate (when $1/D = 0$) at a value of $1/Y_g$ (Figure 4C).

Now whereas the use of double reciprocal plots of yield versus growth rate can provide measurements of both the so-called maintenance energy requirement and the 'true' (maximum) yield value, they suffer the disadvantage that heavy emphasis is placed on the substrate consumption rate expressed at low growth rates. This can be readily appreciated by considering the fact that the difference between the reciprocals of $D = 0.1$ and $0.2\,h^{-1}$ is 5.0 whereas the difference between the reciprocals of $D = 0.4$ and $0.5\,h^{-1}$ is only 0.5. Moreover, an evaluation of both q_m and Y_g can be obtained from a direct plot of substrate uptake rate versus growth rate since the slope of this line (that is $ds/d\mu$) is itself the reciprocal of the maximum yield value, and the intersect with the ordinate (when $D = 0$) is the maintenance rate (q_m). The precise relationship between the plots of q versus μ (or D) and $1/Y$ versus $1/D$ can be best understood by considering the general equation for a straight line — that is:

$$Y = ax + b.$$

In the case of a plot of q against D,

$$q = aD + b.$$

Now b is the value of q when $D = 0$ and has been termed (above) the 'maintenance' rate (q_m). Further, a is the slope of the line (that is dq/dD) and has the same units as reciprocal yield: it is a constant that is numerically identical with $1/Y_g$. It follows, therefore, that the above equation can be written:

$$q = (D \times 1/Y_g) + q_m ,$$

and dividing this equation by D transforms it into equation (10): that is,

$$q/D = 1/Y_g + (q_m \times 1/D),$$

or

$$1/Y = 1/Y_g + q_m /D.$$

This latter derivation of the growth yield equation reveals the uncertain nature of the constant Y_g. It is, in fact, the incremental increase in substrate consumption rate required to support an incremental increase in growth rate. However, when comparing populations of organisms growing at different rates one is not comparing like with like since, compositionally and functionally, organisms vary markedly with growth rates. In some circumstances this may not create serious problems of interpretation, but in others it does. Thus, populations growing in the presence of an excess of carbon substrate, and with growth limited by the availability of, say, phosphate, express a high rate of carbon substrate uptake (that is a low yield value) but the incremental increase in substrate consumption rate with growth rate $(1/Y_g)$ is much decreased, giving the false impression that these organisms are expressing a vastly improved 'true' growth yield (Y_g) as compared with carbon substrate-limited cultures (Figure 4B). The absurdity of this conclusion becomes apparent when one considers the situation at or near μ_{max} (that is, close to D_C). Here $\bar{s}$ is approaching S_R and the culture is only nominally limited by the chosen growth-limiting substrate. In effect, all such cultures must be virtually identical, irrespective of their nominal limitation, and must express the same substrate uptake rate and yield value. Hence they must be growing with the same real efficiency: Their maintenance requirements (at μ_{max}) must be closely similar, even though they may be vastly different at low growth rates, and hence their 'true' yield values for carbon substrate also must be similar. An erroneous interpretation of the data obtained with carbon-sufficient chemostat cultures arises from the assumption that the 'maintenance rate' is a constant, which clearly it is not. Indeed, the fact that a linear relationship generally has been found between the specific rate of substrate consumption and the growth rate has been used to support the conclusion that the maintenance rate *must* be constant and independent of growth rate. But clearly this need not necessarily be so, though the fact that the substrate consumption rate increases linearly with growth rate suggests that, should the main-tenance rate vary, then in all probability it also would do so as a linear function of the growth rate. It is therefore sensible to modify equation (10) to allow for the possibility of a varying maintenance rate, and the simplest way of doing this is to assume, as stated above, that should the maintenance rate vary (positively or negatively), it will do so as a linear function of the growth rate. Then:

$$1/Y = 1/Y_g + q_m (1 \pm cD)/D$$

where c is the slope of the line of q_m versus D. Rearrangement of this equation reveals that the main effect on any progressive change in q_m, with dilution rate, will be to change the value of Y_g:

$$1/Y = (1/Y_g \pm cq_m) + q_m /D. \tag{11}$$

This modification allows of the possibility (indeed probability) that variously-limited chemostat cultures may express markedly different relationships between q and D and yet be closely similar in their strictly growth-associated substrate requirements.

The importance of yield values (particularly those for carbon substrate and oxygen) resides in the fact that they are indicators of the energetic efficiency with which microbes grow. Thus, as shown in Figure 5, the carbon substrate is taken up and metabolized to provide both intermediates and reducing equivalents required for cell synthesis, and further reducing equivalents necessary to provide, aerobically, substrate for the ATP-generating reactions of respiration. Clearly, some balance must be established between the substrate-catabolizing reactions associated with the generation of biologically useful energy, and the substrate- and ATP-consuming reactions of anabolism. This balance is reflected in the yield value.

Since the major currency of biologically useful energy is ATP, it might be argued that a more meaningful comparison of the efficiencies of growth of different organisms on some specific substrate (or of a specific microbial species on a range of different carbon substrates) can best be made on the basis of Y_{ATP}, that is, g organisms synthesized per mol of ATP *consumed*. The problem here is that, generally speaking, it is only the generated ATP that can be measured, and then only with a degree of imprecision. It can be assumed, of course, that all the ATP that is generated is turned over in energy-consuming reactions, since the adenine nucleotide 'pool' is of a relatively constant size and composition, and only a small amount of the ATP synthesized is actually incorporated into cell substance (e.g. as RNA and DNA). However, what bedevils any realistic evaluation of Y_{ATP} is the seeming absence of direct coupling (in the strict sense of the word) between ATP

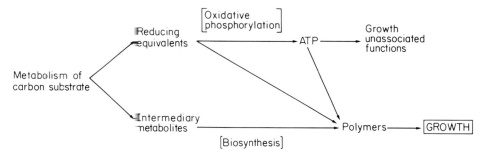

Figure 5. Pathways of carbon and energy flow in the growth of microorganisms. Schematic representation.

turnover and biosynthesis. In other words, although biosynthesis causes the ATP pool to turn over, this can equally well happen in the absence of biosynthesis. Indeed, this is patently obvious from the fact that washed suspensions of organisms can oxidize substrates like glucose at a high rate under conditions where polymer synthesis is grossly impeded. It follows, therefore, that any comparisons of the efficiency of utilization of ATP in biosynthesis are likely to be misleading, unless methods are found for assessing that portion of the ATP that is turned over by processes that are not associated with biosynthesis.

Notwithstanding the above strictures, evaluations of Y_{ATP} (as the ratio of cells synthesized per mol ATP *synthesized*) have been made for a number of microbial species, and comparisons drawn. The least equivocal situation is found with cultures of strictly anaerobic bacteria, particularly with those species that have no electron transport chain activity. In such species, ATP is generated by substrate-level phosphory-lation reactions, and therefore ATP formation from ADP and P_i is a coupled process (i.e. catabolism cannot occur in the absence of ATP synthesis). The actual rate of ATP synthesis can be assessed from the rate of substrate consumption and/or by the rate of accumulation of some product whose synthesis involved substrate-level phosphorylation reactions. In addition, the use of substrates that cannot be assimilated into cell substance, either directly or after partial catabolism, further aids in eliminating 'imponderables' from the final assessment. With those organisms fulfilling most of the above conditions, ratios in the region of 10 (for g organisms synthesized per mol ATP formed) have been obtained. Later estimates, which attempted to correct for the maintenance energy requirement, were in the region of 14.

With organisms that can generate ATP by electron transport chain (oxidative phosphorylation) reactions (i.e. aerobes and anaerobes), so many pitfalls are encountered that a reliable evaluation of yield per mol ATP synthesized is seemingly impossible. With organisms growing aerobically, the major stumbling block is the assessment of the efficiency of oxidative phosphorylation. In this connection, it is known that the complete oxidation of 1 mol of glucose requires 6 mol of oxygen and can lead to the synthesis of, maximally, 38 mol of ATP (from ADP and P_i). Hence, since the rate at which oxygen is consumed by growing organisms can be accurately determined, then by assuming a ratio of $q_{ATP}:q_O$ of 38/12 (that is, 3.17), the Y_{ATP} (in terms of g organisms synthesized/mol ATP *generated*) can be calculated from the equation:

$$Y_{ATP} = Y_O/3.17,$$

where Y_O is the g organisms synthesized per g-atom oxygen consumed.

In practice, this gives values well below 10 (e.g. 7—8 with *Klebsiella aerogenes* growing on glucose as the sole carbon and energy source) which is uncomfortably low — though possibly realistic. On the other hand, if the oxidation of 1 mol of glucose generated less than 38 mol of ATP (that is, if the efficiency of oxidative phosphorylation was less than that found with, say, mitochondria) then the actual Y_{ATP} value may be considerably higher — that is, much nearer to the value found with organisms growing anaerobically. As yet, however, there is seemingly no way of deciding unequivocally the number of sites of energy conservation on the respiratory chains of prokaryotic organisms, or, indeed, whether the transfer of electrons from NADH to oxygen is coupled *stoichiometrically* to the phosphorylation of ADP. All that can be stated is that the rate of ATP synthesis (q_{ATP}) will be related to the rate of oxygen consumption and to the efficiency of oxidative phosphorylation by the generalized equation:

$$q_{ATP} - q_{ATP}^{s.l.} = 2q_{O_2} \times P/O$$

where $q_{ATP}^{s.l.}$ is the rate of ATP synthesis by substrate-level phosphorylation reactions, and P/O the ratio of the number of ADP molecules phosphorylated per atom of oxygen that is reduced. Assuming that, aerobically, substrate-level phosphorylation reactions contribute only to a minor extent to the overall rate of ATP synthesis, then, by dividing the above equation by the dilution rate D, one obtains the relationship:

$$1/Y_{ATP} \approx (P/O) \times 1/Y_O$$

or,

$$Y_{ATP} \approx Y_O /(P/O). \tag{12}$$

Again it should be emphasized that Y_{ATP}, as defined by equation (12), is not a genuine yield value but simply a comparison of two rates of synthesis (cell synthesis and ATP synthesis). Yield in the sense of cells synthesized per mol of ATP *consumed* in growth-associated processes may be something very different.

The use of facultatively anaerobic organisms for studies of the energetics of cell synthesis would seem to offer some advantages since, by comparing organisms growing at a fixed dilution rate, first aerobically and then anaerobically, one might think that it would be possible to derive corresponding values for both q_{ATP} (anaerobically) and q_{O_2} (aerobically), from which the efficiency of oxidative phosphorylation could be evaluated directly. However, the obvious snags with this procedure are, firstly, that it is exceedingly difficult to ensure that absolutely no oxygen enters a stirred fermenter, for should it do so, even in minute traces, a false evaluation of q_{ATP} will be made.

Secondly, anaerobic cultures may be fundamentally different from aerobic cultures in that the former will, in all probability, be energy-limited, whereas it is conceivable (indeed likely) that the latter will be carbon-limited. Therefore the rate of aerobic ATP synthesis may be substantially greater than the anaerobic synthesis with cultures of organisms *growing at the same dilution rate*.

In view of the uncertainties surrounding the assessment of q_{ATP} values, and the multiplicity of processes (both growth-associated and growth-unassociated) that cause the ATP pool to turn over, it would seem inappropriate at the present time to formulate more detailed equations relating growth to the synthesis and utilization of ATP.

D. Transient-state Phenomena

Although a chemostat culture is inherently self-balancing and tends always to move towards the establishment of a steady state, much can be learned about the properties of microorganisms by either perturbing the steady state or else changing it, and then following the events occurring during the transitional stages leading to the establishment of a new steady state. Indeed, since chemostat cultures provide fiercely selective conditions in which mutant organisms may rapidly replace the parent organisms in the culture (see following section), analysis of the kinetics of transient states may provide the most direct method of determining whether different physiological properties, expressed by organisms growing in different environments, are the result of phenotypic variation or mutant selection. A specific example serves to make the point.

Bacillus subtilis possesses a wall that contains substantial amounts of teichoic acid (a phosphorus-containing polymer). When grown under conditions of phosphate limitation, however, synthesis of this wall-bound teichoic acid is seemingly inhibited and, in its place, a non-phosphorus-containing polymer (teichuronic acid) is formed. The basic question that has to be answered, therefore, is whether the change in wall composition is due to regulation of the syntheses of teichoic acid and teichuronic acid, or whether it is, in fact, due to the selection of some mutant organisms that were present as a small proportion of the initial population. Further, if the former is the case, then it would be important to determine whether the replacement of one wall-bound polymer by some other polymer involved turnover of the wall or simply a progressive build-up associated with the synthesis of new wall material. Both these questions can be answered by analysing the kinetics of the change-over in wall polymer composition during the transitional stages between one type of growth limitation and the other, as detailed below.

Supposing that the organisms exhibited some property p (e.g. teichoic acid) that was present in high concentration in the initial population, but virtually absent from the steady-state population following a change in the growth condition. Then, if at time t_0 the cells ceased to synthesize this component (but did not destroy that already present), its concentration in the culture would decline at a rate such that:

$$p_t/p_0 = e^{-Dt} \qquad (13)$$

where p_0 is the concentration of the substance in the culture at time t_0 and p_t is its concentration after time t. Similarly, if some change occurred in the reverse direction (that is, if at time t_0 the organisms started to synthesize some component that was not previously synthesized, and at a rate proportional to the growth rate), then its concentration would increase at a rate such that:

$$p_t/p_s = 1 - e^{-Dt}$$

or

$$\ln(1 - p_t/p_s) = -Dt \qquad (14)$$

where p_s represents the final steady state concentration of the property, and p_t its concentration after time t.

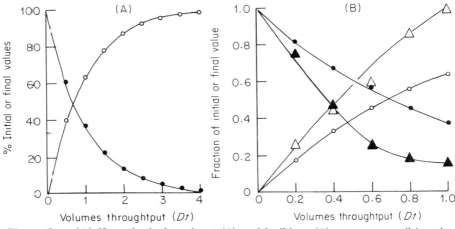

Figure 6. (A) Hypothetical washout ($\bullet$) and build-up ($\circ$) curves prescribing the rates of change in some property that either ceases, or starts to be synthesized, at time zero, and whose synthesis is strictly growth associated. Reproduced by permission of Academic Press Inc. (London) Ltd. (B) Changes in the culture content of ($\blacktriangle$) teichoic acid and ($\triangle$) teichuronic acid following the change from magnesium limitation to phosphate limitation with a chemostat culture of *Bacillus subtilis* var. *niger*: lines indicated by ($\circ$) and ($\bullet$) indicate the theoretical rates of change assuming no turn-over of wall material.

It follows, therefore, that if the change in the composition of the wall of *B. subtilis*, following a change in its environmental conditions, was due to some phenotypic change, then the *minimum* rate at which this should occur would be defined by the exponential wash-out rate (equation 13). A faster rate of change would indicate actual turn over of the existing wall material, whereas a substantially lower rate of change-over would be indicative of mutant selection (Figure 6A). The results that were actually obtained are shown in Figure 6B. These clearly indicate that this organism does possess the capacity to change its wall composition phenotypically, and that this may, indeed, involve turn over of the wall material.

IV. Growth of Mixed Microbial Populations

A landmark in the history of microbiology was the development of techniques for isolating individual microorganisms and for culturing them free from the 'contaminating' organisms with which they were naturally associated. And although, in recent years, interest has tended to move towards the behaviour of organisms in mixed cultures, nevertheless, microbiologists still invest heavily in studies of axenic cultures. Therefore it is important to understand the environmental forces which may act upon some contaminant organism that enters a batch or a chemostat culture, and to predict its fate.

Assuming no interaction between different organisms in a culture except competition for the available nutrient, it is clear that, with a closed system (batch culture), organisms of each species would accumulate at rates related to their exponential growth rate. This can be readily appreciated from a consideration of the basic growth equation:

$$N_t = N_0 2^n = N_0 2^{t/t_d}$$

assuming there to be present two different microbial species ('a' and 'b') initially at concentrations N_0^a and N_0^b, and further assuming that each species grew without exhibiting any lag period up to a point where further growth was impeded due to the exhaustion of an essential nutrient. At that time t (the onset of the stationary phase), the concentrations of organisms 'a' and organisms 'b' would be, respectively:

$$N^a = N_0^a 2^{n^a}$$

and

$$N^b = N_0^b 2^{n^b}$$

where n^a and n^b are the number of doublings of the initial populations (N_0^a and N_0^b) over the course of the exponential growth phase. It follows, therefore, that the ratio of organisms 'a' to organisms 'b' at the onset of the stationary phase would be:

$$N^a/N^b = 2^{(n^a - n^b)} N_0^a/N_0^b. \tag{15}$$

Now the number of doublings (n) is inversely proportional to the population doubling time ($= t/t_d$) and directly proportional to the specific growth rate (since $\mu = 0.693/t_d$). In fact,

$$n = t/t_d = \mu t/0.693$$

Therefore the rate of change in the proportion of organisms 'a' to organisms 'b', in a batch culture, will depend solely on the difference in the specific growth rates of the two species. The extent to which one species outgrows the other will also depend on the number of doublings that the environmental conditions can support. Repeated subculturing will allow progressive enrichment of one species over the other to a point where, for all practical purposes, the culture becomes axenic. This condition applies not only to a culture that becomes contaminated with a different microbial species, but also to one which becomes self-contaminated (so to speak) with mutant organisms. In either case there will be selection of one cell type providing that there is some difference in the growth rate (that is μ_{max}) and that the culture is well mixed. The affinity of the different organisms for the nutrients present in the medium will have almost no effect on the selection process since, except for a brief period of time prior to the onset of the stationary phase, all nutrients will be present in cell-saturating concentrations (see Figure 1).

A very different situation obtains in a chemostat culture by virtue of the fact that organisms must compete for some essential nutrient that is present in a growth-rate-limiting concentration. Moreover, since a chemostat is an 'open' system, organisms that exhibit a growth rate lower than the dilution rate will be progressively washed out from the fermenter culture (see p. 7.00). The above arguments assume that Michaelis–Menten kinetics apply to the growth of organisms in nutrient-limited cultures. That is:

$$\mu = \mu_{max} \left(\frac{s}{K_s + s} \right)$$

If this is indeed the case, and that μ depends critically on s, then the outcome of a deliberate or accidental contamination of the culture can be predicted from the relationships that exist between μ and s for both

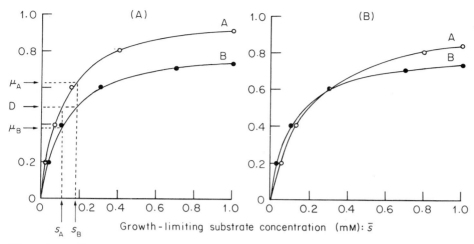

Figure 7.   Theoretical curves for two organisms (A and B) showing the relationship between growth-limiting substrate concentration ($\bar{s}$) and specific growth rate
($\mu = D$) assuming Monod kinetics apply. In the first graph (A) the organisms have
widely different affinities for the substrate, and different maximum growth rate
values. Reproduced by permission of Academic Press Inc. (London) Ltd. In the
second graph (B), organism A still possesses a capacity to grow faster than B at high
substrate concentrations, but organism B has a greater affinity for the growth-
limiting substrate.

the initial culture of organisms and for the contaminant organisms
(Figure 7).

   Figure 7 depicts graphically the relationship defined by equation (3)
for two different species of organism (A and B) that exhibit different
maximum growth rate values and have different affinities for the
growth-limiting substrate. Let us assume that initially the chemostat
contains a pure culture of organism B, growing at a dilution rate of
$D \, h^{-1}$, and that this becomes contaminated with organism A. The
growth-limiting substrate concentration in the culture at the time that
the contamination occurred would be s_B, and at this concentration the
contaminant organisms would grow at a rate equal to μ_A. Since this
growth rate is greater than the dilution rate D, the concentration of
contaminant organisms in the culture must increase (equation 4) but
this will cause the growth-limiting substrate concentration to decrease
correspondingly until μ_A equals D. At this time, the growth-limiting
substrate concentration in the culture will have decreased to s_A which
will permit organism B to grow only at a rate of μ_B. However, since μ_B
is less than the fixed dilution rate D the concentration of B type
organisms in the chemostat culture must diminish (ultimately to zero)
as they are progressively washed from the growth vessel. Hence, with a

chemostat culture, as opposed to a batch culture, the selective pressure is not the maximum growth rate value (μ_{max}) but the affinity of the organisms for the growth-limiting substrate *at the dilution rate at which the culture is being grown*. Of course, Figure 7 represents an ideal situation in which the saturation curves are well separated. It is possible that the two saturation curves may cross, in which case (as indicated above) the outcome would depend on the dilution rate at which the culture was being grown.

It must be emphasized that the fate of the contaminant organism will be influenced by many factors and the situation described above assumes that (i) the dilution rate is maintained constant, (ii) the culture is in a steady state at the moment that contamination occurs, (iii) growth rate is limited by the availability of a single nutrient substance that is also essential for the growth of the contaminant organism, (iv) there is no interaction between the different organisms except for the competition for the single growth-limiting nutrient, (v) the culture is perfectly mixed and homogeneous (that is, there is no accretion of organisms on the walls and other parts of the culture vessel), and (vi) growth rates of the organisms adjust themselves to changes in substrate concentration without appreciable lag. These conditions may be adequately realized in many real chemostat cultures (as seen in Figure 7B) but it is not uncommon to find with chemostat enrichment cultures the establishment of a stable, mixed microbial population growing on a single growth-limiting, carbon-containing nutrient.

V. Concluding Remarks

When one considers the vast range of prokaryotic and eukaryotic microbes present in the biosphere, and the variety of environmental conditions that act upon these creatures in natural ecosystems, then it is clear that the half-dozen or so topics included in this chapter represent only a small fraction of those that properly could have been included under the title 'Dynamics of Microbial Growth'. However, restrictions of space necessitated a rigorous selection of the subject matter and, whilst this selection was by no means arbitrary, the omission of certain topics is not easy to justify. In particular, four important aspects of the subject had to be either omitted or treated superficially, and all that usefully can be done by way of compensation is to point to key publications where further information may be obtained.

Firstly it should be emphasized that, whereas growth has been considered principally at the level of the population, primarily it occurs at the level of the cell. In this connection, much information has accumulated over the last decade on the mechanisms by which the events occurring in the cell division cycle are timed and coordinately regulated. And, although such information relates largely to cells growing at, or near, their potentially maximum rate, it forms a solid foundation for studies, yet to be undertaken, of organisms growing at the more ecologically relevant rates.

Secondly, although the rates at which microorganisms are likely to grow in natural ecosystems is far below those exhibited by most laboratory cultures, the consequences of growth at exceedingly low rates was not considered, nor was any analysis made of the influence of viability changes on the growth kinetics of the viable portion of a microbial population.

Thirdly, in considering the growth kinetics of mixed microbial populations, the simplifying assumption was made that no interaction occurred between cells other than competition for essential nutrient substances. However, although this condition may hold with many artificial mixtures of laboratory cultures, it is clear that, in natural ecosystems, other interactions (such as commensalism, host—parasite and prey—predator relationships) frequently occur.

Finally, no account was given of the dynamics of fungal growth. This is a subject of considerable interest, not only academically (by virtue of their particular growth mode) but also industrially, since these organisms are cultured extensively in the production of antibiotics and other economically important metabolites.

References to papers dealing with each of these four subjects are therefore included under 'Further Reading' below.

VI. Further Reading

1. Microbial Growth (General)

Monod, J. (1949). The growth of bacterial cultures. *Annual Review of Microbiology*, 3, 371—394.
Novick, A. (1955). Growth of bacteria. *Annual Review of Microbiology*, 9, 97—110.
Meadow, P. M. and Pirt, S. J. (1969). Microbial Growth. *19th Symposium of the Society for General Microbiology* Cambridge: Cambridge University Press.
Pirt, S. J. (1975). *Principles of Microbe and Cell Cultivation*. Oxford: Blackwell.

2. Microbial Growth (Continuous Culture)

Herbert, D., Elsworth, R., and Telling, R. C. (1956). The continuous culture of bacteria; theoretical and experimental study. *Journal of General Microbiology*, 14, 601—622.

Malek, I. and Fencl, Z. (1966). *Theoretical and Methodological Basis of Continuous Culture of Microorganisms*. Prague: Czechoslovak Academy of Sciences.

Tempest, D. W. (1970). The continuous cultivation of micro-organisms: 1. Theory of the chemostat. In: *Methods in Microbiology*, 2, 259–276. London and New York: Academic Press.

Kubitschek, H. E. (1970). *Introduction to Research with Continuous Cultures*. Englewood Cliffs, New Jersey: Prentice-Hall, Inc.

Veldkamp, H. (1976). *Continuous Culture in Microbial Physiology and Ecology*. Durham, England: Meadowfield Press Ltd.

3. The Cell Cycle

Marr, A. G., Painter, P. R. and Nilson, E. H. (1969). Growth and division of individual bacteria. In: *Microbial Growth, 19th Symposium of the Society for General Microbiology*, pp. 237–261. Cambridge: Cambridge University Press.

Cooper, S. and Helmstetter, C. E. (1968). Chromosome replication and the division cycle of *Escherichia coli* B/r. *Journal of Molecular Biology*, 31, 519–540.

Donachie, W. D., Jones, N. C., and Teather, R. (1973). The bacterial cell cycle. In: *Microbial Differentiation, 23rd Symposium of the Society for General Microbiology*, pp. 9–44. Cambridge: Cambridge University Press.

4. Other Aspects

Bull, A. T. and Trinci, A. P. J. (1977). The physiology and metabolic control of fungal growth. *Advances in Microbial Physiology*, 15, 1–84.

Meers, J. L. (1973). Growth of bacteria in mixed cultures. *CRC Critical Reviews in Microbiology*, 1, 139–184.

Koch, A. L. (1971). The adaptive response of *Escherichia coli* to feast and famine existence. *Advances in Microbial Physiology*, 6, 147–217.

Powell, E. O. (1972). Hypertrophic growth. In: *Environmental Control of Cell Synthesis and Function, 5th International Symposium on the Continuous Culture of Micro-organisms*, pp. 71–78. London: Academic Press.

Intermediary Metabolism

PATRICIA H. CLARKE

University College London, Gower St, London WC1E 6BT

I. Growth

Most of the metabolic activities of microorganisms are directed towards growth. Bacteria increase in size, then divide into two and increase in number. The bacilli and clostridia form spores which allow them to survive for long periods when the environmental conditions are unfavourable for growth. The budding bacteria go through a growth cycle with distinct morphological forms. More complex differentiation, with the formation of mycelia and aerial spores, is characteristic of the *Streptomycetes*. The *eukaryotic* microorganisms, including yeasts and more complex fungal species, differ from the *prokaryotes* in their growth cycles, their gross morphological features and also in details of cell structure. For example, the mitochondria of eukaryotic cells are the centres for metabolic activities which in bacteria are located in the cytoplasmic membrane or associated with it. The deoxyribonucleic acid of the eukaryotic cell is found in the nucleus contained within a membrane, while in prokaryotic cells it is intricately folded but not separated from the cytoplasm. The genetic material of prokaryotes is therefore not physically separated from the metabolic activities of the cell. The structural polymers of eukaryotes and prokaryotes differ but there are also many variations in detail between these compounds in different species of prokaryotes.

All microbial growth and differentiation depends on the synthesis of the macromolecular components of the cells and of the very many low molecular weight compounds which are required for cellular activity. The remarkable thing about the biochemistry of microorganisms is the variety of compounds which can be used to support microbial life and growth. Many species of bacteria and fungi are able to grow if supplied with a solution containing a few inorganic salts and any one of a very large number of different carbon compounds. It is this ability which enables microorganisms to act as the biological scavengers of the world and to recycle all natural products. Some species of prokaryotes can make use of atmospheric nitrogen and very many can synthesize all the nitrogen compounds of the cell from ammonium salts or nitrates. Carbon dioxide can provide all the carbon needed for growth for the species able to carry out *photosynthesis*, and for those able to obtain energy by the oxidation of inorganic compounds for *chemolitho-synthesis*.

Intermediary metabolism is concerned with the reactions by means of which all carbon and nitrogen compounds entering the cell are transformed either into new cell material or converted into products which are excreted.

II. Metabolism, Energy, and Building Blocks

The synthesis of chemical compounds is an *endergonic* or energy-demanding process. The photosynthetic organisms capture *light energy* to drive the chemical reactions which convert carbon dioxide into cell material. The carbon metabolism of photosynthetic organisms starts with the reactions of carbon dioxide assimilation which are only found in organisms which are *autotrophs* and able to use carbon dioxide as the major carbon source for growth. Most of the subsequent reactions are similar to those to be found in heterotrophs. *Heterotrophs* obtain energy by the breakdown of organic compounds. The *chemical energy* of the compounds used as growth substrates is released in such a way that it can be used for the synthesis of cell material and for the maintenance of such cellular activities as transport of nutrients across the membranes and motility. The organic compounds which are metabolized by heterotrophs to provide energy are also the sources of the molecular building blocks for biosynthesis. The *biosynthetic* pathways for essential compounds, such as amino acids, are similar in autotrophic and heterotrophic microorganisms, and are also very similar in prokaryotic and eukaryotic organisms. The degradative or *catabolic* pathways of heterotrophs, on the other hand, exhibit a variety of different patterns but these too converge on certain key compounds which occupy central positions in the complex web of metabolism.

III. Enzymes and Genes

Within microbial cells are enzymes designed to catalyse many kinds of biochemical transformations. In any one cell there will be a thousand or more different enzymes. However, at any one time the enzymes present in an organism will depend on the environment in which it is growing and its hereditary characteristics. Different species of microorganisms can be distinguished from one another by their *metabolic potential*. This describes the *total range of enzymes* an organism can produce and depends on the genetic constitution. *At any one time*, the *actual enzyme content* will reflect the way in which the organism reacts to the environment at that stage of its growth and life cycle. Metabolic maps of enzyme reactions, linking together the compounds found in living cells, are drawn up by biochemists from the results of studies on many kinds of plants, animals, and microorganisms. When these reactions are assembled together they give a vast array of interlocking enzyme transformations. At first sight a metabolic map of this sort is so daunting that it seems unlikely that it will be possible to drive any clear routes through. Yet, the elementary units of all organisms are the same, biosynthetic reactions always follow similar pathways, and catabolic or degradative pathways are found to fit into well-defined groups.

A heterotrophic bacterium, capable of growing in an inorganic salt medium with ammonium salts providing the nitrogen source and a single organic compound the carbon source for growth, produces some enzymes which are so *essential* for growth that they will invariably be present. Examples of these are the enzymes required for nucleic acid synthesis. Certain biosynthetic enzymes may not be needed if some of the end products or metabolic intermediates are present in the growth medium. Bacteria which are able to grow in both minimal salt media and complex media, need fewer biosynthetic enzymes when they are growing in complex media than when they are growing in a simple salt medium. If an enzyme can no longer be made, because mutations have occurred in the gene which determines its structure, then the mutant bacterium will be unable to grow in the minimal salt medium but may continue to grow in complex medium. The wild type *Escherichia coli* is able to grow in complex medium and in minimal medium with ammonium salts and glucose to supply the nitrogen and carbon requirements. Mutations in genes determining the structure or regulation of enzymes required for the biosynthesis of amino acids produce *auxotrophs,* which will only grow in minimal salt medium if the amino acid, or a suitable intermediate of that biosynthetic pathway, is added to the medium.

During evolution many groups of bacteria have become adapted to environments rich in organic nutrients and these have frequently lost the ability to synthesize many essential metabolites. Although exacting organisms require to be provided with a number of complex organic molecules which they are unable to synthesize for themselves, their other metabolic reactions are found to be identical to those of the *prototrophs,* or non-exacting bacteria. The evolution of organisms with complex nutritional requirements does not necessarily result in the total disappearance of the genes for the enzymes of a pathway. In the Lactobacilli the amino acid requirements, at least in some instances, results from the loss of active gene product. Single-step mutations can restore the ability to synthesize one of the amino acids normally required and this can be repeated to allow the organism to become independent of the need for several different amino acids. This suggests that the relevant genes have become silent but not lost.

Some species of bacteria and fungi can grow in simple salt medium with an unusually large range of organic compounds as the sole carbon source. *Pseudomonas putida* strains are known which grow on octane, camphor, or naphthalene, as the carbon source in a minimal salt medium, and each of these compounds is metabolized by enzymes which are not needed for the other two compounds. Specialized enzymes of this class may be *dispensable.* The genes which determine the enzymes required for the first few steps in the metabolism of these compounds are not carried on the chromosome of *P. putida* but are carried by small genetic elements known as plasmids. The plasmids may be lost from most of the bacterial cells when there is no demand for the enzymes. When octane, camphor or naphthalene appears in the environment, then the appropriate plasmid may spread rapidly through the bacterial population by cell to cell transfer.

A *structural gene* determines the amino acid sequence of an enzyme and mutations in these genes result in altered enzymes or the absence of gene products. A *regulator gene* controls gene expression either by preventing gene transcription, *negative control,* or by assisting gene transcription, *positive control.* The regulator gene product may have no function other than controlling the expression of a gene by affecting transcription, and in that case it acts solely as a *regulator protein.* On the other hand some proteins may have *dual functions,* acting as enzymes and also as regulator proteins by controlling gene expression. A regulator protein may control the expression of a single enzyme gene or of a group of genes linked together in an *operon.* A special case of this type of control of gene expression is when a regulator protein controls the rate of its own synthesis and this is described as *autogenous regulation.* An example is glutamine synthetase.

IV. Enzymes and Coenzymes

Enzymes catalyse reactions which are chemically feasible but allow them to take place at temperatures which are compatible with cellular viability. Thermophilic bacteria have enzymes which, not unexpectedly, are very resistant to heat whereas the psychrophils have enzymes which are still relatively active at very low temperatures. These qualities are not necessarily exclusive and some bacterial enzymes, such as the histidase of *Pseudomonas putida,* are active over a wide range of temperatures. Histidase of this species is most active between 20 and 35 °C but is active over the range 0 to 60 °C and the activity at 1.5 °C is 30 per cent of the maximum. In most cases the activity increases about twofold for each 10 degree rise in temperature but this is counteracted at higher temperatures by the inactivation of the enzyme. There are no clear indications of the special features of protein structure that confer thermostability on enzymes, but it is frequently the case that mutations in the amino acid sequence result in enzymes which are still active but more thermolabile than the parent. When this occurs the mutant enzyme may limit the upper temperature at which the mutant strain can grow. Temperature-sensitive mutants have been very useful in analysing the physiological roles of particular enzymes. The occurrence of such thermolabile mutant enzymes also illustrates the way in which the enzymes of different species which carry out the same biochemical reactions may have acquired their different thermal stabilities.

Although all enzyme reactions are theoretically reversible it is found that *in vivo* many are essentially unidirectional. This is particularly important in the control of rates of synthesis and breakdown of molecules since there are steps at which the same chemical conversion occurs in both the biosynthetic and catabolic pathways. There are often two enzymes at such steps, one for biosynthesis and one for catabolism, whose activities and rates of synthesis are independently regulated.

In addition to the specific enzyme, for many reactions it is also essential to have present a low-molecular weight reactant or coenzyme. The coenzymes participate in the reaction and may act as hydrogen acceptors as do the nicotinamide adenine dinucleotide coenzymes, NAD and NADP. Others are concerned with the transfer of particular functional groups. Thiamin pyrophosphate transfers aldehyde groups. Tetrahydrofolic acid, with one-carbon transfer, and Coenzyme A, with acyl group transfer, are other examples of reactions requiring co-enzymes. Metal ions may be essential for some enzyme reactions.

V. Classes of Enzymes

As enzymes were discovered they were given names which could be associated with the reactions catalysed. It is convenient to have a simple or trivial name to recall an enzyme and its reaction but it is also useful to group enzymes together according to the class of reaction. The current classification arranges enzymes into six main groups.

A. Hydrolases

Urease is an example of a hydrolase. It catalyses the hydrolysis of urea to carbon dioxide and ammonia. This is a very common enzyme among microorganisms and enables urea, which is the end product of nitrogen metabolism of animals, to be converted to ammonia and then to be used for microbial growth.

Urease

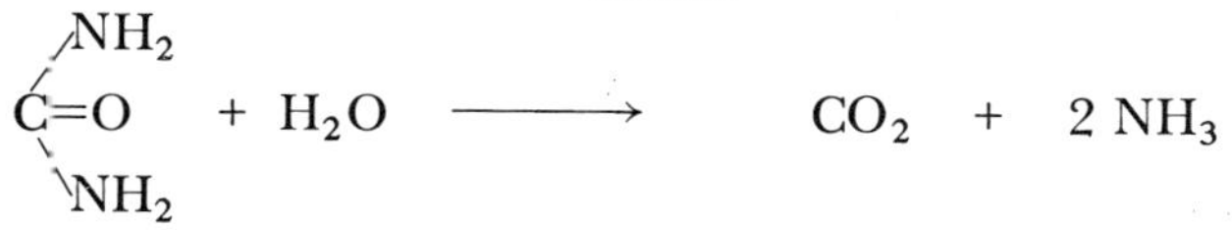

$$\begin{array}{c} NH_2 \\ | \\ C=O \\ | \\ NH_2 \end{array} + H_2O \longrightarrow CO_2 + 2 NH_3$$

Urea + Water $\longrightarrow$ Carbon dioxide + Ammonia   (1)

Another example of a hydrolase is *glucose-6-phosphatase*. The phosphate esters of sugars are of great importance in metabolism and *phosphatases* are of common occurrence.

Glucose-6-phosphatase

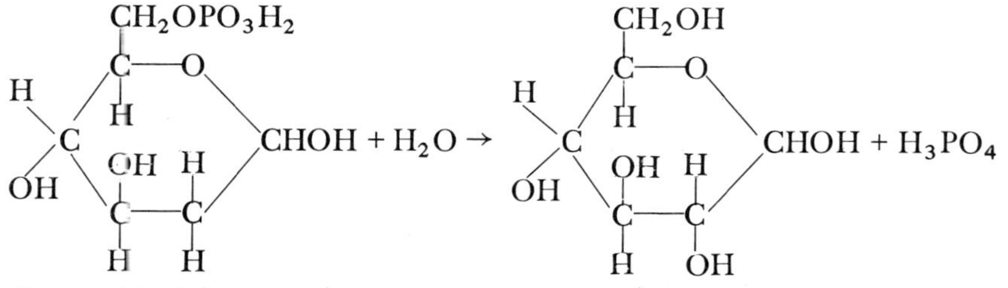

Glucose-6-phosphate + Water $\longrightarrow$ Glucose + Orthophosphate   (2)

B. Lyases

A second class of splitting enzymes are the *lyases*. In these reactions a compound is split into two components. An example of a lyase is *aldolase,* an enzyme of glucose metabolism. The diphosphate ester of fructose is split into two triose phosphates.

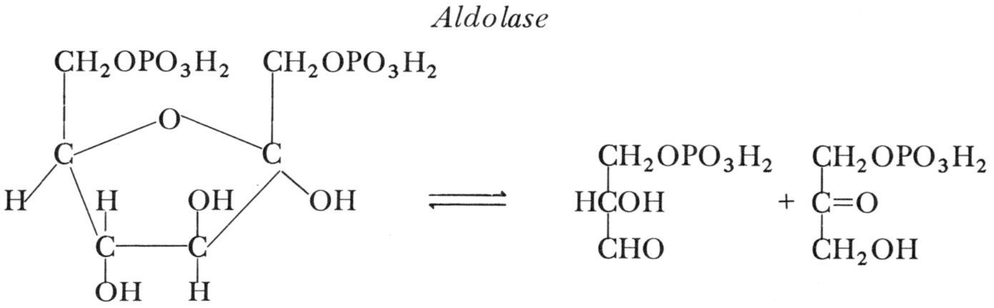

Aldolase

$$CH_2OPO_3H_2 \quad CH_2OPO_3H_2 \quad \rightleftharpoons \quad HCOH \atop CHO \quad CH_2OPO_3H_2 + C=O \atop CH_2OH$$

Fructose-1,6-diphosphate $\rightleftharpoons$ Glyceraldehyde-3-phosphate $\quad$ (3)

$\qquad$ + Dihydroxy acetone phosphate

Some lyases require coenzymes for activity. Among these are the *amino acid decarboxylases* which require the coenzyme pyridoxal phosphate, which also takes part in other amino acid reactions.

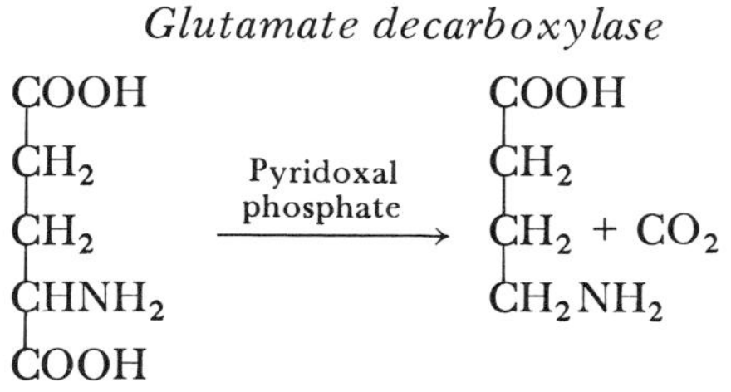

Glutamate decarboxylase

$$COOH \atop CH_2 \atop CH_2 \atop CHNH_2 \atop COOH \xrightarrow[\text{phosphate}]{\text{Pyridoxal}} COOH \atop CH_2 \atop CH_2 + CO_2 \atop CH_2NH_2$$

Glutamate $\longrightarrow$ 4-Aminobutyrate + Carbon dioxide $\qquad$ (4)

C. Isomerases

The sequential reactions of metabolic pathways frequently include steps for which the only change is a molecular rearrangement of the atoms within the molecule. Coenzymes may be essential for these reactions but there are no net changes in the reactants. A most important member of this class is *triose phosphate isomerase,* another enzyme of glucose metabolism. The two triose phosphates are formed by the action of aldolase (equation 3) and are interconverted by the isomerase. The removal of glyceraldehyde-3-phosphate in subsequent reactions results in the entire hexose molecule being metabolized via glyceraldehyde-3-phosphate.

Triose phosphate isomerase

$$\begin{array}{ccc}
\mathrm{CH_2OPO_3H_2} & & \mathrm{CH_2OPO_3H_2} \\
\mathrm{CHOH} & \rightleftharpoons & \mathrm{C{=}O} \\
\mathrm{CHO} & & \mathrm{CH_2OH}
\end{array}$$

Glyceraldehyde-3-phosphate $\rightleftharpoons$ Dihydroxyacetone phosphate   (5)

Racemases also belong to the class of isomerases and enable the interconversion of optical isomers. An organism may be able to metabolize both L and D forms of compounds, for example *Pseudomonas putida* utilizes both L- and D-mandelate and possesses a mandelate racemase. D-Alanine is needed for mucopeptide synthesis and can be made from L-alanine by most bacteria. Alanine racemase requires *pyridoxal phosphate* (PP).

Alanine racemase

$$\begin{array}{ccc}
\mathrm{COOH} & & \mathrm{COOH} \\
\mathrm{H_2N{-}C{-}H} & \underset{}{\overset{PP}{\rightleftharpoons}} & \mathrm{H{-}C{-}NH_2} \\
\mathrm{CH_3} & & \mathrm{CH_3}
\end{array}$$

(6)

L-Alanine $\rightleftharpoons$ D-Alanine

Since there are no large free energy changes involved in the reactions of isomerases the same enzyme may be able to take part in both biosynthetic and catabolic pathways, provided that one of the reactants is removed by appropriate enzymes. This is also true for some lyases and aldolase is an enzyme which functions in the biosynthesis of glucose and other sugars as well as in the major catabolic pathway. Hydrolases are potentially reversible but play no synthetic roles *in vivo*. The synthesis of hexose phosphate esters is achieved by specific enzymes and phosphatases act unidirectionally in releasing the free sugars from their esters. Polysaccharides, such as starch and cellulose, are broken down into the monomers by hydrolytic enzymes but synthesis involves other enzymes and requires energy.

D. Transferases

Transferases catalyse the transfer of a functional group from one compound to another. In amino acid metabolism an important group of enzymes are the *aminotransferases* or *transaminases* which transfer the amino group of one amino acid to an α-oxoacid thus producing a different pair of amino acids and α-oxoacids. These are freely reversible reactions and, as for many other amino acid transformations, the coenzyme is *pyridoxal phosphate* which forms an intermediate with the amino group.

Glutamate: Aspartate aminotransferase

$$
\begin{array}{cc}
COOH & COOH \\
| & | \\
CH_2 & CH_2 \\
| & \\
CH_2 & + C=O \\
| & | \\
CHNH_2 & COOH \\
| & \\
COOH &
\end{array}
\quad \overset{PP}{\rightleftharpoons} \quad
\begin{array}{cc}
COOH & COOH \\
| & | \\
CH_2 & CH_2 \\
| & | \\
CHNH_2 & + CH_2 \\
| & | \\
COOH & C=O \\
& | \\
& COOH
\end{array}
$$

Glutamate + Oxalacetate $\rightleftharpoons$ Aspartate + α-Oxoglutarate   (7)

The *kinases* are important enzymes which transfer phosphate groups from nucleotides to acceptor molecules. The first step in the catabolism of many sugars is the formation of a phosphate ester by the transfer of a phosphate group from ATP (adenosine triphosphate). Examples are *hexokinase* which produces glucose-6-phosphate from glucose and *phosphofructokinase* which produces fructose-1,6-diphosphate from fructose-6-phosphate (see equations 2 and 3).

Another important transferase is the enzyme which transfers the acetyl group from acetyl-CoA (Coenzyme A) to oxalacetate to produce citrate. This enzyme is known as *citrate synthase*, or the condensing enzyme, and is one of the key enzymes of metabolism. This reaction is not reversible.

Citrate synthase

$$
\begin{array}{cc}
& COOH \\
& | \\
CH_3 & CH_2 \\
| & + | \\
CO-S-CoA & C=O \\
& | \\
& COOH
\end{array}
\quad \longrightarrow \quad
\begin{array}{c}
CH_2COOH \\
| \\
HO-C-COOH \quad + CoASH \\
| \\
CH_2COOH
\end{array}
$$

Acetyl–CoA + Oxalacetate $\longrightarrow$ Citrate + CoASH   (8)

E. Ligases

Ligases or synthetases are joining enzymes which form new bonds between molecules. Specific ligases join sections of DNA (deoxyribonucleic acid) together and others add additional units to a polysaccharide chain which is being built up. Some ligases join two small molecules together. One of the latter is *glutamine synthetase* which produces glutamine from ammonia and glutamate and requires ATP which is converted to ADP and inorganic phosphate in the reaction.

Glutamine synthetase

$$\begin{array}{c} \text{COOH} \\ | \\ \text{CH}_2 \\ | \\ \text{CH}_2 \\ | \\ \text{CHNH}_2 \\ | \\ \text{COOH} \end{array} + \text{NH}_3 \quad \longrightarrow \quad \begin{array}{c} \text{CONH}_2 \\ | \\ \text{CH}_2 \\ | \\ \text{CH}_2 \\ | \\ \text{CHNH}_2 \\ | \\ \text{COOH} \end{array} + \text{ADP} + \text{H}_3\text{PO}_4$$

$$\text{Glutamate} + \text{ATP} + \text{Ammonia} \longrightarrow \text{Glutamine} + \text{ADP}$$
$$+ \text{Inorganic phosphate} \quad (9)$$

Glutamine synthetase is required by some organisms, including *Klebsiella aerogenes*, for the assimilation of ammonia during nitrogen-limited growth. Glutamine synthetase has regulatory properties and controls its own synthesis by autogenous regulation and in addition it plays a part in the regulation of the synthesis of other enzymes of nitrogen metabolism.

F. Oxidoreductases

Enzymes which catalyse the oxidation or reduction of metabolites are the oxidoreductases. These include the *oxygenases* which use molecular oxygen directly as a substrate. Among these are the enzymes which introduce oxygen atoms into hydrocarbons such as octane and the enzymes which add oxygen to diphenols and open up the ring structures of aromatic compounds. Oxygenases are found only in aerobic organisms.

Most oxidations take place, not by the addition of oxygen, but by the removal of hydrogen atoms or electrons. *Dehydrogenases* include many enzymes which require as coenzymes either NAD (nicotinamide-adenine dinucleotide) or NADP (nicotinamide-adenine dinucleotide phosphate). The coenzyme is reduced in the reaction *in* which the substrate is oxidized. The dehydrogenase reactions often act *in vivo* in either direction but may function predominantly in one direction. Lactic acid is produced as a fermentation product by the reduction of pyruvate by lactate dehydrogenase. The lactobacilli include the genus *Streptococcus* which produces L-lactic acid and *Leuconostoc* which produces D-lactic acid. Other lactobacilli produce both optical isomers.

Lactate dehydrogenase

$$\text{CH}_3\text{CHOHCOOH} + \text{NAD}^+ \rightleftharpoons \text{CH}_3\text{COCOOH} + \text{NADH} + \text{H}^+$$

$$\text{Lactate} + \text{NAD}^+ \rightleftharpoons \text{Pyruvate} + \text{NADH} + \text{H}^+ \quad (10)$$

Glyceraldehyde phosphate dehydrogenase catalyses the oxidation step on the major glucose catabolic pathway. This is a complex reaction and results in the formation of 1,3-phosphoglyceric acid from the aldehyde phosphate.

$$Glyceraldehyde\ phosphate\ dehydrogenase$$

$$
\begin{array}{l}
CH_2OPO_3H_2 \\
HCOH \\
CHO
\end{array}
+ NAD^+ + H_3PO_4
\rightleftharpoons
\begin{array}{l}
CH_2OPO_3H_2 \\
HCOH \\
COOPO_3H_2
\end{array}
+ NADH + H^+
$$

D-Glyceraldehyde-3-phosphate + NAD$^+$ + Inorganic phosphate

$$\rightleftharpoons 1,3\text{-Phosphoglyceric acid} + NADH + H^+ \qquad (11)$$

The reduced coenzyme can be reoxidized at the expense of the reduction of another cell metabolite. This may be a later intermediate of the same pathway. In anaerobic growth, reduced coenzymes are reoxidized by this means and the reduced products are excreted from the cells. Typical fermentation products which are formed in this way are lactic acid (see equation 10) and ethanol produced by yeasts and bacteria from acetaldehyde by the action of *alcohol dehydrogenase*. These reactions enable the reduced coenzyme to be reoxidized and to continue the dehydrogenation reactions of the catabolic pathway.

The nicotinamide coenzymes take part in many metabolic reactions and although some enzymes are active with both NAD and NADP, other enzymes are specific in their coenzyme requirement. The need for a supply of reduced coenzymes is most obvious for the reduction reactions which occur in biosynthetic pathways and in most cases the dehydrogenases of biosynthesis require NADPH.

VI. The Role of ATP

ATP (adenosine triphosphate) occupies a central position in intermediary metabolism. Many biosynthetic reactions, such as that catalysed by glutamine synthetase, are dependent on ATP for the reaction to occur. Again, the first reaction in the catabolism of many compounds is an activation step as in the hexokinase reaction which produces glucose-6-phosphate from glucose and ATP. The synthesis of ATP takes place as a result of several different biochemical events, all of which involve energy capture. The fermentation of glucose by microorganisms which can grow in the absence of oxygen leads to the synthesis of ATP by *substrate level phosphorylation* and the release of chemical energy from the splitting of the glucose molecule is coupled to the synthesis of ATP from ADP and inorganic phosphate. In the

Embden—Meyerhof pathway of glucose catabolism there are two stages at which this can take place. Substrate level phosphorylation occurs when a phosphate group is transferred from 1,3-phosphoglycerate (see equation 11) by *phosphoglycerate kinase* and when a phosphate group is transferred from phosphoenol pyruvate by *pyruvate kinase*.

Phosphoglycerate kinase

$$
\begin{array}{l}
CH_2OPO_3H_2 \\
HCOH \\
COOPO_3H_2
\end{array}
+ ADP \rightleftharpoons
\begin{array}{l}
CH_2OPO_3H_2 \\
HCOH \\
COOH
\end{array}
+ ATP
$$

1,3-Diphosphoglycerate + ADP $\rightleftharpoons$ 3-Phosphoglycerate + ATP   (12)

Pyruvate kinase

$$
\begin{array}{l}
CH_2 \\
COPO_3H_2 + ADP \\
COOH
\end{array}
\longrightarrow
\begin{array}{l}
CH_3 \\
C{=}O \\
COOH
\end{array}
+ ATP
$$

Phosphoenol pyruvate + ADP $\longrightarrow$ Pyruvate + ATP   (13)

Reaction (12) can occur in the reverse direction for the synthesis of glucose from low molecular weight growth substrates such as acetate and pyruvate. Reaction (13) is essentially unidirectional.

During aerobic growth the reduced coenzymes are reoxidized by a series of reactions which eventually transfer the hydrogen atoms to molecular oxygen with the formation of water. Hydrogen atoms and electrons are transferred to oxidation—reduction carriers which make up the electron transport or *cytochrome chain*. During this transfer process ATP is synthesized from ADP and inorganic phosphate by *oxidative phosphorylation*. In eukaryotes the mitochondrion is the site of oxidative phosphorylation but in prokaryotes the reactions are associated with the cytoplasmic membrane. In photosynthetic organisms ATP is also synthesized by *photophosphorylation* which is dependent on the capture of light energy by the photosynthetic pigments followed by the transfer of electrons through a series of carriers which include specialized cytochromes.

The main components of the respiratory chain are the nucleotide coenzymes, flavoproteins, and cytochromes. Cytochrome a_3, or cytochrome oxidase, carries out the final step involving molecular oxygen. The complete oxidation of one molecule of reduced NAD can lead to the synthesis of three molecules of ATP and coupling steps have been located at three regions of the electron transport chain. Certain bacteria can grow anaerobically with nitrate or sulphate as the terminal electron acceptors and these organisms synthesize ATP by oxidative phosphorylation with the inorganic salt substituting for oxygen at the end of the chain. The following reactions show a simplified version of the electron transport chain with the sites of coupling of ATP synthesis. Some substrates are oxidized by flavoproteins and slot into the carrier chain between sites 1 and 2.

$$Cytochrome\ chain$$

$$NADH \dashrightarrow \overset{1}{FP} \dashrightarrow Cyt\ b \overset{2}{\dashrightarrow} Cyt\ c \dashrightarrow Cyt\ a \dashrightarrow O_2$$

$$3\ ADP + Inorganic\ phosphate \longrightarrow 3\ ATP \qquad (14)$$

VII. Metabolic Pools

Living organisms consist mostly of water and when this is removed it is found that most of the *dry weight* is contributed by the cell polymers. In bacteria the cell walls, membranes, intracellular proteins, and nucleic acids make up about 90 per cent of the dry weight. The rest is made up of low molecular weight compounds, amino acids, monosaccharide derivatives, coenzymes, nucleotides, inorganic salts, and a variety of metabolic intermediates. These constitute the *metabolic pools*. The individual components of the pools are in a state of flux. Compounds are withdrawn for biosynthesis and may enter the cells from the external environment or be produced endogenously by the reactions of the metabolic pathways. The concentrations of compounds in the metabolic pools is normally low and may be difficult to measure, but it is very easy to observe the effects of altering the concentrations of some of the compounds by adding substances to the growth medium.

A. Biosynthetic Enzymes

When *Escherichia coli* is growing in an ammonium salt medium the amino acids are synthesized from ammonia and the intermediates produced from the catabolism of the carbon compounds of the growth medium. If an amino acid, such as tryptophan is added to the medium the immediate effect is to inhibit the first enzyme of the tryptophan biosynthetic pathway. The addition of the amino acid increases the size of the tryptophan pool and this results in a rapid adjustment of the rate of synthesis of tryptophan by the inhibition by tryptophan of the first enzyme of the pathway. Regulation of the rate of synthesis of an amino acid by inhibition of an early enzyme of the pathway is a general method of regulation known as *feedback inhibition*. The inhibitor may be the end product of the pathway or an intermediate and this activity may be mimicked by amino acid analogues which thereby inhibit growth in minimal medium. If the exogenous supply of tryptophan is maintained for some time it will affect the rates of synthesis of the enzymes of the pathway. This longer term effect is known as *repression of enzyme synthesis*. These two regulatory mechanisms allow *E. coli* to avoid the unnecessary expenditure of energy in making tryptophan from its precursors, and the unnecessary expenditure of making the tryptophan enzymes, when sufficient tryptophan is available exogenously. In its natural environment in the human gut *E. coli* is exposed to fluctuating levels of tryptophan and these regulatory mechanisms allow rapid adjustments to be made. The patterns of regulation have been studied in most detail in *E. coli* and other enteric bacteria but many other species have been investigated and it has been found that regulation of biosynthetic pathways may take somewhat different forms in organisms living mainly in soil and water.

B. Catabolic Enzymes

The catabolic pathways give rise to the intermediates which provide the precursors for biosynthetic pathways as well as releasing chemical energy which is trapped as ATP. Some growth substrates allow much higher growth rates than others and this means that the enzymes have evolved in such a way that the overall pathways of catabolism and biosynthesis function more effectively with the substrates which support the faster growth rates. Glucose supports rapid growth of *Escherichia coli* but although *Pseudomonas aeruginosa* can grow with glucose as a carbon source, it grows much faster in succinate medium. If two carbon compounds are present at the same time they may be metabolized together, but frequently only one of the pair of potential growth substrates is used during the first growth period and when this has disappeared the other compound is then utilized for growth. Many catabolic enzymes are *inducible* and are synthesized only when their substrates are present in the medium. Some enzymes are constitutive and are synthesized at the same levels whether or not the potential growth substrate is present. In *E. coli* the enzymes of glucose metabolism are constitutive and glucose is generally the preferred growth substrate if mixtures of substrates including glucose are present. Glucose exerts several effects on the catabolism of other compounds and may prevent the synthesis of inducible enzymes even in the presence of the inducing substrate. This effect of glucose in preventing enzyme synthesis is known as *catabolite repression*. An inducible enzyme which is subject to catabolite repression by glucose is the β-galactosidase of *E. coli* which is essential for this organism to grow on lactose.

$$\text{Lactose} \xrightarrow{\ \beta\text{-}Galactosidase\ } \text{Galactose} + \text{Glucose} \tag{15}$$

When lactose is present in the growth medium as the sole carbon source the enzyme β-galactosidase is induced and the bacteria grow. The enzyme can also be induced by analogues of lactose, various thio-galactosides, which act as inducers although they are not substrates of the enzyme as is lactose itself. When both glucose and lactose are present the synthesis of β-galactosidase is repressed. After all the glucose has been used up there is a brief growth lag during which β-galactosidase is synthesized and then growth starts again at the expense of the lactose. This preferential use of one carbon compound rather than another is another regulatory device which economizes on the synthesis of enzymes if a more suitable growth substrate is available. In *E. coli* glucose can shut down the synthesis of β-galacto-sidase in bacteria which had previously been induced and were actively synthesizing the enzyme. This effect is mainly due to a lowering of the concentration of cyclic AMP ($3',5'$-cyclic adenosine monophosphate) in the metabolic pool. This nucleotide is essential for the synthesis of many inducible catabolic enzymes. It interacts with an activator protein whose function is to stimulate the expression of the genes for these enzymes of catabolic pathways. When the level of cyclic-AMP in the metabolic pool falls below a threshold value then those enzymes whose synthesis is controlled in this way will not be synthesized.

The inducer of the enzymes of a catabolic pathway may be the first substrate of the pathway and may control the synthesis of a group of enzymes which catalyse sequential reactions. Genes controlled together by a single regulatory arrangement form a unit of regulatory control called an *operon*. All the enzymes determined by the genes of a single operon will be synthesized in the same relative amounts and their rates of synthesis will be *coordinately* controlled. Some enzymes of a catabolic pathway are not induced by the first substrate but by a later intermediate. When the concentration of this intermediate in the metabolic pool reaches a sufficiently high level, then the later enzymes are induced. This is known as *sequential* induction. It has an obvious advantage in cell economy since it allows for the induction of synthesis of all the enzymes of the pathway if the first substrate is provided, and the induction of the synthesis of only the later enzymes of the pathway if an intermediate is available as an exogenous growth substrate. Another variation on induction of catabolic enzymes is found in pathways where the enzyme is induced by its product and not by its substrate. This control by *product induction* occurs in the pathways for the catabolism of histidine by *Klebsiella aerogenes, Pseudomonas aeruginosa,* and *Pseudomonas putida.* Histidine is synthesized and a certain amount must be available in the metabolic pool for incor-poration into protein. It would be undesirable for histidine, or other amino acids, to be too readily drawn away by catabolic enzymes. Only when the metabolic pool of histidine reaches high levels does the histidine catabolic pathway begin to operate. A very low level of histidase, the first enzyme of the pathway, is present and converts histidine into urocanate. The actual inducer of the histidine catabolic pathway enzymes is urocanate and only when the concentration of urocanate in the metabolic pool reaches the threshold value for induction can the enzyme of the pathway be synthesized.

Catabolic enzymes may be regulated by induction by substrate or end-products by the cyclic-AMP activator protein and also by activation by the enzyme glutamine synthetase in its role as regulator protein. All these regulatory mechanisms are thought to act at the stage of transcription of the DNA of the gene into the messenger RNA complementary copy from which the enzyme proteins are then translated. The regulatory processes may be either *negative control* when the gene transcription is switched off by the regulator protein, or *positive control* when the gene transcription is switched on by the regulator protein. In each case the concentration of low molecular weight compounds in the metabolic pools is very important in switching enzyme synthesis off or on. Only a few of the catabolic pathways are known to be controlled by all three systems, enzyme induction, catabolite repression, and glutamine synthetase, but many are controlled by both induction and catabolite repression. The mechanism of catabolite repression by glucose exerted by changes in the cyclic AMP levels of the cell shows how one growth substrate may influence the metabolism of another compound by altering the internal pool of essential metabolites.

VIII. Transport of Metabolites

Cells are not normally permeable to polar and charged molecules and specific transport systems, sometimes described as permeases, are required for the uptake of amino acids, sugars, and carboxylic acids into bacterial and fungal cells. *Active transport* requires energy and the substrate of the transport system is concentrated within the cells without modification of structure. Specific transport proteins are present and these frequently have very high affinities for the substance transported so that the organism is able to take up nutrients from very dilute solutions. This could clearly be of advantage to organisms living in aqueous environments when the concentrations of organic compounds would be likely to be very low. Many transport systems are present under most conditions at low constitutive levels and can be induced to much higher levels in the presence of the substrate. This enables the cells to take up the potential growth substrate and to increase the total activity of the specific transport system when it is required. The permeases for substrates of catabolic enzymes may be induced coordinately with the enzymes. This occurs in the case of β-galactoside permease which is induced together with the enzyme β-galactosidase when *Escherichia coli* is grown with lactose as the carbon source.

Group translocation is also energy requiring and leads to a concentration of compounds above that of the external environment but in this case the substrate of the transport system is chemically modified. Many sugars are transported by the phosphotransferase system which requires phosphoenol pyruvate and several protein components, one of which is specific for the sugar being transported. The other proteins function in the transport of all sugars taken up by this route. The initial step is the phosphorylation of a protein known as HPr, a low-molecular weight protein with histidine at the active site. This phosphorylation is carried out by the transferase action of Enzyme I with phosphoenol pyruvate as the phosphate donor. As can be seen from the kinase reaction shown in equation (13) this is equivalent in terms of energy expenditure to activation by ATP. The second reaction is the phosphorylation of the sugar by Enzyme II. The sugar phosphate is released in the cell and is available for metabolism.

(i) Phosphoenol pyruvate + HPr $\xrightarrow{\text{Enzyme I}}$ HPr-P + Pyruvate

(ii) HPr-P + Sugar $\xrightarrow{\text{Enzyme II}}$ HPr + Sugar-phosphate   (16)

Facilitated diffusion also involves specific binding proteins but no energy is required and the substrate is not concentrated in the cells at a higher level than the outside.

Passive diffusion depends only on the compounds concerned being able to diffuse across the cytoplasmic membrane. This is presumed to occur for many uncharged molecules including the hydrocarbons which are growth substrates for many species. When the catabolic enzymes are present in the cell the incoming molecules will be rapidly metabolized and thus make way for the entry of more substrate. Entry of substrate in this manner, or by facilitated diffusion, will not be rate-limiting for growth if the necessary catabolic enzymes have been induced in the organism. The entry by passive diffusion appears to be adequate to obtain inducing levels of these uncharged potential growth substrates. On the other hand although charged molecules may diffuse slowly across the membranes this is usually not at a rate sufficient to support growth in the absence of the specific permease system. Loss of activity of the specific transport system is often the reason why some bacterial strains are unable to utilize amino acids used as growth substrates by the wild type parent. Loss of β-galactoside permease makes *Escherichia coli* unable to grow on lactose. Such strains which are able to produce the necessary catabolic enzymes but which are unable to take up the compounds from the medium are said to be *cryptic*.

IX. Metabolic Pathways

If we take a common microbial growth substrate such as glucose, it is possible to trace the origins of the biosynthetic pathways and to follow the steps by which compounds including amino acids, purines and pyrimidines, carbohydrates and lipids are constructed from intermediates of the catabolic pathways. Figure 1 shows a simplified picture with some of the key metabolites and the connections between the pathways. Many other amino acids are derived from intermediates of the tricarboxylic acid cycle, shown in more detail in Figure 5. During aerobic growth the tricarboxylic acid cycle (also known as the Krebs' cycle after its discoverer) provides a series of reactions which convert the two-carbon acetyl fragment derived from glucose breakdown into carbon dioxide and water. The reduced coenzymes formed during these reactions are reoxidized by the cytochrome chain (see equation 14) with the formation of ATP by oxidative phosphorylation. During both aerobic and anaerobic growth the tricarboxylic acid cycle provides the amino acid precursors.

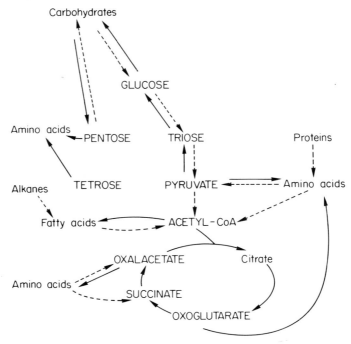

Figure 1. Key metabolites of catabolic and biosynthetic pathways. The dashed lines indicate the direction of catabolism and the solid lines the direction of biosynthesis. All the compounds in capital letters are precursors of biosynthetic pathways. The catabolic role of the tricarboxylic acid (TCA) cycle is shown in Figure 5

The glucose metabolite which enters the tricarboxylic acid cycle reactions is acetyl-coenzyme A and this compound is also the starting point for the biosynthesis of fatty acids and other lipids. Other intermediates of glucose catabolism are precursors of amino acids, purines, and pyrimidines. Growth on two-carbon and three-carbon compounds may require additional enzymes to replenish the intermediates withdrawn for biosynthesis and will demand pathways for converting these compounds into hexoses required by the cells. Some of these will be described later.

A. Glucose Metabolism

1. Embden–Meyerhof Pathway

The most common pathway of glucose catabolism is the Embden–Meyerhof pathway of glycolysis which was first recognized in yeasts but occurs very widely and is found in animals, plants, and micro-organisms. The main steps of the pathway are shown in Figure 2 and several of the enzymes were described earlier. During aerobic growth glucose is degraded completely to carbon dioxide and water except for the intermediates which have been withdrawn for biosynthesis. Strict anaerobes which grow at the expense of glucose are unable to carry out oxidation and derive all their energy by fermentation reactions leading to substrate level phosphorylation. The facultative anaerobes are able to grow both aerobically and anaerobically and their metabolism is fermentative during anaerobic growth and mainly oxidative during aerobic growth. Many different fermentation products are known and these are formed by specific enzymes. The main function of these reactions is to accomplish the reoxidation of the reduced coenzyme NADH and in some cases this is accompanied by additional phosphorylations. Ethanol is the major yeast fermentation product, the homofermentative lactobacilli produce only lactic acid while other bacteria such as *Escherichia coli*, the heterofermentative lactobacilli and the anaerobic clostridia produce a mixture of compounds.

Many of the reactions of the Embden–Meyerhof pathway are freely reversible and can be used for the synthesis of hexose as well as for degradation. Three of the reactions are irreversible and *gluconeogenesis* requires a different enzyme from glycolysis at each of these steps. In the biosynthetic direction phosphoenol pyruvate is synthesized from pyruvate and ATP by the action of *phosphoenol pyruvate synthase* (there are also alternative routes which lead to the same product). Fructose-1,6-diphosphate and glucose-6-phosphate are hydrolysed by specific phosphatases (see equation 2). The enzymes of the degradative pathway at these steps are kinases and require ATP.

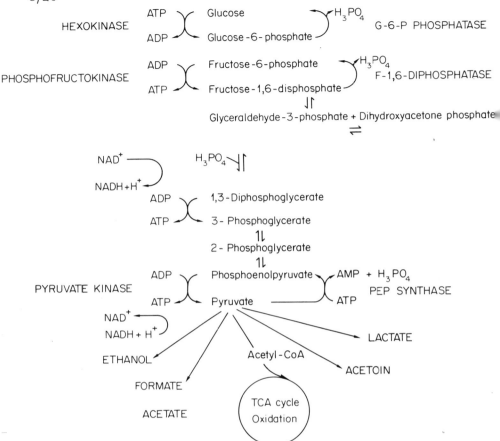

Figure 2.   The Embden–Meyerhof pathway of glucose catabolism. During aerobic growth the endproducts of metabolism are carbon dioxide and water produced by the tricarboxylic acid (TCA) cycle reactions. Anaerobically pyruvate may be reduced to give various fermentation products of which examples are shown above

2. Entner–Doudoroff and Pentose Pathways

A few bacteria, including *Pseudomonas* species, do not metabolize glucose by the Embden–Meyerhof pathway. Instead, glucose-6 -phosphate is oxidized to 6-phosphogluconate. The Entner–Doudoroff reactions (Figure 3) involve dehydration to give 2-deoxy-3-oxo-6-phosphogluconate which is split by aldolase cleavage to glyceraldehyde-3-phosphate and pyruvate. In the aerobic pseudomonas species the catabolism is completed via acetyl-CoA and the tricarboxylic acid cycle. The fermentative bacterium *Zymomonas lindneri* uses this pathway to ferment glucose, producing two molecules of ethanol and two of carbon dioxide from each glucose molecule. Although the products are the same as those of yeast fermentation the distribution of labelled

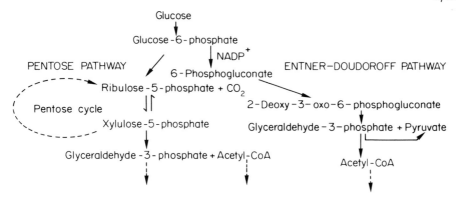

Figure 3.   The Entner—Doudoroff and pentose pathways of glucose catabolism.

carbon atoms in the products formed from glucose with radioactively labelled carbon atoms shows very clearly that the catabolic pathways differ.

The sugar acid 6-phosphogluconate may be oxidatively decarboxylated to produce ribulose-5-phosphate which can be converted to ribose-5-phosphate and xylulose-5-phosphate. These reactions are important in the provision of ribose and deoxyribose for nucleic acids. The enzymes *transaldolase* and *transketolase* transfer two-carbon fragments between the various sugars producing pentoses, hexoses, tetroses, and trioses by a series of reactions known as the pentose cycle (Figure 4). These reactions can account for the complete oxidation of glucose but the most important role of the pentose cycle in aerobic organisms as well as in anaerobic organisms is the provision of intermediates for biosynthesis. One of the intermediates of the pentose cycle is the four-carbon compound erythrose-4-phosphate which is the precursor of the aromatic amino acids. A few anaerobic bacteria including *Leuconostoc mesenteroides,* ferment glucose by these reactions splitting xylulose-5-phosphate by an aldolase cleavage and producing lactate and ethanol as fermentation products. Similar pentose cleavage reactions occur in pathways of pentose utilization as sole carbon source.

Many microorganisms are able to grow on sugars, polyalcohols, and polysaccharides. The initial attack on the insoluble polysaccharides is usually by the action of extracellular enzymes which hydrolyse them to disaccharides and monosaccharides. *Amylases* hydrolysing starch, and *cellulases,* hydrolysing cellulose, are produced by many soil microorganisms. The fermentation of carbohydrates is used by bacteriologists to identify bacterial species. In general the main reactions for the metabolism of hexoses and the disaccharides are the same. A few initial enzyme reactions convert hexoses, such as galactose, into glucose or a derivative of glucose which is then metabolized by one of the pathways which have been described. A disaccharide, such as

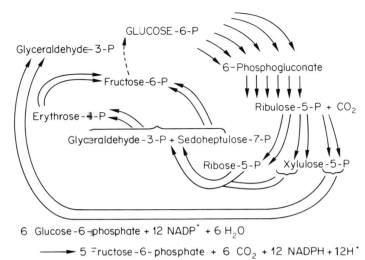

6 Glucose-6-phosphate + 12 NADP$^+$ + 6 H$_2$O
$\longrightarrow$ 5 Fructose-6-phosphate + 6 CO$_2$ + 12 NADPH + 12H$^+$

Figure 4. The reactions of the oxidative pentose cycle can account for the complete oxidation of glucose but may be of more importance for the provision of intermediates for biosynthesis. Some of these reactions are used in pathways for the catabolism of pentoses and pentitols

lactose, is hydrolysed by a specific enzyme, in this case β-galactosidase to give glucose and galactose (see equation 15). The difference between two isolates which are positive and negative respectively for the fermentation of a particular sugar may be due to the absence of a single enzyme required for the initial step in catabolism. Occasionally it may be due to a difference in the uptake of the sugar by the two strains. Pentoses are used as the carbon source for growth by many species. If the pentose can be transported into the cell and converted into a phosphate ester it is usually metabolized by pentose cycle or pentose cleavage reactions.

3. The Pentose Cycle and Photosynthesis

When carbon dioxide is the sole carbon source for growth then the two primary demands on intermediary metabolism are for reactions to convert the one-carbon compound into organic molecules containing a larger number of carbon atoms, and for reactions to reduce carbon dioxide to the oxidation state of the carbon compounds of the cell. These reactions require energy and the ATP and the reduced coenzyme NADPH needed are both produced by the photosynthetic reactions. The photosynthetic pigments, chlorophylls and associated carotenoids, capture light energy. The excitation of these molecules is followed by an electron flow through the photosystems which is coupled to the synthesis of ATP by photophosphorylation and also results in the reduction of $NADP^+$. These are then available for the assimilation of carbon dioxide by reactions which do not require light. Many carboxylating reactions are known in different pathways which involve the uptake of carbon dioxide but the primary carboxylation of photosynthesis is unique and occurs only when carbon dioxide is the sole carbon source.

Carbon dioxide is taken up by the enzyme *ribulose diphosphate carboxylase* and added to ribulose diphosphate which is cleaved into two molecules of 3-phosphoglycerate. This compound can be reduced to glyceraldehyde phosphate and after partial isomerization to dihydroxy-acetone phosphate, the two triose phosphates can be converted to fructose-1,6-diphosphate by the action of aldolase. Apart from the first reaction, the enzymes for these assimilatory reactions are the same as those occurring in the Embden–Meyerhof glycolytic pathway. It is also necessary, for assimilation of carbon dioxide to continue, to have reactions to provide a constant supply of ribulose diphosphate and this is accomplished by the reactions of the Calvin cycle involving pentose pathway reactions. The reaction producing ribulose diphosphate is also specific for the assimilation of carbon dioxide and requires the enzyme phosphoribulokinase.

Ribulose diphosphate carboxylase

$$CO_2 + \text{Ribulose-1,5-diphosphate} \longrightarrow 2(\text{3-Phosphoglycerate}) \qquad (17)$$

Phosphoribulokinase

$$\text{Ribulose-5-phosphate} + ATP \longrightarrow \text{Ribulose-1,5-diphosphate} + ADP \quad (18)$$

The overall equation for the production of one molecule of hexose from six molecules of carbon dioxide requires six molecules of ribulose-1,5-diphosphate which are regenerated by the pentose cycle reactions. Three molecules of ATP are needed for each molecule of CO_2 fixed and two molecules of NADPH. These reactions occur in photosynthetic microorganisms and also in chemosynthetic bacteria fixing carbon dioxide at the expense of oxidizing inorganic compounds.

Overall equation for the fixation of carbon dioxide in photosynthesis

6 Ribulose-1,5-diphosphate + 6 CO_2 + 18 ATP + 12 NADPH
+ 12 H^+ + 12 H_2O $\longrightarrow$

6 Ribulose-1,5-diphosphate + Hexose + 18 ADP + 12 $NADP^+$
+ 18 Phosphate

$$CO_2 + 3ATP + 2\,NADPH + H^+ \longrightarrow 1/6\ \text{Hexose} \qquad (19)$$

X. The Tricarboxylic Acid and Glyoxylate Cycles

The reactions of the tricarboxylic acid cycle shown in Figure 5 allow the complete oxidation of acetyl groups produced from the breakdown of glucose and other carbon compounds. Many catabolic pathways converge on the tricarboxylic acid cycle giving rise to acetate, acetyl-CoA, or some of the cycle intermediates. The cycle of reactions provides the main channel of terminal respiration in aerobic micro-organisms. Isocitrate, α-oxoglutarate, succinate and malate are oxidized by dehydrogenases and the reduced coenzymes are reoxidized by the reactions of the cytochrome chain (see equation 14). The yield of ATP from the terminal oxidation of glucose by the cytochrome chain is much greater than that from the earlier stages of glucose catabolism. One molecule of glucose can give a net yield of *two molecules of ATP* from the substrate level phosphorylations of the Embden—Meyerhof pathway. The oxidative phosphorylation which is coupled to electron transport along the cytochrome chain gives a theoretical yield of 12 molecules of ATP for each acetyl group oxidized by the tricarboxylic acid cycle reactions. The complete oxidation of a molecule of glucose could give rise to *38 molecules of ATP* from oxidative phosphorylation associated with the oxidation of acetyl-CoA and its formation from pyruvate, together with the substrate level phosphorylations. This is a theoretical value and the rate of catabolism and the synthesis of ATP will depend on many factors. ATP is formed as a result of the operations of catabolic pathways and used up in biosynthesis and membrane transport but there is no accumulation of ATP in the metabolic pool. Fluctuations in the ratios of the adenosine nucleo-tides, ATP/ADP/AMP have feedback effects in controlling the rates of some of the key enzymes.

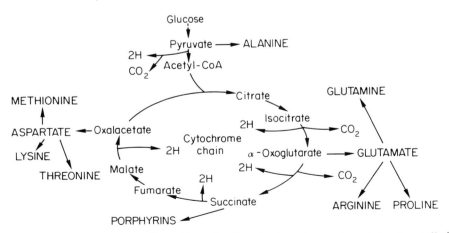

Figure 5.   The tricarboxylic acid cycle. Terminal oxidation is mainly channelled through these reactions. Intermediates are withdrawn for biosynthesis

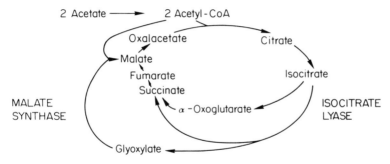

Figure 6.   The glyoxylate cycle. These reactions allow the replenishment of the pool of tricarboxylic acid cycle intermediates

Some of the intermediates of the tricarboxylic acid cycle are very important for biosynthesis and these include α-oxoglutarate and oxalacetate which are amino acid precursors. During growth in ammonium salt medium these compounds are continually being withdrawn from the metabolic pool for the amino acid biosynthetic pathways. If the major growth substrate is broken down to give an intermediate of the tricarboxylic acid cycle then any intermediates withdrawn for biosynthesis are replaced by the normal catabolic reactions. If acetate is the sole carbon source it is essential to have additional enzyme reactions to replenish the pool of tricarboxylic acid cycle intermediates. This is accomplished by two enzymes, *isocitrate lyase* and *malate synthase* which are synthesized during growth on acetate or compounds which give rise to acetyl-CoA as the result of catabolic reactions. These enzymes provide a bypass around some of the tricarboxylic acid cycle reactions and the net effect is the synthesis of one molecule of succinate from two molecules of acetate. Acetyl-CoA enters the glyoxylate cycle at two points. It undergoes the condensation reaction with oxalacetate to give citrate which is the entry point for the tricarboxylic acid cycle and the next reactions lead to the formation of isocitrate. *Isocitrate lyase* is a splitting enzyme producing succinate and glyoxylate. The second molecule of acetyl-CoA is condensed with glyoxylate to give malate by the action of *malate synthase*. Figure 6 shows the way in which the two enzymes fit together with enzymes of the tricarboxylic acid cycle to form the glyoxylate cycle. Enzymes which carry out replenishment reactions in this way are known as *anaplerotic* enzymes, their function being to maintain the pool of essential intermediates for biosynthesis. Reactions of this type are also important for growth on pyruvate and other C_3 compounds since this will also require enzymes to synthesize the four-carbon intermediates of the tricarboxylic acid cycle. Pyruvate is converted directly to oxalacetate by the ATP-dependent *pyruvate carboxylase* in *Pseudomonas* species but *Escherichia coli* and other microorganisms employ an indirect route and form phosphoenolpyruvate which is then carboxylated to produce oxalacetate.

The importance of the anaplerotic reactions is clear from the properties of mutants lacking these enzymes. *Pseudomonas aeruginosa* is able to grow with acetate as the carbon source and also with fatty acids of longer chain length. Mutants lacking isocitrate lyase are unable to grow on acetate and they are also unable to grow on fatty acids containing an even number of carbon atoms. The degradative pathway for fatty acids involves oxidation at the β-carbon atom and the successive cleavage to give acetyl-CoA. If the fatty acid contains an even number of carbon atoms then the only product is the two-carbon fragment. However, fatty acids containing an odd number of carbon atoms generate one three-carbon fragment for each molecule of the fatty acid. Isocitrate lyase-negative mutants are able to grow on odd-numbered fatty acids although they cannot grow on the even numbered.

The enzymes of the tricarboxylic acid cycle are not synthesized at the same rates in all growth media. The highest levels of the enzymes in a facultative anaerobe such as *Escherichia coli* occur when it is growing aerobically in a minimal salt medium with a simple carbon source such as pyruvate. Under these conditions the cycle is important both for biosynthesis and for oxidation of the growth substrate. The addition of amino acids reduces some of the biosynthetic requirements and lowers the enzyme levels. Anaerobic growth reduces the need for some of the steps of the cycle and again lowers the enzyme levels. The addition of glucose also reduces enzyme levels and this can be accounted for by the substrate level phosphorylations providing part of the ATP requirement. Under some growth conditions *E. coli* synthesizes all the enzymes of the cycle with the exception of α-oxoglutarate dehydrogenase. This incomplete cycle is sufficient for biosynthetic needs but cannot of course allow the normal cyclic terminal respiration to take place.

XI. Nitrogen Metabolism

The nitrogen-fixing species include the aerobic *Azotobacter*, the anaerobic clostridia, the facultative anaerobe *Klebsiella*, species of cyanobacteria (blue-green algae) and the symbiotic rhizobia. Nitrogen fixation is of great economic importance and the nitrogen-fixing microorganisms enable the nitrogen of the air to be harvested as food. For a long time the occurrence of nitrogen-fixing systems in bacteria as diverse as the highly aerobic *Azotobacter* and the highly anaerobic clostridia was puzzling. It is now known that the *nitrogenase* of all nitrogen-fixing organisms is very sensitive to oxygen and is so readily inactivated that it must be kept in a reduced state. Both the anaerobic metabolism of the clostridia and the high rate of respiration of *Azotobacter* are able to maintain the low oxidation—reduction potential of the micro-environment of their respective nitrogenase systems. The first identifiable product is ammonia, and the energy and reducing power for the reactions is supplied by catabolism of growth substrates or by photophosphorylation in the case of the photosynthetic organisms. If fixed nitrogen is available in the form of amino acids, or ammonium salts, the synthesis of nitrogenase is repressed. Repression of synthesis of nitrogenase in growth conditions for which it is not essential for growth is comparable to regulatory systems which control the synthesis of catabolic enzymes. Recent experiments indicate that the enzyme glutamine synthetase acts as an activator for the transcription of the genes determining the enzymes required for nitrogen fixation. In *Klebsiella aerogenes* these genes, known as the *nif* genes for *ni*trogen *f*ixation, are carried on plasmids which can be transferred from cell to cell. It has been possible to construct a hybrid plasmid which can carry the *nif* genes from *K. aerogenes* to *Escherichia coli* which does not normally fix nitrogen.

Many bacteria and fungi can use nitrate and nitrite as nitrogen sources for growth. Nitrate and nitrite reductases reduce these compounds to ammonia which can be assimilated. Some bacteria, including species which are otherwise strictly aerobic in growth habitat, can use nitrate as the terminal oxidant for catabolism. The reduction of nitrate and nitrite does not always lead to ammonia, and oxides of nitrogen or molecular nitrogen can be produced by some species. The release of molecular nitrogen in these reactions results in the loss of fixed nitrogen and impoverishes the soil in some regions. Ammonia is released from organic combination by the action of microorganisms on plant and animal residues and may be taken up again into new microbial life. The nitrifying bacteria, *Nitrosomonas* and *Nitrobacter,* have a special use for ammonia and obtain energy by oxidizing it to nitrite and nitrate. Taken together, these microorganisms account for the constant recycling of nitrogen atoms through inorganic and organic compounds.

When ammonium salts supply the nitrogen for growth ammonia is combined with organic carbon compounds by amination reactions. Several of these are known and the most widespread is that which leads to the production of glutamate by the amination of α-oxoglutarate obtained as an intermediate of the tricarboxylic acid cycle.

Glutamate dehydrogenase

$$\alpha\text{-Oxoglutarate} + NADPH + H^+ + NH_3 \rightleftharpoons$$

$$Glutamate + NADP^+ + H_2O \quad (20)$$

This reaction is known to be important *in vivo* for the growth of many microorganisms in minimal salt media. There is however another route which operates in some organisms when the ammonium ion concentration is very low or when growth is nitrogen limited by other nitrogen sources. This has been examined in most detail in *Klebsiella aerogenes* and it was found that the affinity of glutamate dehydrogenase for ammonia was too low for it to be effective when the internal pool was at a low level. Growth in minimal salt media containing less than 1 mM ammonium ions requires two enzymes and is more energy demanding since ATP is needed in addition to the reduced coenzyme. The enzymes are (i) *glutamine synthetase* which adds ammonia to glutamate to form glutamine and requires ATP, and (ii) *glutamate synthase* or GOGAT (glutamine:oxoglutarate aminotransferase) which transfers the amide group of glutamine to α-oxoglutarate to give two molecules of glutamate.

Glutamine synthetase

$$\text{Glutamate} + \text{ATP} + NH_3 \longrightarrow \text{Glutamine} + \text{ADP} + \text{Phosphate} + H_2O \quad (21)$$

Glutamate synthase

$$\text{Glutamine} + \alpha\text{-Oxoglutarate} + \text{NADPH} + H^+ \rightleftharpoons 2\,\text{Glutamate} + NADP^+ + H_2O \quad (22)$$

Glutamine synthetase activity is regulated by some of the metabolites which are formed from glutamine. The enzyme is remarkable in that it acts as a regulator protein for its own synthesis, for the nitrogen-fixing enzymes, and for other enzymes of nitrogen metabolism. The central role of glutamine in nitrogen metabolism is indicated in Figure 7 which shows some of the compounds containing heterocyclic rings or other nitrogen groups which are synthesized by pathways requiring glutamine as the donor of amino groups.

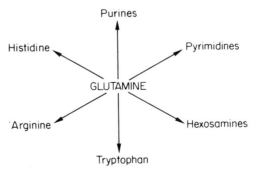

Figure 7.   Outline of biosynthetic pathways requiring glutamine

XII. Biosynthesis of Amino Acids

The detailed steps of the pathways for the biosynthesis of amino acids were mostly established by studying the properties of mutants requiring amino acids for growth. A mutant with a defect in an enzyme in the middle of a biosynthetic pathway will be able to accept later intermediates (if they can enter the cell) but intermediates earlier than the enzyme block will not be able to substitute for the end-product. In some cases the intermediate before the enzyme block will continue to be synthesized and may be excreted into the medium. In addition to *auxotrophic* mutants which require amino acids, it is often possible to isolate mutants which are resistant to growth inhibition by amino acid analogues. Mutants which are resistant to amino acid analogues may be altered in regulation of activity or in regulation of synthesis and some may be altered in permeability and exclude the analogue from the cell. *Regulatory* mutants have been important in understanding the various control systems which have evolved for biosynthetic pathways.

Figures 1 and 5 indicated some of the carbon compounds of the major catabolic pathways which provide the starting points for amino acid biosynthesis. Arginine and proline are synthesized via glutamate produced from α-oxoglutarate by one of the amination reactions. Lysine, methionine, threonine, and isoleucine are produced via aspartate from oxalacetate. Pentose-pathway intermediates are the precursors of the more complex amino acid side-chains and erythrose-4-phosphate with phosphoenolpyruvate provides the starting point for the biosynthesis of the aromatic amino acids, tryptophan, phenylalanine, and tyrosine. The biosynthetic pathways for the aromatic amino acids can be taken as an example of the types of regulation which may be found for branched pathways leading to more than one product. If the first enzyme were to be totally inhibited by one of the end-products this would prevent the synthesis of the other two end-products as well and growth would be inhibited. Two main types of controls have evolved. Among the enterobacteria it is normal to find multiple enzymes with the same catalytic functions at control points of the pathway. Figure 8 indicates this for the regulation of the first enzyme of the aromatic pathway in *Escherichia coli*. This step is carried out by three isoenzymes. One is regulated by feedback inhibition and repression of synthesis by phenylalanine, a second is regulated by feedback inhibition and repression by tyrosine, and the third is not subject to feedback inhibition and is controlled by repression of synthesis by tryptophan. In other microorganisms there is a single enzyme for this step in biosynthesis which is controlled by the end-products acting in a concerted way.

The last steps in the biosynthesis of tryptophan are specific for that amino acid. In *Escherichia coli* the genes for this group of enzymes are

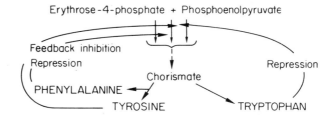

Figure 8. Outline of the regulation of synthesis of the aromatic amino acids in *Escherichia coli*

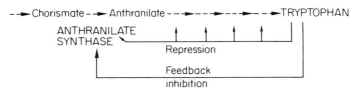

Figure 9. Regulation of tryptophan biosynthesis in *Escherichia coli*. The first enzyme anthranilate synthase is inhibited by tryptophan. The synthesis of all the enzymes of the pathway is repressed by tryptophan.

found closely linked together on the chromosome and in the wild type strain the addition of tryptophan to the growth medium results in the cessation of synthesis of all five enzymes of the pathway. The *trp* (tryptophan biosynthetic genes) form an operon under common control regulated by repression by tryptophan. The flow of intermediates through the pathway is regulated by feedback inhibition of the first enzyme, *anthranilate synthase.*

The clustering of biosynthetic genes in this way is uncommon in other groups of microorganisms. In other bacterial genera the tryptophan genes may be arranged in two or more different clusters and the regulation of gene expression may be different.

Although the arrangements of genes for biosynthetic pathways may not be the same in the various microbial groups and the details of the feedback inhibition and controls of enzyme synthesis may vary considerably the final result is the same. The rates of synthesis of all the amino acids are so regulated as to provide a balanced mixture of amino acids for protein synthesis. Only when mutations have been introduced is the balance upset and excess amino acids excreted from the cells.

XIII. Catabolism of Amino Acids

Many natural environments contain amino acids in sufficient quantities to support microbial growth and many species have evolved catabolic pathways for the fermentation of various amino acids. The clostridia include species producing extracellular proteases which hydrolyse proteins to their constituent amino acids and these are then fermented to provide energy and carbon and nitrogen compounds for growth. An unusual fermentation occurs in some species of clostridia which ferment two amino acids, oxidizing one and reducing the other. Among the pairs of amino acids which can take part in the mutual oxidation—reduction, first described by Stickland, are alanine which is oxidized and glycine which is reduced, both being converted into acetic acid. The overall reaction is the following

Stickland reaction

Alanine + 2 Glycine + 2 H_2O $\longrightarrow$

$$3 \text{ Acetic acid} + 3 \text{ NH}_3 + \text{CO}_2 \quad (23)$$

The aerobic bacteria and fungi break down amino acids by a variety of catabolic pathways. The amino acid biosynthetic pathways are essentially similar in all microorganisms although the regulatory systems show considerable diversity. The catabolic pathways on the other hand exhibit a wide variation in the degradation routes found in different organisms. Of the many pathways known, that for histidine utilization in *Klebsiella, Pseudomonas,* and *Bacillus* species has very interesting characteristics. The enzymes of the pathway are induced by the product of the first enzyme and not by the substrate histidine. It was mentioned earlier that this regulatory device could ensure that endogenous histidine would not trigger off the synthesis of enzymes which would destroy it. Synthesis of the histidine enzymes requires the presence of the inducer, urocanate, which is produced from histidine by the enzyme histidase which is present at low levels in the non-induced culture. In addition to the operon-specific induction it is essential for the expression of the genes for enzymes of the histidine utilization pathway that one of the activator systems should be effective. If alternative carbon compounds are not available the metabolic pool of cyclic AMP will allow the general catabolic gene activator protein to act and facilitate transcription. This can operate irrespective of the ammonia concentration. However, if the amount of an alternative carbon source is fairly high it may still be possible to express the histidine utilization genes. Glutamine synthetase is present in an active form during growth in nitrogen-limiting media, and as was explained earlier, can play a dual role as regulator protein as well as enzyme. When histidine is the only nitrogen source although glucose or other carbon compounds may be present the histidine utilization enzymes can be expressed with glutamine synthetase acting as the regulator protein.

XIV. Further Reading

Accounts of intermediary metabolism of microorganisms will be found in textbooks covering general aspects of metabolism as well as in textbooks of general microbiology. Many more texts are available on the metabolism of bacteria than of other groups of microorganisms. Recent advances in microbial genetics have included studies on genes and enzymes of metabolic pathways and on regulation of gene expression and some texts on microbial genetics include descriptions of intermediary metabolism. For detailed accounts of enzyme properties and the mechanisms of enzyme action an advanced textbook of biochemistry would provide the best source. The following titles are suggested for further reading:

Stanier, R. Y., Doudoroff, M., and Adelberg, E. A. (1971). *General Microbiology,* 3rd ed. London: Macmillan.
Sokatch, J. R. and Ferretti, J. J. (1976). *Basic Bacteriology and Genetics.* Chicago: Year Book Medical Publishers Inc.
Mandelstam, J. and McQuillen, K. (1973). 2nd ed. *Bacterial Growth,* Oxford: Blackwell.
Conn, E. E. and Stumpf, P. K. (1977). *Outlines of Biochemistry,* 4th ed. New York: John Wiley,
Lehninger, A. L. (1975). *Biochemistry,* 2nd ed. New York: Worth Publishers Inc.
Lewin, B. (1974). *Gene Expression-1.* London: John Wiley.
Maas, W. K. and McFall, E. (1964). Genetic Aspects of Metabolic Control. *Annual Review Microbiology,* 18, 95.
Cohen, G. N. (1965). Regulation of Enzyme Activity in Microorganisms. *Annual Review Microbiology,* 19, 105.
Datta, P. (1969). Regulation of branched biosynthetic pathways in bacteria. *Science* 165, 556.
Englesberg, E. and Wilcox, G. (1974). Regulation: Positive Control. *Annual Review Genetics,* 8, 219.
Goldberger, R. F. (1974). Autogenous control of gene expression. *Science,* 183, 810.

Classification of Microorganisms

P. H. A. SNEATH

Department of Microbiology,
Leicester University, Leicester

I. Introduction

A. The Importance of Classification

Classification is an activity essential to all scientific work. It would be impossible to make any generalizations about microorganisms and their role in nature if we could only refer to each strain by a different and arbitrary name. We must arrange microorganisms into groups that share common properties, so that we can talk about sets of strains that have properties in common, e.g. the typhoid bacillus, the foot-and-mouth virus, brewers' yeast *Saccharomyces cerevisiae*. Classification is the activity of grouping things together.

We know only a minority of the kinds of microbes that occur in nature. Lack of knowledge of properties and groupings of microbes is holdings up many areas of microbiology. Perhaps half the bacterial colonies on an agar plate inoculated with river water cannot be identified to species level even by experts. We know little more about most natural habitats. Classification of microorganisms is more difficult than that of higher organisms, so the methods used are here discussed at some length.

The present chapter is complementary to the one on identification (diagnosis) of microorganisms. Identification is distinguished from classification because it requires a previously-made classification to whose classes one can assign unknown organisms. Although each chapter is self-contained there are aspects that touch on both areas, so they are best read in conjunction with each other.

B. Some Definitions

Classification is the process of recognizing and describing groups of living organisms. It is part of the wider study of systematics, as shown below.

Systematics is the scientific study of the diversity of organisms and their relationships. All the following, (a) to (e), are included in systematics in this broad sense:

(a) *Classification* is the ordering of organisms into groups.
(b) *Taxonomy* is often used synonymously with classification, but it sometimes means the theory of classification. The *products* of the activity of classification, or of taxonomy, are often referred to as classifications or taxonomies (which mean essentially the same thing). *Numerical taxonomy* is the grouping of organisms on the basis of numerical methods: it is sometimes called *taxometrics* or *numerical classification*.
(c) *Identification* is the assignment of unidentified organisms to a particular class in a previously made classification.

(d) *Nomenclature* is the subject that deals with the correct international scientific names of organisms.
(e) *Phylogeny* is the study of the evolutionary history of organisms. The study of the genealogy — the branching pattern of the lineages — is called *cladistics*, and a *clade* is a monophyletic group. Phylogeny usually connotes the study of long-term evolutionary change, whereas short-term change is generally now studied under *population genetics*.

Relationship. The term relationship is one that is used in many different ways. We may distinguish two broad kinds of relationship:

(a) *Similarity*, or *resemblance*, in the observed properties of organisms. This refers to attributes that organisms possess, without consideration of how those attributes arose. Similarity is expressed as proportions of existing similarities and differences, (including genetic ones) and is called *phenetic relationship*.
(b) Relationship by ancestry, or evolutionary relationship. This refers to the phylogeny of organisms, and not necessarily to their present attributes. Evolutionary relationship is expressed as the time to the common ancestor, or the amount of change that has occurred in some evolutionary lineage. It is not expressed as a proportion of similar attributes, although it may sometimes be *deduced* from similarity *on the assumption* that evolution has proceeded in some orderly and defined way.

Taxonomic categories and rank. In order to marshal the diversity of organisms into some workable scheme, the taxonomist sets up a taxonomic *hierarchy*, in which a taxonomic group (*taxon*, plural *taxa*) occupies a position in a nested scheme, for example:

```
Class     Bacteria
          Order 1        Pseudomonadales
              Family  Pseudomonadaceae
                      Genus 1 Pseudomonas
                              Species 1  Pseudomonas aeruginosa
                              Species 2  Pseudomonas fluorescens
                              etc.
                      Genus 2 Xanthomonas
                              Species 1  Xanthomonas campestris
                              etc.
          Order 2  Spirochaetales
              Family  Spirochaetaceae
                      Genus   Spirochaeta
                      etc.
```

The hierarchy is intended to express the fact that species of a genus are very similar to one another — more similar than they are to species of other genera. Likewise, genera of one family are more similar than those from different families, and so on. The most diverse groups are the highest (*Class*). The names Class, Order, etc., are taxonomic *categories*, and they possess a relative position in the hierarchy called taxonomic *rank*. The Class has the highest rank in the example above.

The species category. The basic category is the *species*, composed of individuals (in microbiology these are usually isolates, clones, or strains). Rather uncommonly microbial species are divided into *subspecies*, but many minor classes are often recognized as *infra-subspecific forms* that usually refer to some minor variation pattern (defined, for example by serology). The term species, however, has been used in so many senses that unless it is further qualified it becomes a source of great confusion. It is useful to distinguish three terms:

(a) *Taxospecies*. A group of individuals of high mutual similarity that are separated by a phenetic gap from nearby taxospecies. We shall see that taxospecies emerge as distinct *clusters* in numerical taxonomic analyses.
(b) *Genospecies*. A group of individuals capable of exchanging genetic material (either by hybridization or by gene transfer, transduction, transformation, conjugation, etc.). Members of different taxo-species can sometimes hybridize (so they are a single genospecies), or there may be barriers to gene exchange within a taxospecies so that it contains many genospecies (e.g. races of *Paramecium aurelia*).
(c) *Nomenspecies*. A taxonomic group that carries a binominal name, whatever its status in other respects. Thus it is not clear whether *Xanthomonas hederae* is a variant of *X. campestris*, or a taxospecies or a genospecies, so for the present it can be treated as a nomenspecies.

C. Natural Classifications

There have been many attempts to place classifications on a rigorous philosophical basis, reaching back to the ancient Greek philosopher Aristotle. The advent of scientific classification must be credited to Linnaeus, although he made no great contributions to philosophy or theory. However, since his time it has been increasingly realized that an entirely logical basis for classification is difficult to achieve. Micro-organisms pose yet greater problems than higher organisms. This is because (a) the study of their evolution is exceedingly difficult, so that a phylogenetic basis for classification is scarcely attainable; (b) they have many different and unexpected systems of gene transfer, so that genetic bases are seldom firm, and (c) many of them have simple morphological structure so that the traditional methods of taxonomy of higher organisms (which rests heavily on the ability of the eye and mind to integrate complex structural detail) are not readily applied. The earliest work was based on a few readily observable properties (like cell shape) which were given undue significance: we now realize that a much wider range of properties must be studied.

The microbial world is very diverse: some groups, e.g. algae and protozoa, are classified largely by the methods used in higher organisms, based mostly on morphology. Others, like microfungi, are commonly studied by a combination of morphological and chemical methods, and in some (e.g. yeasts) the work is mainly chemical or physiological, as it is with bacteria. Viruses are a special case, and systematic methods are only now being devised for them. There is heavy reliance on the fine structure and symmetry of the virus particle, backed up by serology and tests of pathogenicity to whole organisms or tissue cultures. In order to provide a common conceptual framework for microbial classification, the present chapter uses the logical outline that has developed from numerical taxonomy. The calculations are, in practice, done by computers, but the basic principles and arithmetic are simple to describe.

The naturalness of a classification can be measured by the amount of common information in its groupings. Hence, the group Birds implies many common properties found in all birds, whereas the group Pests implies only that pests are noxious in some way. They are often distinguished as *general purpose* and *special purpose* groupings respectively. They are formed in different ways, as shown below.

	Organisms	Attributes		
		X	Y	Z
Group (a)	1	+	−	−
	2	+	−	−
	3	+	−	−
Group (b)	4	+	+	−
	5	+	−	+
	6	−	+	+

A group formed by choosing a single defining character, as in (a) above, is called *monothetic* (from the Greek, 'one arrangement'). Those formed by the greatest number of shared characters, are called *polythetic* (many arrangements), as in (b) above. Polythetic groups permit of a limited number of exceptional characters, and this is important because there always seem to be some exceptions to any character used to define a taxonomic group. Monothetic groupings are only commonly used for infrasubspecific variants like patterns of resistance to bacteriophages.

D. Logical Steps in Classification

A résumé of the logical steps in classification is given below:

(a) Collection of data. The *organisms* to be classified are collected, and they are examined for numerous properties (*taxonomic characters*). The results are tabulated in a table of organisms versus characters.

(b) The results are *coded* and *scaled* into a suitable form for numerical analysis. Commonly in microbiology the characters are presence—absence ones, but quantitative characters can also be used.

(c) The *similarity*, or *resemblance*, between each organism and every other is then calculated, to give a table of *Similarity Coefficients* between all pairs of organisms.

(d) The similarities are now analysed for *taxonomic structure*, which is intended to discover the polythetic groups that are present. The organisms are then arranged into these groups (*phenons*) which are broadly equated with taxa.

(e) The results now require evaluation, both by statistical methods and by comparison with other kinds of available information. This leads to the formal description of the taxa (either the establishment of new taxa or the revision of existing classifications).

(f) The properties of the phenons can now be tabulated for further study or generalizations. One important aspect is to set up *identification systems* by choosing the best characters (*diagnostic characters*) by which to identify new specimens.

Planning a taxonomic study requires a clear grasp of these logical steps, whether it is to be done by numerical or other methods. A properly designed scheme of work, to obtain complete comparative data under well-standardized conditions, can greatly assist in mastering what may be very difficult problems.

II. Data Used in Classification

A. Organisms and Their Characters

The data needed for classification consist of the properties of the organisms that are to be studied. These data should be as complete as possible and of sufficient quantity. Further, they must be of high quality, because observational errors are an important cause of poor taxonomic results, and microbiological data are very susceptible to slight differences in techniques. Data from earlier publications are commonly not of the quality required, so that frequently the microorganisms will have to be studied anew.

B. Operational Taxonomic Units

One may wish to classify a great variety of entities — species, strains, specimens, local populations, genera — for which no common term is available. The entities to be classified are therefore called Operational Taxonomic Units (OTUs). Each OTU can be represented as a column in a table of t OTUs versus n characters. The OTUs should be chosen so as to represent the known diversity of the organisms that are being studied. The exact composition of the set of OTUs usually makes little difference to the end result, although it is obviously important that adequate samples are available. For most purposes OTUs are considered to have a single value for any given character, but if there is much variation several representatives should be chosen so as to cover this variation.

It is best to throw the net fairly widely in choosing OTUs. Computer programs, however, are especially limited by the number of OTUs they can handle (usually a few hundred) so the choice of OTUs must take this into account. It is particularly important to obtain authentic representatives of the species that have been previously described, ideally type cultures. Sources of these can be found in publications listed in the section on further reading. A small proportion of strains should be put through the analysis in duplicate in order to keep a check on test errors (see Section IV. E).

C. Characters, Character Complexes, and Unit Characters

A *character* is defined as any property that can vary between OTUs, and the possible values that it can assume are called *character states*. Thus 'fermentation of glucose' is a character, with the states 'positive' and 'negative'. Similarly, 'length of spore' is a character, and '1.5 micrometres' is one of its states.

When organisms are compared it is obviously important to compare like with like, e.g. length of spore in this OTU with length in that OTU. The recognition that two structures are spores, and that a particular dimension is the length (rather than the breadth) is called the determination of *homology*. Homology poses many problems in complex organisms but, fortunately, in microbial taxonomy these are seldom serious. There are, however difficulties in being sure of the homologies of the organs and organelles in some groups, particularly fungi and protozoa.

A single character treated as independent from others is called a *unit character*. Patterns of reaction are *character complexes*, e.g. the end-products of metabolism permit recognition of many unit characters.

D. Kinds of Characters

Characters are said to be *taxonomic characters* if they convey the sort of information that is useful in taxonomy. It is difficult to give a watertight definition of this, but clearly the serial number on a culture label is not a taxonomic character of the culture. Such characters are therefore inadmissible. Other inadmissible characters are obvious duplicates of those previously recorded, or those that are logically completely correlated. Hence, for a circular structure it would be inadmissible to score both radius and diameter. However, characters that are observed as being correlated should be all included if there is no logical reason to think they are inevitable expressions of one underlying property.

Characters should be chosen to cover the broadest range of properties of taxonomic interest. The usual aim is to represent the entire phenome (the phenotype and genotype), and data from morphology, chemistry, physiology, and the other aspects should be considered.

In microorganisms the most important classes of characters are as follows:

Morphological	e.g. number of flagella; shape of spores.
Physiological	e.g. ability to grow anaerobically.
Biochemical	e.g. oxidase activity; acid production from galactose.
Chemical constituents	e.g. presence of lysine in the cell wall.
Cultural	e.g. visual appearance of colonies on a defined medium.
Nutritional	e.g. ability to grow on acetate as sole carbon-energy source; requirement for thiamine.
Drug sensitivities	e.g. sensitivity to benzyl penicillin.
Serological	e.g. agglutination by an antiserum to a reference culture; presence of a specific precipitin band in a gel-diffusion experiment.
Genetic	e.g. percent GC in DNA; ability to be transduced by a given bacteriophage preparation; extent of pairing with a reference sample of DNA.

It should be noted that serological and genetic data are of two kinds. In one, the reaction indicates the presence or absence of some defined property (e.g. an antigen, or a phage receptor). These can be scored like any other character. In the other, the cross-reaction between organisms is measured, and this is analogous to a similarity value, which is discussed further in Section III.C.

There are some other classes of characters, of less general use than those cited, but which are occasionally employed. They include chemicals detected by liquid or gas chromatographic analyses of whole cells or fermentation products, pattern of proteins in gel electrophoresis, and so on. The amino acid sequences of the same protein in various organisms can also be used as a set of characters.

E. Numbers of Characters

Clearly we could not get a very informative similarity value from a single character. Although it is not possible to say that characters form a random population, yet in most biological applications they behave as quasi-random variables. This leads to the conclusion that the sampling error of similarity coefficients will roughly follow the well-known statistical square-root law, that is the standard error of estimate will be roughly proportional to $1/\sqrt{n}$. This is found to be sufficiently true in practice. In order to obtain enough discriminative power, at least 50 and preferably several hundred characters are desirable; with higher numbers the gain in accuracy falls off disproportionately to the labour of obtaining the data.

F. Technical Methods

Great strides are now being made in automation of microbiological tests. Numerous micromethods are also available. Many of these are used with semi-automatic replication of inocula. Typical techniques include replica plating of numerous cultures from a master plate on to a series of plates containing different carbon-energy sources, or the inoculation of one culture into an array of different fermentable carbohydrates.

Microbiological tests are subject to much experimental error due to difficulties in standardizing the test conditions. For example, there may be unsuspected differences in temperature on different shelves of an incubator. Even within one laboratory it is difficult to get better than 97—98 per cent agreement in test results between duplicate subcultures. Results from different laboratories may have only 80—90 per cent reproducibility. Some tests are particularly prone to such difficulties, but the problem cannot be circumvented by retaining only a few highly reproducible tests. If this is done the sampling error may become too large because n is so small. One can only afford to exclude a few of the least reproducible tests. The solution, then, must lie in good test standardization. The effect of poor experimental data is discussed in Section IV.E.

III. Data Processing

A. Steps in Data Processing for Taxonomy

The steps in data processing follow the logical steps listed in section I.D, which may be recapitulated as follows:

(a) Observation of the character states of the OTUs.

(b) Coding and scaling of the data for computation.
(c) Calculation of a table of similarity coefficients between the OTUs.
(d) Examination of taxonomic structure and grouping of OTUs into phenons.

These are then followed by evaluation of the results and drawing appropriate generalizations about the phenons.

These steps will be illustrated as diagrams with an example on a selection of bacteria belonging to the genus *Serratia*.

B. Coding and Scaling of Characters

The data are accumulated as a table of OTUs (t in number) versus characters (n in number). They must be converted from the form in which they are recorded (often in words) into a form suitable for computation. Figure 1 shows a page from a laboratory record book recording test results on strains. If information is missing, because a test

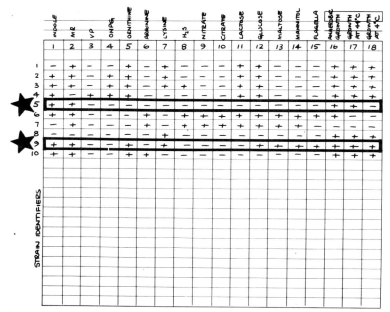

	INDOLE	MR	VP	ONPG	ORNITHINE	ARGININE	LYSINE	H₂S	NITRATE	CITRATE	LACTOSE	GLUCOSE	MALTOSE	MANNITOL	FLAGELLA	ANAEROBIC GROWTH	GROWTH AT 44°C	GROWTH AT 4°C
	1	2	3	4	5	6	7	8	9	10	11	12	13	14	15	16	17	18
1	−	+	−	−	+	−	+	−	−	−	+	+	−	−	−	+	+	+
2	+	+	−	+	+	−	+	−	−	−	+	+	−	−	−	+	+	+
3	+	+	−	+	+	−	+	+	−	−	+	+	−	−	−	+	+	+
4	+	−	+	+	+	−	−	−	−	−	+	+	−	−	−	+	+	+
5	+	+	−	−	−	−	−	−	−	−	−	−	−	−	−	+	−	−
6	+	+	−	−	−	+	−	+	+	+	+	+	+	+	+	+	−	−
7	−	+	−	−	−	+	−	+	+	+	+	+	+	+	−	−	−	−
8	−	−	−	−	+	−	−	−	−	−	−	−	−	−	−	+	+	+
9	+	+	−	−	+	−	+	−	−	−	−	+	+	+	+	+	+	+
10	+	+	−	−	+	+	−	−	−	−	−	−	−	−	−	+	+	+

STRAIN IDENTIFIERS

Figure 1. A page of a laboratory record book with some of the test results upon strains of *Serratia* and its allies. The characters (tests) and OTUs (strains) have been numbered sequentially. The stars mark two strains which are being compared numerically, and the corresponding rows of test results are boxed in with heavy lines. All the tests in this figure are recorded as + or −

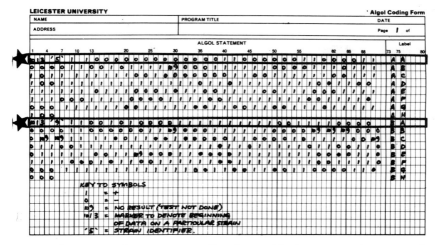

Figure 2. A page of a computer coding-form with the tests on the two strains coded ready for punching on to cards. The form also has entries for the strain numbers, and symbols to indicate the beginning of data for each strain. The letters at the right are to identify the separate cards, because eight cards are needed per strain (there are in all 244 presence-absence characters). The symbol for 'no comparison' is 10^9

has not been performed, or because it could not be recorded (e.g. spore shape in a non-sporing organism) the symbol NC is entered standing for 'no comparison is to be made with this entry'. Some suitable computer code is used for NC, e.g. 10^9, as in Figure 2, which shows the coded data for two strains.

The coded data table is called the *data matrix* or $n \times t$ matrix. Its entries are symbolized as X. X_{ij} is the value of the ith character in OTU j. Characters can be quantitative, binary, or qualitative multistate (as shown below, with some common coding schemes).

1. Coding of characters

(a) Quantitative characters, e.g. length of spore in micrometers.

Character	States		
Spore length	0.9	1.1	2.3

The actual measurements can be used, or one can group them into a small number of classes that can be distinguished with confidence, e.g.:

Character	States		
Spore length	0–0.99	1.0–1.99	2.0–2.99

They could then be coded 0, 1, 2, for example.

(b) Binary characters or qualitative two-state characters. These are often called presence–absence characters, and have only two states. They may be +, – or A, B, but are usually coded as 1 and 0, e.g.:

Character	State	
Lipid inclusions	Present	Absent
	1	0

It is wise to score only definite positive values as 1, and doubtful results as 0. With some characters the choice of 1 and 0 is arbitrary, and does not denote presence or absence (e.g. mating type in fungi).

(c) Qualitative multistate characters are those where there are more than two mutually exclusive alternatives, which however cannot be arranged along a quantitative scale. States are either different or the same, and the degree of difference is identical for any pair. Hence the symbols *, † ‡ would be states of this kind. An example is:

Character	States		
Colony colour	Red	Yellow	Blue

These can be coded in several forms, which depend essentially on the computer program used, e.g.:

Character	States		
Colony colour	1	2	3

or else

Character	States		
Colony colour	A	B	C

Sometimes it is desired to convert quantitative characters into binary ones. This involves breaking the quantitative characters into as many classes as can be confidently distinguished and allocating binary characters to the classes. Two methods are useful:

(i) Additive coding. Differences between states retain magnitude. An example is:

Character	Binary characters		
Amount of indole	1	2	3
Less than 1 (undetectable or –)	0	0	0
1.0–1.9 (+)	1	0	0
2.0–2.9 (+ +)	1	1	0
3.0–3.9 (+ + +)	1	1	1

(ii) Non-additive coding. We may wish to ignore the magnitude of any differences, and consequently can code a character differently, e.g.:

Character	Binary characters		
Amount of indole	1	2	3
Less than 1 (undetectable or –)	0	NC	NC
1.0–1.9 (+)	1	1	0
2.0–2.9 (+ +)	1	NC	1
3.0–3.9 (+ + +)	1	NC	NC

2. Scaling and weighting of characters

It is usual to give equal weight to each character, largely because of the difficulty in deciding logically on what weights should be given. It should be noted that different weights may properly be given to different characters in an identification scheme, but this is not what is being made at this juncture. Some scaling and weighting is introduced by coding characters as described above, but more explicit changes must be considered. The question of whether characters should have unit weight depends largely on what is meant by a character and how much information it carries. If a character is broken down into unit characters as mentioned above it is reasonable to give unit weight to each unit character. For binary and qualitative characters the problem is not acute. The major difficulty is with quantitative ones, because the scale of measurement affects the results (e.g. whether a structure is measured in μm, mm, or cm). From a logical viewpoint one should perhaps convert all quantitative characters into a coded form of several classes and then give equal weight to a difference of one class. In practice the averaging that occurs in calculating a similarity coefficient has a pronounced equalizing effect on weights. Two convenient scaling methods are commonly used, ranging and standardization.

(a) *Ranging of values.* In this the smallest value of character i in any OTU (i.e. $X_{i\min}$) is subtracted from each X_i value, and divided by the observed range. The transformed value, X' is given as

$$X'_{ij} = \frac{(X_{ij} - X_{i\min})}{(X_{i\max} - X_{i\min})}$$

For example, values of 3.2, 4.0, 7.2, 9.6 would be ranged by subtracting 3.2, and dividing by $(9.6 - 3.2)$, giving 0, 0.125, 0.625, 1.0. $X_{i\min}$ always becomes 0 and $X_{i\max}$ becomes 1.

(b) *Standardization of values.* By this the values of i are converted to a new scale with mean of zero and standard deviation of 1. The mean, $\bar{X}_i$, and standard deviation, s_i, of the X_i values over the t OTUs are first calculated:

$$\bar{X}_i = \frac{1}{t} \Sigma X_{ij}$$

$$s_i = \left[\frac{1}{t-1} \Sigma (\bar{X}_i - X_{ij})^2 \right]^{\frac{1}{2}}$$

Then each X_{ij} value is standardized as

$$X'_{ij} = (X_{ij} - \bar{X}_i)/s_i$$

For the example in (a) above we get a mean of $24/4 = 6.0$, and

$s_i = \sqrt{[\frac{1}{3} 26.24]} = 2.9575$. Then the standardized values become $(3.2 - 6)/2.9575 = -0.9467$, and so on. The new values are $-0.9467, -0.6762, 0.4057$, and 1.2172.

In many applications standardization gives very similar results to those obtained using the unstandardized data. If size effects are pronounced it is probably better to use unstandardized character states and use a shape coefficient (discussed in Section III.C). Otherwise standardization (which does not remove size effects) may be preferable. Standardization of binary characters is usually undesirable.

Sometimes the states of a character are recorded for only a small proportion of the OTUs. Although a few such gaps are tolerable, it is as well to omit characters that are not scored for at least half the OTUs. Such characters are said to have *low relevance*.

3. Preparation of data for computing

It is important to avoid unnecessary handling of data, because this wastes effort and introduces errors. If possible the form of data for the available computer program should be found out early on. The worker should find out if scaling or standardizing can be done by the computer, or whether he must do it himself on a desk machine. A sample of the data input is valuable. This usually requires information on the number of OTUs, characters, etc., and options for similarity coefficients and cluster methods. The character values after coding should then be copied onto the computer coding sheets for punching, as an $n \times t$ data matrix. This should be thoroughly checked for missing or reduplicated values (coloured lines can be placed after every fifth column, for example), and then the cards punched. If the data is too big, two computer runs should be made, and then a third one using a few OTUs from each cluster so as to match up the phenons in the two runs.

C. Coefficients of Resemblance or Similarity

1. Phenetic hyperspaces

It is convenient to think of OTUs as distributed in a phenetic hyperspace of numerous dimensions, one for each character. One can then express the resemblance between a pair of OTUs in one of two ways: as *similarity* (e.g. percentage of characters in which they agree), or as *dissimilarity* (e.g. the percentage in which they disagree). Dissimilarity is particularly useful because it can be expressed as the *taxonomic distance* between the positions of the OTUs in the phenetic hyperspace. This permits the construction of *taxonomic maps* or *taxonomic models* in which similar organisms form *clusters* or *phenons*. One aim of numerical taxonomy is to find these clusters or phenons, but first the similarities or dissimilarities must be calculated. The principle is simple; the two OTUs below are scored for five characters.

Characters	1	2	3	4	5
OTU *j*	+	−	−	+	−
OTU *k*	+	−	+	−	−

they agree on the first, second, and fifth, so their similarity is 3/5 = 60 per cent. Their dissimilarity is therefore 40 per cent.

2. *Some commonly-used coefficients of resemblance*

These coefficients are used with different kinds of characters, as noted below. The character states of a pair of OTUs *j* and *k* are compared over the *n* characters. If for any character an NC symbol is given for one or both OTU's, that character is not considered, and 1 is subtracted from *n*.

For qualitative characters the numbers of the four possible different combinations of matches (agreements) and mismatches (disagreements) of the character states of two OTUs can be shown as a 2 x 2 table. The numbers of the four combinations are *a*, *b*, *c*, and *d* and the total $a + b + c + d = n$.

		OTU *j*		
		+ or 1	− or 0	Totals
OTU *k*	+ or 1	*a*	*b*	*a + b*
	− or 0	*c*	*d*	*c + d*
	Totals	*a + c*	*b + d*	Grand total $a + b + c + d = n$

A numerical example, with 20 characters is

		OTU *j*		
		+ or 1	− or 0	Totals
OTU *k*	+ or 1	12	3	15
	− or 0	1	4	5
	Totals	13	7	20

The most commonly used coefficients are given below.

The Simple Matching Coefficient S_{SM}:

$$S_{SM} = (a + d)/n$$

This is related to taxonomic distance (see below). The example gives $S_{SM} = (12 + 4)/20 = 0.80$

The Jaccard Coefficient S_J:

$$S_J = a/(a + b + c)$$

This takes no account of negative matches (the number *d*, the number of characters where both OTUs are negative). It therefore requires decisions on which state is the negative state (e.g. whether antibiotic resistance or antibiotic sensitivity is positive). The example gives $S_J = 12/16 = 0.75$.

For quantitative characters the most commonly-used coefficients are those listed below.

Taxonomic Distance, d:

$$d_{jk} = \sqrt{\left[\sum^{n} (X_{ij} - X_{ik})^2 / n \right]}$$

where the character state of OTU j for character i is X_{ij} and that of OTU k is X_{ik}. The symbol Σ^n indicates the sum over n characters, $1, 2, 3, \ldots, i, \ldots, n$. The value of d is the distance in a phenetic space divided by $\sqrt{n}$. The distance with 1, 0, characters is related to S_{SM} as follows:

$$d = \sqrt{(1 - S_{SM})}$$

The Correlation Coefficient, r:

$$r_{jk} = \frac{\sum^{n} [(X_{ij} - \bar{X}_j)(X_{ik} - \bar{X}_k)]}{\sqrt{\left\{ \left[\sum^{n} (X_{ij} - \bar{X}_j)^2 \right] \left[\sum^{n} (X_{ik} - \bar{X}_k)^2 \right] \right\}}}$$

Here the symbols $\bar{X}_j$ and $\bar{X}_k$ are the average values of X over all n characters for OTUs j and k respectively.

For mixed qualitative and quantitative characters the most useful coefficient is a form of the Gower Coefficient, S_G:

$$S_G = \frac{\sum^{n} s_{ijk}}{n}$$

where s_{ijk} is a score on character i for the comparison of OTUs j and k such that $s_{ijk} = 1 - (|X_{ij} - X_{ik}|/R_i)$. Here R_i is the range of character i over all the t OTUs. If all characters are qualitative (1, 0) then S_G becomes S_{SM}. This coefficient excludes characters that have invariant character states for all OTUs.

3. Size, Shape, Vigour, and Pattern

We know that two organisms can differ in shape but be similar in size (for example rats and guinea-pigs) or differ in size but be similar in shape (e.g. rats and mice). Most resemblance coefficients combine an element of both size difference and shape difference. In many applications it is shape difference that is chiefly relevant to taxonomy.

When the organisms under study differ greatly in size, therefore, a shape coefficient should be used. The correlation coefficient is an almost pure shape coefficient. An alternative is to scale all measurements by some standard such as maximum length before calculating the resemblance.

For qualitative characters an analogue of shape difference can be obtained as follows, called the Pattern Difference, D_P:

$$D_P = 2\sqrt{b\ c}/n$$

To convert into a similarity one can use Pattern Similarity, $S_P = (1 - D_P)$. For the 2×2 table used earlier, $D_P = 2\sqrt{3}/20 = 0.173$. This coefficient is useful for situations where some strains grow much more slowly, or are metabolically much less active, than others (provided that not too many characters of an OTU are positive or negative, when D_P becomes unreliable).

4. Resemblance in other guises

The amount of serological cross-reaction, or the percentage of DNA pairing, between two organisms is an analogue of a similarity coefficient. We can think of the former as a similarity measure between the antigens which is calculated, so to speak, by the immunological mechanism of the animal in which the antiserum is prepared. In the latter the physico-chemical properties of the DNA molecules reflect the similarity in the nucleotide sequences.

There are some problems in producing a similarity matrix by such methods. The relative contributions of different genes or gene-products may not be the same in different situations. The cross-reactions may not be the same for OTU j compared with OTU k as for k compared with j: the two values may need to be averaged. Also, it may be too laborious to obtain a complete resemblance matrix between all OTUs. This leads to the use of a few reference OTUs against which to compare the rest. If each phenon is represented by a reference OTU the strategy works well, but it is difficult to ensure this before doing all comparisons (which is impracticable). Therefore, the major phenons are often first delineated by a numerical taxonomic study, and the other techniques are used to amplify and confirm the findings.

D. The Discovery of Taxonomic Structure

Taxonomic structure refers mainly to the separate groups of organisms that are present, but it can relate to other patterns such as geographic trends. The search for structure starts with a *Similarity Matrix* (*S* matrix) of the similarities (or dissimilarities) between all pairs of OTUs. This matrix is square, of size $t \times t$, but (except for some

1. Clustering of a set of OTU's uses the similarities between all pairs of the OTU's, e.g.

(a)

1	1.0					
2	.90	1.0				
3	.80	.75	1.0			
4	.29	.30	.28	1.0		
5	.42	.40	.44	.31	1.0	
6	.39	.41	.43	.33	.70	1.0
	1	2	3	4	5	6

OTU-NAMES MATRIX:

SIMILARITY MATRIX 3 SIG. FIG. OUTPUT:

(b)

1	5	1.00																	
34	16	.972	1.00																
19	293	.975	.958	1.00															
8	207	.971	.959	.962	1.00														
7	60	.950	.847	.912	.889	1.00													
14	261	.922	.914	.925	.906	.943	1.00												
35	57	.933	.921	.916	.905	.959	.967	1.00											
33	12	.900	.895	.899	.897	.913	.934	.933	1.00										
2	9	.801	.797	.793	.792	.780	.786	.725	.813	1.00									
9	221	.806	.793	.794	.790	.802	.794	.808	.813	.966	1.00								
21	319	.780	.792	.792	.772	.780	.784	.782	.795	.957	.954	1.00							
16	283	.791	.803	.794	.774	.786	.794	.792	.813	.957	.946	.950	1.00						
13	239	.797	.800	.784	.772	.793	.801	.807	.805	.944	.942	.941	.946	1.00					
15	278	.760	.772	.772	.763	.759	.793	.779	.799	.940	.921	.933	.921	.904	1.00				
28	507	.731	.781	.777	.765	.753	.778	.775	.764	.945	.926	.954	.942	.925	.963	1.00			
27	503	.757	.753	.755	.744	.723	.740	.745	.754	.855	.854	.875	.880	.858	.895	.871	1.00		
12	238	.723	.719	.716	.716	.737	.765	.758	.734	.838	.839	.833	.816	.837	.851	.847	.822	1.00	
18	289	.747	.751	.743	.743	.730	.755	.762	.753	.808	.813	.807	.792	.815	.816	.808	.791	.896	1.00
24	392	.744	.731	.739	.731	.736	.760	.763	.742	.876	.834	.820	.805	.824	.822	.817	.808	.930	.942
20	299	.760	.763	.764	.764	.752	.777	.778	.767	.842	.834	.828	.813	.824	.837	.834	.870	.894	.946
22	325	.736	.731	.739	.731	.744	.752	.754	.742	.835	.846	.837	.813	.821	.830	.831	.796	.880	.936
26	410	.725	.714	.713	.702	.715	.773	.736	.704	.813	.805	.799	.788	.788	.796	.806	.792	.851	.900
23	329	.696	.692	.699	.701	.697	.710	.706	.736	.777	.775	.769	.746	.786	.774	.767	.757	.846	.878
25	517	.716	.703	.716	.692	.700	.700	.714	.694	.765	.754	.752	.746	.752	.748	.758	.770	.839	.868
3	22	.715	.711	.727	.728	.683	.700	.700	.701	.719	.727	.742	.702	.733	.725	.736	.739	.731	.742
6	39	.714	.710	.717	.715	.690	.707	.707	.709	.718	.726	.741	.701	.732	.725	.734	.733	.740	.753
5	37	.720	.715	.724	.717	.697	.713	.714	.715	.725	.733	.734	.708	.739	.722	.733	.731	.737	.747
10	224	.712	.701	.715	.717	.699	.705	.705	.707	.716	.724	.726	.708	.730	.722	.733	.723	.737	.747
11	231	.707	.702	.710	.704	.643	.691	.692	.693	.702	.711	.725	.696	.717	.710	.719	.718	.724	.737
4	36	.695	.691	.699	.693	.690	.689	.689	.690	.708	.724	.730	.700	.722	.722	.733	.719	.745	.755
36	513	.711	.707	.706	.706	.671	.695	.696	.697	.712	.719	.712	.702	.726	.718	.728	.710	.715	.729
17	288	.675	.671	.678	.672	.650	.668	.668	.669	.709	.704	.714	.696	.702	.689	.704	.724	.721	.716
31	C 1	.631	.627	.633	.607	.653	.620	.629	.633	.731	.718	.720	.716	.707	.699	.710	.737	.718	.703
32	A 1	.636	.624	.655	.632	.651	.653	.642	.675	.664	.668	.653	.650	.674	.664	.668	.688	.692	.671
29	E20	.607	.603	.605	.584	.638	.630	.617	.627	.668	.653	.698	.682	.646	.650	.692	.647	.686	.646
30	E29	.436	.423	.437	.413	.471	.434	.444	.438	.568	.568	.573	.552	.548	.556	.568	.593	.606	.586
	OTU:	1	34	19	8	7	14	35	33	2	9	21	16	13	15	28	27	12	18

24	392	1.00																	
20	299	.917	1.00																
22	325	.934	.954	1.00															
26	410	.888	.921	.905	1.00														
23	329	.871	.887	.900	.855	1.00													
25	517	.866	.851	.851	.864	.863	1.00												
3	22	.755	.784	.768	.730	.775	.733	1.00											
6	39	.763	.796	.771	.742	.783	.745	.979	1.00										
5	37	.760	.798	.781	.744	.786	.738	.963	.971	1.00									
10	224	.760	.798	.791	.736	.780	.738	.963	.975	.975	1.00								
11	231	.748	.793	.763	.734	.760	.729	.959	.975	.971	.963	1.00							
4	36	.769	.798	.791	.769	.788	.730	.942	.946	.959	.959	.947	1.00						
36	513	.739	.789	.752	.723	.754	.712	.926	.942	.955	.963	.942	.938	1.00					
17	288	.736	.778	.753	.770	.777	.726	.890	.950	.892	.884	.875	.880	.879	1.00				
31	C 1	.708	.704	.692	.742	.674	.728	.647	.658	.569	.561	.660	.669	.656	.686	1.00			
32	A 1	.702	.667	.660	.658	.675	.577	.726	.714	.719	.719	.715	.719	.722	.695	.796	1.00		
29	E20	.676	.672	.656	.705	.633	.720	.628	.643	.642	.642	.636	.650	.636	.654	.731	.676	1.00	
30	E29	.579	.596	.583	.646	.559	.668	.510	.521	.517	.508	.506	.517	.494	.577	.717	.571	.730	1.00
	OTU:	24	20	22	26	23	25	3	6	5	10	11	4	36	17	31	32	29	30

Figure 3. (a) A small similarity matrix to illustrate the clustering method shown in Figures 4 and 5. (b) The full similarity matrix produced by the computer on 36 strains of *Serratia* and allied bateria. This is a sorted matrix, with the strains arranged in the order given by the cluster analysis illustrated by Figure 6. Figure 8 is this matrix in shaded form.The entries are values of S_{SM}. The phenons are represented by triangles of high values. At the left and along the bottom are the serial numbers of the 36 OTUs. On the extreme left are also the strain catalogue numbers in the culture collection from which they were chosen

```
Similarity        OTU's
                  1 2 3 5 6 4
   1.0            | | | | | |
                  | | | | | |
   0.9            1-2 | | | |
                  1-2 | | | |
   0.8            1-2-3 | | |
                  1-2-3 | | |
   0.7            1-2-3 5-6 |
                  1-2-3 5-6 |
   0.6            1-2-3 5-6 |
                  1-2-3 5-6 |
   0.5            1-2-3 5-6 |
                  1-2-3-5-6 |
   0.4            1-2-3-5-6 |
                  1-2-3-5-6-4
   0.3            1-2-3-5-6-4
```

Figure 4. Single Linkage clustering. One simple way of clustering is to list the links between OTUs in falling order of similarity as shown. As a start the closest OTU's (here OTU's 1 and 2 with 0.9 similarity) are combined into a *group* or *cluster*. The two closest groups now join (single OTUs are treated as single-membered groups) here (1 + 2) and 3, and similarities involving the new groups are calculated. The process continues until all the OTUs have joined to form one group. The last join is of OTU 4 to all the rest

exceptions noted above) since the similarity of *j* to *k* is the same as that of *k* to *j*, only one half is ordinarily filled in. An example of a small matrix is shown in Figure 3 to illustrate the clustering in Figures 4 and 5. Before the search for structure the matrix will have the OTUs in the original order, perhaps a haphazard one, but afterwards they can be reordered to produce a *sorted matrix* where similar OTUs have been arranged together to form phenetic groups.

1. Cluster analysis

There are a number of ways of finding the compact clusters in the phenetic hyperspace, collectively called *cluster analysis*. One of the simplest is called the Single Linkage method which is illustrated in the example in Figures 4 and 5. In this the OTUs that are mutually most similar are joined into pairs, and then the OTUs most similar to these pairs are added on at successively lower similarity levels, until all the OTUs have been joined together.

With Single Linkage the link between a candidate for joining a group and the existing group members is formed at the highest similarity level between the candidate and the existing members. This method can therefore find long straggly clusters if they are present, because only small distances may separate successive OTUs in a chain. Rather more laborious to calculate, but somewhat more satisfactory, is Average Linkage, where the average of similarities between a candidate and existing cluster members is used for linkage. A common type is UPGMA (Unweighted Pair Group Method with Averages, Figure 6). Average Linkage gives more compact clusters than Single Linkage.

2. Phenograms

The results of cluster analysis can be expressed as a tree-like diagram (*dendrogram*) which because it is based on phenetic evidence is distinguished from other tree diagrams as a *phenogram*. Compact

OTUs

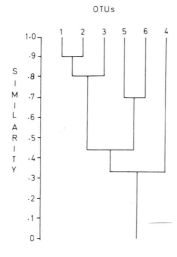

Figure 5. The Single Linkage phenogram derived from Figure 4. The vertical scale is that of the similarity coefficients. The horizontal scale is only to separate the OTUs, and the exact order is without significance

S—SM

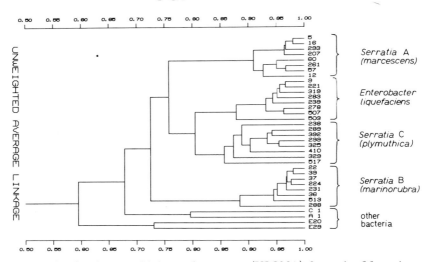

Figure 6. The Average Linkage phenogram (UPGMA) from the 36 strains of *Serratia* and allied bacteria. The phenogram is drawn sideways by the computer to permit strain numbers to be conveniently added. Four main phenons can be seen: (1) *Serratia* A, which is the well-known species *Serratia marcescens*; (2a) *Enterobacter liquefaciens*, a species known to be rather like *Serratia*, and shown by this and other studies to be so closely allied that it would be better named *Serratia liquefaciens* (even though it is non-pigmented); (2b) *Serratia* C, a poorly known group, *Serratia plymuthica*, whose position was uncertain, but is evidently close to *Enterobacter liquefaciens*; (3) *Serratia* B, previously only known from one or two strains, named *Serratia marinorubra*, but clearly a good, separate, species. The other bacteria form at the 75 per cent similarity level the minor groups (4), C1 plus A1; (5) E20 alone; and (6), E29 alone

clusters of OTUs are shown as tightly arranged twigs on long stems (Figure 5 and 6).

3. Ordination methods

The principle of ordination is to reduce the representation in multidimensional phenetic space to a representation in a small number of dimensions, usually two (allowing the OTUs to be plotted on a sheet of paper to give a taxonomic map) or three (allowing a model to be made, giving a 'fish-tank' model). Clusters are *not* formed: they must be made by eye later.

There are many techniques for ordination, ranging from informal ones (where the distances between OTUs are measured and attempts are made to arrange them on paper or to build models with sticks and plasticine), to highly sophisticated methods requiring large computers (based on eigenvalues and eigenvectors, or on iterative fitting and distortion of the original configuration to some criterion of simple structure: see Figure 9).

The basic principle is, however, similar in all. One first seeks for the dimension that expresses the greatest spread or scatter, and arrays the OTUs along this dimension. The dimension is usually a straight axis, but a few methods permit a curved axis. Then the next dimension is found as the one with the next greatest scatter, and so on until enough have been obtained. The basic idea is familiar to everyone: a schematic map of a train route is an ordination in one dimension, simplified from the three-dimensional geography of the railway line.

4. Other grouping procedures

Graphs in graph theory are diagrams consisting of points or *nodes* joined by lines or *internodes*, and they are mainly used in phylogenetic work. The OTUs or common ancestors are the nodes, the evolutionary pathways are internodes. One form of graph is of general value in numerical taxonomy, and that is the *minimum spanning tree*, which consists of internodes connecting the OTUs to give a tree of shortest length. Successive division of links in order of decreasing length will generate a Single Linkage phenogram. But unlike ordination plots, only a very few of the distances between OTUs can be effectually shown (Figure 7). Linkage diagrams are graphs in which different symbolism is used for linkages at different levels. They give a network that effectively shows how clusters form during clustering.

5. Shaded similarity matrices

These are commonly used in microbiology. The order of the OTUs must first be rearranged to the same order as a phenogram, but after

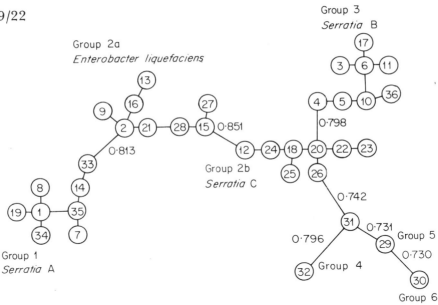

Figure 7. The minimal spanning tree from the *Serratia* example. The strain numbers are in the circles and maximum similarities (equivalent to minimal distances) between the groups of Figure 6 are shown against the appropriate links

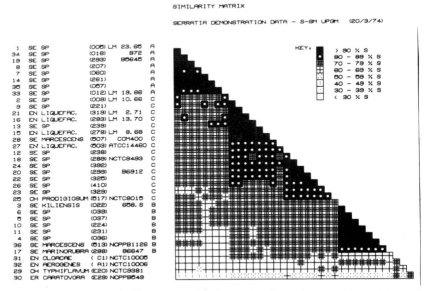

Figure 8. The shaded similarity matrix from the *Serratia* example, with strains recordered according to Figure 6. The phenons are shown by dark triangles. Some of the names under which the strains were received (extreme left) were evidently incorrect: they must have been previously misidentified

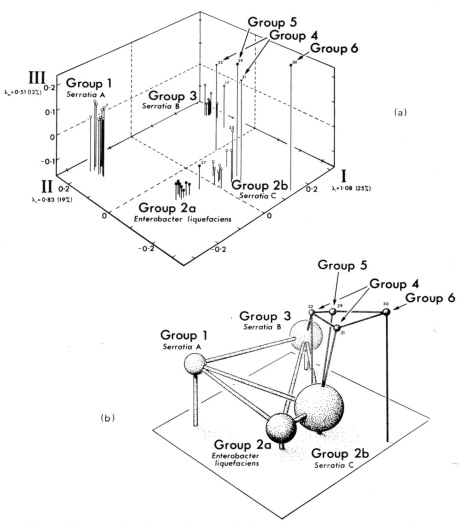

Figure 9. An ordination and a model of the groups of Figure 6. (a) A three-dimensional ordination of the 36 strains of Figure 6, obtained by the computer method called Principal Components Analysis. The new axes I–III are the three largest axes of variation, and the amount of the total variation accounted for by them is shown by their λ values. A vertical view on to the ground-plane is a taxonomic map in the first two dimensions. Note that Groups 4 and 5 would then overlap with Group 2b, though they are quite separate in the third (vertical) dimension. Each strain's position is shown, but only a few have been labelled with the strain number. (b) A taxonomic model derived from Figure 9a. The positions of the individual strains have been replaced by spheres representing the groups of Figure 6. The distances between the spheres are those between the centres of the clusters in Figure 9a, and the diameters of the spheres are proportional to the scatter of the strains of each cluster in that figure

that the clusters of highly similar OTUs show as dark triangles if high similarity values are made dark. This is often useful for getting information on the groups that do not fit very neatly into the system, as they show up as dark or light bars across or down the diagram (Figure 8). About six shades are usually feasible; it is best to cover the whole range of similarity values and not restrict the shading to values higher than some chosen figure.

Ordination plots can be converted into three-dimensional *models* where beads represent the OTUs positioned on three of the factor axes. They can effectively be represented also by stereograms; when these are viewed in a stereoscope they appear to be solid. Often the individual OTUs are not shown but only the groups represented by spheres of diameter proportional to the scatter of the OTUs (Figure 9b).

Many of these procedures for representing taxonomic structure may be useful for research or teaching, even though some (e.g. solid models) are not well suited to publication. Most of these are ancillary to phenograms. It is best to prepare several kinds of representation, and in particular one should not rely entirely on those that are interpreted by eye (e.g. ordination plots, shaded similarity matrices) because of the danger of subjective judgements.

E. Criteria of Goodness of Taxonomic Structure

The results of clustering and the like may not be easy to interpret. In particular one would wish to know if the structure found is significant in some sense, and though this aspect of numerical taxonomy has not been fully explored there are some useful adjuncts that help one to decide.

1. Congruence

One often wishes to compare the results of two different classifications (based, for example, on different sets of characters). This is readily done if the same OTUs are present in both. One calculates the correlation coefficient between the similarity values from one study A and those from the other B, using S_{jkA} and S_{jkB} in place of X_{ij} and X_{ik} in the formula given in Section III.C. Perfect agreement is given by $r = 1.0$, and no correspondence by $r = 0$. This is a test of the *congruence* of the classifications, and is important for evaluating any classification against additional information, provided both can be expressed as similarities (e.g. numerical taxonomy compared with DNA pairing, as below).

Similarity Matrix % S_{SM}

OTUs	a	b	c	d
a	100			
b	93	100		
c	88	90	100	
d	67	71	69	100

% DNA pairing (averaged)

	a	b	c	d
a	100			
b	74	100		
c	82	85	100	
d	25	23	32	100

Then the congruence is expressed by the correlation between the matched pairs (e.g. 93, 74; 88, 82; etc); r is here 0.958, a very good agreement.

2. Cophenetics

If one calculates r between the similarity matrix and the levels on the phenogram derived from that matrix this gives the *cophenetic corre-lation*. This is a measure of how well it is possible to cluster the OTUs, i.e. whether most OTUs do fall into compact clusters. One hopes that the cophenetic r will be above about 0.7. An example is:

	Similarity Matrix % S_{SM}					Matrix from phenogram			
OTUs	OTUs a	b	c	d		a	b	c	d
	a 100					a 100			
	b 93	100				b 93	100		
	c 88	90	100			c 89	89	100	
	d 67	71	69	100		d 69	69	69	100

The cophenetic correlation is that between the matched pairs (93, 93; 88, 89; etc) and it is here 0.993.

There are as yet no good and simple tests for the significance of clusters. The best guide at present is to calculate the standard error of the similarity coefficient (see the references in the further reading), and if the length of the stems separating the lowest members in a phenogram is several times the standard error, the clusters can be considered significantly distinct. A more recent test, the W test, allows a test of whether clusters overlap significantly, and in particular whether they represent simply adjoining regions of a continuous swarm. The mean inter- and intra-group similarities are readily calculated and are useful adjuncts to interpretation as well as giving information that is useful for making taxonomic models (Figure 10).

MEAN INTER- AND INTRA- GROUP SIMILARITIES

GROUP: MATRIX:

1	.929					
2	.758	.853				
3	.699	.737	.939			
4	.639	.692	.690	.796		
5	.614	.669	.642	.704		
6	.437	.584	.519	.644	.730	
GROUP:	1	2	3	4	5	6

Figure 10. The inter- and intra-group average similarities for the groups of Figure 6. To calculate these the computer must be provided either with lists of group members or with a similarity level across the phenogram. The level used here was 75 per cent

IV. Evaluation

A. Comparison of Different Sets of Data

The results of a taxonomic study must be evaluated against other available evidence on the organisms before the groups can be accepted. An important general principle is to compare the groupings obtained from one type of information with those from other types. Thus, a serological classification that was consistent with data from DNA pairing carries much more conviction than either alone. Wherever possible numerical tests of agreement should be used, as described in III.E. above.

B. Evaluation of Genetic Data

Genetical data are mostly of three kinds: (a) hybridization and gene-transfer, (b) GC base ratios, and (c) nucleic acid base pairing.

(a) Hybridization and gene transfer can only occur between fairly closely related organisms. However the degree of closeness varies very widely, with both the group under study and the genetic system used. Thus, with some protozoa only some stocks of a taxospecies can hybridize, so the taxospecies contains many genospecies. In bacteria gene exchange by transformation may be possible only between strains of one taxospecies but gene exchange by plasmid transfer may occur between genera of different families. Lack of gene transfer does *not* imply distant relationship, because it may merely reflect the absence of a gene for mating-type, for example.

(b) GC ratios (the molar percentage of guanine plus cytosine) are important because large differences imply large differences in the DNA. Therefore GC ratios that differ widely imply that the organisms are not closely related. But similarity in GC ratios is *not* evidence of taxonomic similarity. Man and the pneumococcus have almost the same percentage of GC. Strains of one species of microorganism generally have GC ratios within a range of about 5 per cent. The rule for species of one genus is much less clear, though the range is usually less than 15 per cent.

(c) Nucleic acid base pairing gives a physico-chemical estimate of similarity between nucleotide sequence in nucleic acids from two organisms. The commonest method is DNA—DNA pairing, which in bacteria (unlike eukaryotes) is a reliable method because each bacterial cell contains (as a rule) only a single copy of each gene. The DNA preparation therefore is representative of the whole genome. The DNA is dissociated into single strands by heat. The degree of pairing (as a percentage of that for homologous DNA) is estimated either by the binding to a reference DNA on a solid support or by the temperature at which hybrid double helices reform on cooling (*annealing*). Nucleic acid studies are of great theoretical and practical value.

Nucleic acid pairing is sensitive to very small taxonomic differences. It seems that there are often DNA-pairing subgroups within one phenon or taxospecies; the biological significance of this is still uncertain. However, the effect of experimental variation and reproducibility has not been critically examined. DNA preparations from strains of one taxospecies usually pair well (e.g. 70—90 per cent). Preparations from two species of a genus may show such low pairing that it is scarcely above the controls, and commonly that between different genera is scarcely detectable (newer methods may soon overcome this problem).

C. The Significance of Serology

The types of relationship reflected by serology are very varied. Some serological systems are specific for serological varieties of a rank below subspecies, such as flagellar agglutination of *Salmonella*. Others may give useful cross-reactions between genera, e.g. cross-neutralization of enzyme activity in the *Lactobacillaceae*. Each system has to be evaluated for its specificity. In general serological cross-reactions are useful confirmatory evidence, but occasionally a widespread carbohydrate antigen is responsible for cross-reactions that are of little taxonomic significance. For example some microorganisms contain antigens that react with the human ABO blood groups.

D. The Significance of Chemical Constituents

The presence of defined chemical substances in cells, usually small molecules, is often very valuable in taxonomy. Thus the mycolic acids are a family of lipid compounds characteristic of *Mycobacterium,* and another family, the nocardiomycolic acids are found in the genus *Nocardia*. Some of these compounds are almost diagnostic for certain groups, but it should be emphasized that with the great majority of these compounds an occasional strain may lack it (or it may occur in a group where it is not expected). Before such compounds can provide much assistance in evaluating a taxonomy, there has to be a great deal of background work based on a previous classification. There is often a temptation to invest these compounds with an exaggerated significance that it is later realized they do not possess. Chemical compounds can of course be used as characters in a numerical taxonomy, but they should be given only equal weight with other characters.

Proteins and nucleic acids, however, unlike simple chemical compounds, contain much internal information, and are correspondingly more useful. Protein sequence data show that the fine structure of one gene usually reflects surprisingly closely the taxonomic relations based on numerous different genes.

There are many other types of information that occasionally help to evaluate a taxonomy. Electrophoresis patterns, gas liquid chromatography of fermentation products, protein sequences, and many others can afford confirmation of the homogeneity of a taxospecies. In many instances their value is proportional to the extent to which they can be expressed as similarities between OTUs, and some are now being incorporated into numerical analyses.

E. Factors that Disturb Classification

A warning against poor experimental techniques has been made earlier. The result of lack of reproducibility of tests is to degrade seriously the taxonomic structure. If two OTUs have identical states on ten $+$ or $-$ characters, and we introduce 10 per cent of test errors, this will produce on the average two errors among the 20 test results. Probably the errors will be on different tests, so the OTUs will now only have a similarity of 80 per cent instead of 100 per cent. Further, we will not always get exactly two errors, so the new similarities will become uncertain as well as lower. If the average error rate over all tests, p, rises much above 10 per cent the taxonomic structure is lost (Figure 11).

Efforts must be made to keep the error rate low, and one useful way to monitor this is to duplicate a few of the OTUs in the study. The similarity between such duplicates should be 100 per cent, but this will seldom be so. The percent dissimilarity on the average will be about $2p$, so 90 per cent similarity indicates that p is about 5 per cent. If the

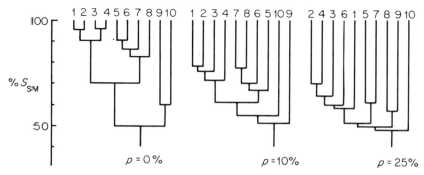

Figure 11. The effect of 10 per cent and 25 per cent test error on a phenogram. With high error rates the groups vanish or become mingled

average between duplicates is less than 80 per cent the data may be almost worthless.

Another source of disturbance is when two OTUs differ considerably in growth rate under the conditions of test, but otherwise give the same reactions. It may not be noticed that one of them gives the same results as the other but only after longer incubation. The Pattern Coefficient is intended to discover such systematic discrepancies.

In recent years it has been shown that many characters are determined by genes borne on plasmids. There is therefore a risk that loss of plasmids, or acquisition of new ones, may make some strains very different from others of that species. Experience is still slight, but it looks as if this source of disturbance is unlikely to be large as a rule. Indeed, if a majority of genes were unstable, it would be difficult to understand how the observed taxospecies have arisen and persisted in nature.

F. Definition of Taxa and Criteria of Rank

At this stage an investigator will be ready to define his taxa, either by describing new ones or improving the descriptions of old ones. He will need to have criteria for their limits and ranks. The simplest way is to draw a line across a phenogram at a suitable similarity level, thus defining phenons by the clusters that are indicated by the line. If the phenons seem acceptable they may be treated as taxa. The lowest branch on a cluster can be used to express the homogeneity of a phenon: thus an 85 per cent phenon is one whose lowest members join at no more than 85 per cent similarity. If one makes the assumption that a cluster is approximately a multivariate normally distributed swarm of OTUs, then one can also define a cluster by two parameters: (a) its position in the phenetic space by its 'centre of gravity' (*centroid*), and (b) its width, by the scatter about the centre (the *taxon radius*, related to the standard deviation in the space). These parameters are particularly useful for identification systems. Another measure of the centre is the *Hypothetical Median Organism*, the imaginary OTU which has the commonest state for each character.

The higher the phenon similarity level, or the smaller the taxon radius, the lower is the taxonomic rank of a taxon. Usually the tightest clusters represent traditional taxonomic species. Higher ranks like genera and families are indicated by major branches on a phenogram and larger volumes in a hyperspace. The criteria for a particular rank (e.g. genus) are at present arbitrary. It is therefore best to choose phenon levels so that existing classifications are not unnecessarily changed. Commonly one will find species at about the 80 per cent level and genera at about 60 per cent, but this is only a rough guide.

G. Description of Taxa

Particularly with a numerical taxonomic study it is simple to arrange for the computer to print out lists of character values for the phenons. The investigator must first give it the OTU membership of each phenon he wishes to recognize. The characters most useful for identification are clearly those that are most constant *within* taxa but vary most *between* taxa. Further discussion of this is given in the chapter on identification

Taxa should be carefully described, listing the OTUs, the characters and the methods with adequate references to laboratory and statistical techniques. Preferably the percentage of positive characters of the taxa (or a similar indication of variability for quantitative characters) should be given in suitable tables. Primary data can be deposited in a museum or institutional library. When describing a new species a typical strain should be nominated as the type strain, and a suitable name chosen in accordance with the provisions of the appropriate Code of Nomenclature.

V. Further Reading

The following are useful for background reading:

Davis, P. H. and Heywood, V. H. (1963). *Principles of Angiosperm Taxonomy.* Edinburgh: Oliver and Boyd.
Lockhart, W. R. and Liston, J. (Eds.) (1970). *Methods for Numerical Taxonomy.* Bethesda: American Society for Microbiology.
Mayr, E. (1969). *Principles of Systematic Zoology.* New York: McGraw-Hill.
Ruse, M. (1973). *The Philosophy of Biology.* London: Hutchinson.

Further details of numerical taxonomy may be found in:

Sneath, P. H. A. and Sokal, R. R. (1973). *Numerical Taxonomy.* San Francisco: W. H. Freeman.
Clifford, H. T. and Stephenson, W. (1975). *An Introduction to Numerical Classification.* New York: Academic Press.
Hope, K. (1968). *Methods of Multivariate Analysis.* London: University of London Press.

Microbiological techniques are covered in detail in the following series, and some articles of special interest are noted in parentheses.

Norris, J. R. and Ribbons, D. W. (Eds.) (1969–1976). *Methods in Microbiology,* 9 vols. London: Academic Press. (Oakley, C. L., 1971, Antigen-antibody reactions in microbiology, Vol. 5A, pp. 173–218; De Ley, J., 1971, Hybridization of DNA, Vol. 5A pp. 311–329; Skidmore, W. D. and Duggan, E. L., 1971, Base composition of nucleic acids, Vol. 5B, pp. 631–639; Sneath, P. H. A., 1972, Computer taxonomy, Vol. 7A, pp. 29–98, also includes information on culture collections and nomenclature).

Reviews include the following:

Colwell, R. R. (1973). Genetic and phenetic classification of bacteria. *Advanc. Appl. Microbiol.*, 16, 137–175.
Jones, D. and Sneath, P. H. A. (1970). Genetic transfer and bacterial taxonomy. *Bacteriol. Rev.*, 34, 40–81.

The W-test for clusters is described by Sneath, P. H. A. (1977). It is a method for testing the distinctness of clusters: a test of the disjunction of two clusters in Euclidean space as measured by their overlap. *Journal of the International Association for Mathematical Geology*, 9, 123–143.

Identification of Microorganisms

P. H. A. SNEATH

Department of Microbiology, Leicester University, Leicester

I. Introduction

A. General Comments

Identification is a branch of systematics. It is, however, different from *classification*, which treats of the way living organisms are grouped together into taxonomic groups or taxa (singular taxon). Identification deals with the process of allocating a new specimen (an 'unknown') to the correct and previously described taxon.

Identification is logically subsequent to classification. An unknown specimen cannot be identified as a member of a taxon until that taxon has been established and named. Therefore, if this chapter is read in conjunction with that upon classification of microorganisms, it is suggested that the chapter on classification should be read first. The chapters are, however, as self-contained as possible, and points especially relevant to both chapters have been included in both.

B. The Importance of Good Identification Systems

Identification is a very important practical activity, which will concern most microbiologists from time to time. This chapter aims to give some idea of *identification systems* (e.g. diagnostic keys), of how to use identification systems, how to find identification systems in the literature, and how a new identification system may be constructed. Because details of different systems in different groups of microorganism are extremely varied, no attempt is made to describe any actual identification systems (except for illustration). Instead, the classes of information, and principles of identification, used in microbiology are briefly considered, but the section on further reading includes some standard works for a number of the more important groups of microorganisms.

Large areas of microbiological work are heavily dependent on good identification (sometimes called *diagnosis*, although the word diagnosis is also used for a concise description of a taxon that serves to distinguish it from its neighbours). Some areas, such as hospital microbiology, are almost entirely concerned with identification and are collectively referred to as diagnostic microbiology. It may be emphasized that numerous kinds of microbiological work, for example in ecology and environmental studies, are seriously hampered by lack of good identification systems, and in addition it should be noted for many groups of microorganisms these systems cannot yet be constructed because the basic classifications have not been made. In these areas poor quality taxonomy is holding up advances in the rest of microbiology.

Identification is a subject that has been studied in its own right relatively little. It consists largely of a hotch-potch of useful techniques without much reasoned design. Very different techniques are used for different groups of microorganisms. In this introductory section, therefore, the subject is divided broadly into the observational bases of identification systems, and the logical bases of such systems, followed by some of the salient points of identification methods for the main microbial groups. In later sections these topics are then expanded.

C. Culture of Microorganisms

Many microorganisms, especially those with simple morphology, must first be isolated in pure culture before they can be identified. The importance of the purity of these cultures cannot be overemphasized: most cultural tests become grossly misleading with mixed cultures. The commonest cause of admixture is isolating a microorganism from a selective medium that inhibits, but does not kill, unwanted species. Isolations from selective media should be re-plated on non-selective medium and new isolations made from this. Usually if all the colonies then have a uniform appearance the culture can be assumed to be pure, but some microbial groups are particularly difficult in this regard, and single-cell isolates under microscopy may have to be made.

The main classes of data used in microbial identification are morphological, chemical, and serological. These are discussed further in Section II. All three may be affected by cultural conditions, so that an important general principle of identification tests is that the technique used must be defined very carefully, that is the test must be *standardized*. An identification test is not intended to give fundamental information about a microorganism, such as whether it possesses a certain metabolic pathway: it is sufficient if one can distinguish taxa because the pathway is active in one taxon but not in another under the standard conditions. For example, production of H_2S can be demonstrated by sensitive methods in most coliform bacteria, but only certain coliforms produce enough to be detected by less-sensitive methods. These less-sensitive methods are therefore appropriate for distinguishing them. The more-sensitive H_2S tests can be used in other bacterial groups. It can therefore be misleading to say a bacterium is H_2S positive without specifying the technique.

Another point in cultural work is that laboratory tests are quite costly, but that identifications are often required very quickly (e.g. in diagnosing infectious diseases). Unnecessary tests are therefore to be avoided, yet enough tests should be done to give a good chance of successful identification without having to do further ones. The choice of appropriate compromises is a major consideration in diagnostic microbiology, together with tests to screen quickly for a few kinds of microorganism of practical importance. Test methods are also being made quicker and cheaper. Because of the great diversity of microorganisms it is often necessary to do a preliminary set of tests that direct the worker to an appropriate subgroup of microbes. Hence, for bacteria, one might first of all determine the following properties: size and shape, Gram stain, motility, presence of endospores, oxygen requirement, and basic type of physiology (chemolithotrophic, heterotrophic, photosynthetic, fermentative or oxidative attack on carbohydrates). The results of these can then direct attention to subgroups of bacteria for which additional tests are prescribed.

D. Symbolism for Identification

An identification system implies the existence of a body of suitable information about a number of taxa, which can in principle always be presented as a table of their properties. This is true even for the highly specific tests used in serology — in which perhaps only one variant of a bacterial species reacts with a highly specific antiserum — because this specificity implies that the reactions against this antiserum for all other bacteria are known (or at least believed) to be negative. In such cases no formal tables may be prepared, but a tabular arrangement is a useful general method for illustrating the logic of identification. In addition, starting from a suitable table one can produce a variety of identification systems such as diagnostic keys, and polyclaves, which are described later. The tables need not be complete, but obviously the fuller they are the easier they are to use, and the more reliable are the results.

Starting with such a table also allows identification to be related to classification. The end result of classification (whereby highly similar organisms are grouped together into taxa according to their characters) readily allows the properties of taxa to be tabulated, and the table can serve as the start for making an identification system. A symbolism similar to that in classification can also be used, and because identification is becoming increasingly quantitative, the symbolism is conveniently related to that in numerical taxonomy, in which organisms or *Operational Taxonomic Units* (OTU's) are grouped into taxa. In identification the OTU's are new specimens for comparing with the known taxa.

The process of identification is the comparison of a new unknown OTU (symbolized as u) with a series of possible taxa to which it might belong (symbolized as A, B, . . . J, K, . . . Q). Capital letters are used for taxa to distinguish them from individual OTU's for which lower-case letters are used (e.g. j,k.). This is because a taxon is treated formally as a *collection* of previously studied OTU's, and is therefore represented by a small table of *several* OTU's versus characters. Also, since not all of the n characters studied in a taxonomy are retained in most identification systems, the smaller number retained is symbolized by m. The number of taxa is indicated by q. The basic matrix from which an identification system is constructed is one of m rows (diagnostic characters) versus q taxa. Each taxon (for example, taxon J) is represented by a number of columns, one for each of the t_J OTU's that were grouped together to form taxon J. Each column therefore contains the character state values of the OTU's that were employed for a numerical taxonomic analysis, but the OTU's have been arranged into taxa. An example of part of an *identification matrix* of this kind is shown below, with taxa separated by vertical lines.

In this example a quantitative character has been included in the form of the rate of attack on gelatin, which is taken as weak but detectable within 9 days in *Proteus morganii*. The other tests are scored 1 for positive and 0 for negative. The general symbol for a character state, X, is used with three subscripts, for the character, the OTU and the taxon respectively.

Taxa (species of *Proteus*)

Characters	A (*P. mirabilis*) $1 \ j \ldots, t_A$			J (*P. morganii*) $1 \ j \ldots, t_J$			Q (*P. vulgaris*) $1 \ j \ldots, t_Q$		
1 Days to hydrolyse gelatin	1	2	1	9	7	9	2	1	7
2 Simmon's citrate test	1	0	1	0	0	0	0	1	0
3 Indole production	0	0	0	1	1	1	1	1	0
⋮									
i		X_{ijA}			X_{ijJ}			X_{ijQ}	
⋮									
m Acid from sucrose	0	0	1	0	0	0	1	1	1

The most useful characters are in general those that *differ between taxa but are constant within taxa*. If the taxa are monothetic in type then one can find characters that are entirely constant within taxa. In the majority of cases the taxa are polythetic, so that there is always the possibility of an exceptional result. In the table above it is seen for example that indole is not invariably positive in *Proteus vulgaris* (only about 90 per cent of strains are positive in this species).

This form of table is usually converted into a more compact one in which the taxa are represented by a single value for each character, with this value standing for some kind of average property of the OTU's in the taxon. At the same time the most useful cutoff point in quantitative characters may be sought. Thus, one could choose gelatin liquefaction within 5 days as a suitable separator for *Proteus* species. The resulting table may be thought of as a *reference library* of diagnostic information.

Often the table is easily expressed as percentages of positive results in the taxa, and this is a very convenient form for further use. The example above would then be turned into a table of the following kind. At the side of the new table we can place a new column for the unknown strain u which it is desired to identify.

	P. mirabilis	P. morganii	P. vulgaris	u
Gelatin hydrolysis in 5 days	95	0	95	1
Simmon's citrate	50	0	5	0
Indole	0	100	90	1
:				
Sucrose acid	20	10	100	1

It can readily be seen that the unknown u (which is coded 1 for a positive test result, 0 for a negative) matches best against *Proteus vulgaris*, and this would be presumed to be its true identity — at least on the basis of these four tests. One can therefore imagine that the process of identification is basically the comparison of the unknown with each taxon in turn, in the search for a single acceptably good matching; or in the case of diagnostic keys the exclusion successively of taxa that disagree on certain characters. If several unknowns, u_1, u_2, etc. are given, they can form a separate table: each is then processed in turn.

Not all the tests may have been performed on an unknown (or some tests may not have been carried out on certain taxa), so that provision must be made to get round the problem of these missing data. Some symbol is thus required equivalent to the 'no comparison' or NC symbol in numerical taxonomy: a blank is often used. More complicated tables (containing statistics like means and standard deviations) can be prepared for special diagnostic problems.

It should be noted that there are four possible outcomes of the identification process:

(a) The unknown is identified with the correct taxon.
(b) The unknown is misidentified, i.e. it is incorrectly attributed to a a wrong taxon.
(c) The unknown is not identified at all, and correctly so because the taxon to which it belongs is not in the matrix.
(d) The unknown is not identified, but should have been identified with a taxon that is in the matrix.

There are therefore two kinds of mistake that can be made, (b) and (d). There are two correct outcomes, (a) and (c), only one of which (a) is a successful identification. The third, (c), is a failure, but not a mistake, but is primarily due to lack of taxonomic information.

E. Data Used with Different Groups of Microorganisms

Some groups of microorganisms are identified from largely morphological observations. These include most algae and protozoa. Electron microscopy and specific staining methods are often used, and occasionally serology (especially with parasitic protozoa).

Bacteria and yeast-like fungi are identified largely on physiological and chemical criteria, but with some assistance from morphology: serology is much employed at and below the level of the species. Most microfungi are identified by morphology with some help from chemical tests.

Virus identification is highly specialized. The main approaches are electron-microscopy of virus particles, observation of pathological changes on isolation in suitable host organisms or tissue cultures, and a large range of serological methods, including the neutralization by specific antisera of their ability to infect a host or tissue culture.

II. Data Used in Identification

A. Morphology

Many microorganisms (e.g. most algae) have sufficiently complex morphology to permit classification and identification to be based largely on microscopic observation. The light microscope is principally used, preferably using phase-contrast. The size and shape of the organisms, and of organelles (such as nuclei, vacuoles, cilia) are important characters. Often the identifications can be made from living organisms. In other instances special stains are used, for example silver impregnation in ciliates, and the Gram stain in bacteria. The morphology of sporing structures is especially important in microfungi.

The electron microscope is also used extensively in virus identification, to determine the size, shape, and symmetry of virus particles, and sometimes the finer structure such as the large regular protein molecules (capsomers) that compose the outer coat of the viruses. For such detail the negative staining technique is often employed, in which the virus particles are embedded in an electron-dense substance in such a way that the virus constituents show as electron-transparent shapes. Electron microscopy is also useful for studying details of cilia or flagella in algae, protozoa, and bacteria.

Colonial appearance is often characteristic if suitable culture media are used. Such media are often combined with a chemical test so that colonies of certain species assume a particular colour or give a visible reaction zone. Hence zones of clearing on milk agar (due to digestion of casein) may be characteristic of some species.

B. Chemistry

Chemical properties are extensively used for microorganisms that are readily cultivated. They may be broadly separated into *physiological tests* and those for *chemical constituents*, but these categories overlap a good deal. Thus physiological tests include such tests as ability to grow anaerobically, or at $40°C$, and production of H_2S from cysteine. Tests for chemical constituents include the percentage of guanine plus cytosine in DNA (the % GC), presence of the pigment prodigiosin in *Serratia marcescens*, and of urease in *Proteus vulgaris*. Some chemical tests are used with the microscope (e.g. histochemical test for chitin in fungal cell walls); but a great many tests, though broadly chemical, fall into an intermediate category. These include those for products of metabolism, such as the production of indole from tryptophan which depends on the action of the enzyme tryptophanase. This could be viewed either as a physiological test or as a test for the enzyme. Most of these are collectively referred to as *biochemical tests*.

Particularly important for identification is the ability of a microorganism to attack only certain specific carbohydrates out of a range of these substances, frequently with the production of acidity that is detected by a pH indicator (*carbohydrate tests*, sometimes, though incorrectly, called fermentation reactions). For example, *Escherichia coli* typically produces acid from glucose and lactose, but not inositol. The ability to grow with specified compounds as sole source of carbon and energy (*CE utilization reactions*) can also produce very useful patterns for identification, provided of course a basal medium is available that contains all the other substances necessary for growth. These compounds may themselves be carbohydrates, but growth rather than change in pH is then determined; e.g. the yeast *Candida krusei* grows on glucose but not on maltose or galactose.

In all these tests the culture medium in which the microorganism is grown must be carefully prescribed, together with the method of inoculation, size of inoculum, and the temperature and conditions of incubation. Hence the medium might be decided as 1 per cent proteose peptone in water plus 0.5 per cent of NaCl and 1 per cent of carbohydrate (e.g. sucrose), with a final concentration of 0.004 per cent of bromocresol purple as pH indicator, adjusted to pH 7.2, dispensed in 10 ml amounts in tubes of 12 mm internal diameter, and sterilized by autoclaving at 115 °C for 20 minutes. The inoculum might be one loopful of an overnight broth culture, followed by incubation for 24 hours at 30 °C in air. Precise details must be given for determining the reaction. Thus in this example a positive result might be a clear yellow colour with no trace of grayish purple. Sometimes a strong positive is diagnostic, but a weak one is not; again, this must be made clear. With other tests one may have to add specified amounts of chemical reagents, and read the results after a specified time. These details together constitute the *standardization* of the test.

It must be emphasized that careful observance of these details of the test must be scrupulously adhered to. Quite small changes in technique may lead to erratic or incorrect results. Indeed, the reliability of microbiological tests is notoriously poor, and this has serious effects on identification (see Section V.C).

In recent years a number of miniature identification tests have been developed commercially, as self-contained identification kits. Once a pure culture has been obtained it is inoculated into all the tests in the kit and the reactions read after the appropriate incubation. These kits are likely to become increasingly used, and they are part of a current trend towards automation in diagnostic microbiology.

C. Serology

The use of serology in identification of microorganisms depends on the high specificity of the *antigen—antibody* reaction. When an antigen is injected into a warm-blooded animal (rabbits are commonly used), it gives rise to *antibodies* (immunoglobulins) which combine only with that antigen (the homologous antigen) or with other antigens (heterologous antigens) that are chemically extremely similar. The serum of the animal is then a *specific antiserum*, and the antigen can be detected by the specific reaction in a number of ways (see below).

A crude antiserum generally needs some treatment before use. Commonly it reacts with unwanted or non-specific components of the test-antigen preparation, and the unwanted immunoglobulin molecules responsible for the non-specific reactions have to be removed. This is often done by absorbing them out by a suitable antigen preparation that lacks the specific antigen component that is to be tested for. For example, a purer antiserum to a flagellar protein can be made by treating the crude serum with a suspension of cells of the strain used for immunizing the rabbit, but a suspension from which all the flagella have been removed. The treated cells are centrifuged down, and carry with them the unwanted antibodies that react with the cell walls, leaving in the supernatant only the antibodies that react with the flagella. In most cases too, a suitable preservative is added to the antiserum.

Some antigens are widely occurring (especially certain carbohydrates), and these may be reponsible for unwanted cross-reactions with different organisms to those which are to be identified. These unexpected reactions can be very misleading, but commonly the reactions are too specific. Serology is excellent for identifying variant forms of one species of microorganism after these have been identified to species level by other methods. Such variants are of great significance in tracing the spread of pathogens. It is sometimes possible to prepare an antiserum that reacts with most, if not all, strains of a particular species. But sera that will identify all strains of a species, or all species of a genus, may be difficult to obtain. Of course, a less specific antigen can be looked for. Also, fluorescence a number of different antisera, each prepared against a different antigenic varient, may be mixed together to give a *polyvalent antiserum*. This is designed to give a reaction (even if a weak one) with all the known antigenic variants. The corollary is of course obvious: such a serum will miss a varient that has not yet been recognized! The most widely-used antisera can be purchased from commercial sources.

Methods for detecting the antigen—antibody reaction are numerous. Only a few of the commoner ones will be mentioned.

1. Agglutination

In this a particulate saline suspension bearing the antigen on the surface of the particles is mixed with antiserum: a positive reaction is shown by clumping (agglutination) of the particles. Most commonly the suspension is a suspension of the microorganism to be tested. Agglutination can be done very quickly in small drops on a microscope slide, but tests in small tubes are preferred for exact or confirmatory studies. The highest dilution of an antiserum that gives a positive reaction is called the *titre*. Sometimes the suspension of the microorganism must react at or near this titre to constitute satisfactory proof of the identification. Agglutinations involving bacterial flagella commonly give characteristic large loose flocks, but otherwise the clumps are small and granular.

2. Precipitation

A soluble antigen will form a precipitate with the specific antiserum, and there are two main forms of *precipitin test*. In one the antigen solution is carefully layered on to the antiserum in a small tube, and a ring of precipitate forms at the junction (precipitin ring test). In the other the soluble antigen and antiserum are diffused against each other in a gel (*gel diffusion precipitin test*): lines of precipitate form where they meet (one line for each reacting antigen). The pattern of lines can indicate whether the antigens are homologous for the antiserum, or are cross-reacting heterologous ones. The technique can be combined with electrophoresis to give *immunoelectrophoresis.* Various methods are used to extract soluble antigens from microorganisms for these tests.

3. Immunofluorescence

If a fluorescent dye is coupled to immunoglobulin molecules, the precipitate with the antigen is fluorescent, and can be seen under a fluorescence microscope. The technique is very sensitive, and is one of the few whereby a microorganism can be identified serologically without isolation in culture, because specific fluorescence can be seen around microorganisms in a smear on a microscope slide. There are two ways of making the antigens fluorescent. In the *direct method* the antiserum is labelled with the fluorescent dye. In the *indirect method* the dye is attached to an *antiserum to rabbit globulin*, and this antiserum precipitates *on top of* the primary precipitate of antigen— antibody to give a *second, fluorescent, layer.* The indirect method only requires one antiserum to be labelled with dye, an 'antirabbit serum', because *unlabelled* antisera (provided they were made in rabbits) can be used to detect any desired antigen.

4. Other Methods

The most commonly used other methods are *complement-fixation tests,* and the production of *hypersensitivity reactions* in the skin of guinea-pigs. These are rather complex systems which would take much space to describe, but both of them allow antigens to be detected by specific antisera through their immunological effects.

D. Bacteriophage

Viruses of bacteria (bacteriophages) are known for most species of bacteria, and they usually attack only strains that are closely related to those in which they are propagated. A few bacteriophages will lyse almost all strains of one bacterial species and none of other species. But such bacteriophages are not easy to find, and bacteriophage lysis is used mainly for characterizing different variants within a species. These are useful for tracing the spread of infections caused by one such variant, in the same way that serological variants can assist in epidemiology. For this purpose a battery of *typing phages* is chosen, and a particular strain of the species is spread as a uniform lawn on a nutrient plate, and spotted with the set of bacteriophages. The pattern of clear spots of lysis (where the bacteria were killed by some but not others of the phages) can be used to characterize a variant of the species. When the same variant is isolated again it can be recognized by the same phage pattern.

E. Other types of information

There are numerous other types of test which are used for rather special purposes or for special microbial groups. Many are still too laborious or unreliable for general use, or else the necessary basis of information from enough microbial species is not available. For these reasons they are unlikely to be used widely in the near future. In *gel electrophoresis* the different proteins in an extract of a microorganism are separated in a gel to give a pattern of bands which may be so characteristic of that species that it can be used to identify unknown strains. This can be combined with tests for enzymes, because certain enzymes (e.g. esterases) have characteristic mobilities in different species. Ultraviolet and infrared spectrometry of microorganisms or their extracts can be studied. Pyrolysis gas-liquid chromatography of microorganisms, and chromatography of their cell walls, or metabolic products, can give characteristic patterns for different species. The % GC of DNA can help in diagnostic work, and DNA base-pairing shows some promise. Occasionally genetic phenomena (e.g. hybridization with appropriate mating types) has been used. The pattern of sensitivity to antibiotics can help identification, but its value is much reduced by the emergence of antibiotic resistance in many groups. In a few instances, notably in viruses, the behaviour of a suitable host organism to infection by a microorganism may be used for identification.

III. Diagnostic Characters

A. Character Weighting

As has been noted, ideally the characters used for identification would be constant within taxa. In microbiology there are few characters which can be treated as if they were entirely constant. This is not only because occasional exceptional strains are likely to occur (although this is the commonest reason). In addition, there are quite likely to be exceptional species within a genus, and similarly there may be exceptions in the higher categories. But another reason is the danger of observational or experimental error. There may be little practical difference whether 1 per cent of strains of a species are truly urease-negative or 1 per cent of strains are falsely recorded as urease-negative because the test is carelessly performed; unless something else arouses the worker's suspicions there is a high chance that one strain in a hundred will be effectively scored as urease-negative. And, as has been noted, microbiological tests are exceptionally difficult to standardize, even within one laboratory.

For these reasons it is wise to assume that some exceptions will occur in practice and therefore tests will range from those that vary greatly (e.g. with half the strains of a species positive) to those that are almost but not quite constant (e.g. 1 per cent or 99 per cent of positives). In principle the more constant characters should be given greater emphasis in identification, so that unlike classification, it would be permissible to weight certain characters more than others when using them for identification. Practical identification schemes seldom do this numerically, but in several schemes this is done implicitly, as for example when an entry in a table is marked 'variable', which at least implies it should not be given too much emphasis. In some statistical methods described later differential weights are in effect used. An allied consideration is whether one should assume that every taxon is equally likely to turn up among the unknowns. One could give a taxon a weight that reflected how common it was thought to be (in mathematical language, its 'prior probability'), but it is considered safest to assume that an unknown is equally likely to belong to any of the taxa. Typhoid is rare in Britain, but it would be foolish to give the typhoid bacillus such a low probability that it was never identified even if it did occur.

B. Choice of Diagnostic Characters

Where characters are expressed as positive or negative one can readily calculate the overall usefulness of each one. A method that is widely used is to consider that where the percentage of positives is 85 or more the character is taken as positive; where it is 15 per cent or less it is taken as negative, and otherwise it is considered variable. If the number of taxa that are positive is multiplied by the number of taxa that are negative, a *separation index* is obtained, which is greatest for the most discriminatory characters. For example, if among the four species of *Proteus* the hydrolysis of gelatin at 5 days is positive in two species and negative in two, the index is $2 \times 2 = 4$. This test scores higher than indole which is positive in three and negative in one (index $3 \times 1 = 3$), and higher than Simmon's citrate (positive in one, negative in two, variable in one, index 1 x 2 = 2). In this way the more useful characters can quickly be found, though if they give similar patterns in the taxa they may not all be useful (it is better to consider the separation index as a way of screening out the less useful characters).

The minimum number of characters that can be used for a reliable identification system is usually about the same as the number of taxa that have to be distinguished (although in theory it may be possible to use many fewer). There should always be some extra characters over and above the minimum, because not all laboratories can conveniently do all the recommended tests, and some additional ones are useful for confirmation of the diagnosis. The characters that are simplest to observe, and the tests that are most reliable, should be preferred.

IV. Identification Systems

A. Types of Identification System

There are several different methods by which identification can be carried out, and these give rise to different *identification systems*. It has been noted above (Section I.D) that one must first have a 'reference library' of diagnostic information about the taxa that are to be covered. This reference library is not necessarily the identification system itself, though one can build one from it.

There are two main approaches to such systems. They can either employ characters in a sequence, or else a number of characters can be considered simultaneously. In the sequential methods the possible alternatives are successively narrowed down by considering more and more tests until only one possibility remains. In the simultaneous methods some sort of matching is made between the unknown and the taxa using all the characters at once, in order to find the best match. The systems that are entirely specific for a single taxon (such as the identification of a single antigenic variant by a completely specific antiserum) are matching methods with only one correct possible match. Some combination of sequential and simultaneous methods can also be employed, as when a preliminary screening is done to obtain the most likely higher taxon, followed by a matching method for taxa contained within it.

Of the identification systems described here, two are sequential, *diagnostic keys* and *polyclaves*; two are simultaneous, *taxon-radius models* and *discriminant analysis*; and one, the *diagnostic tables*, can be used in either way.

B. Diagnostic Tables

A part of a diagnostic table has been illustrated earlier, but it may be noted that the entries can be numerical or not. Hence the example given in Section I.D might be represented as follows:

Taxa: Species of *Proteus*

Characters	*P. mirabilis*	*P. morganii*	*P. rettgeri*	*P. vulgaris*
Gelatin hydrolysis within 5 days	(+)	—	—	(+)
Simmon's citrate	v	—	(+)	(−)
Indole	—	+	+	(+)
Hydrogen sulphide in TSI medium	+	—	—	(+)
Ornithine decarboxylase	+	+	—	—
Gas from glucose	+	+	(−)	+
Mannitol acidified	—	—	+	—
Sucrose acidified	v	(−)	v	+

+ = positive (more than 95% of strains positive)
(+) = usually positive (85 to 95% positive)
v = variable (16 to 84% positive)
(−) = usually negative (5 to 15% positive)
— = negative (0 to 4% positive)

Note that the percentages corresponding to the symbols are shown at the foot; where these are not given the user should be sceptical of the reliability of the table unless the accompanying text makes it clear that exceptions are rare or unknown. Many diagnostic tables given in the microbiological literature are superficially attractive but very unreliable. A table of percentages is far better than the form shown above (see Section I.D). Commonly the symbol *d* is used to indicate variability.

These tables are used by comparing the results of the unknown u with the taxa in turn to see where it fits best. This is usually done by eye (preferably after writing the test results of the unknown on to a strip of paper). One can see if the unknown agrees well with one taxon, and whether there is an occasional exceptional test result. When the comparison is done by mechanical or mathematical methods it leads to the polyclave and taxon-radius systems described later. When a properly prepared diagnostic table is used with care it is a very powerful identification device; the main disadvantages are that it becomes cumbersome with large numbers of taxa or characters, and it is difficult to appreciate quickly the overall significance of more than a few exceptional findings. It is ideally suited for publication in reference books.

C. Diagnostic Keys

Diagnostic keys are familiar systems for identification, but preparing them is often difficult. They are best arranged as *dichotomous* keys, i.e. each branch point has only two alternatives, called *couplets*, that describe possible alternative character states. Each of the contrasting statements is called a *lead*. An example is:

Key to species of Proteus

1	Ornithine decarboxylase positive	2
	Ornithine decarboxylase negative	3
2(1)	Indole positive	*P. morganii*
	Indole negative	*P. mirabilis*
3(1)	Mannitol acidified	*P. rettgeri*
	Mannitol not acidified	*P. vulgaris*

The key is used by answering each couplet in turn from the observations on the unknown and then moving to the numbered couplet that is indicated thereby. This is a 'bracketed' key, and the numbers in parentheses refer to the lead that brought one to the couplet in question (i.e. lead 2 was reached from lead 1), so that the key can be retraced if one thinks one has taken a false step. Another arrangement called an 'indented key', is sometimes used, but it is much less convenient.

Clearly, the couplets should be contrasting and mutually exclusive. This is not always so, even in reputable texts. One such contained the couplet

'Gelatin and/or glucose gelatin may or may not be liquefied'
versus
'Action on gelatin and glucose gelatin not recorded'.

It would seem doubtful whether even the author of the key could decide which path to take if the test on an unknown was either positive, negative or undetermined.

The characters best suited to a key are those that have the highest consistency within taxa, and that separate the set of taxa under consideration into approximately equal halves. The character 'Ornithine decarboxylase' is an example: the percentages of positives within the taxa are less than 5 or greater than 95; it divides the four species into two sets of two. This makes for a short, reliable key. The consistency is the most important thing, particularly for the first divisions, because mistakes early on will lead the user a long way from the correct identification and makes it much more difficult to retrace the steps and check them. A secondary consideration is the ease of determining the character. If provided with a diagnostic table, a computer can prepare and print a key, which can then be used apart from the computer.

Sometimes more than one character is mentioned in a couplet. This has the advantage that the key can be worked if only one of the characters in the couplet has been observed, thus sometimes permitting identification from partially examined material, or material in only one growth phase (e.g. asexual stages in fungi). Also, useful confirmation can be obtained that one is on the correct path. The disadvantage is that when the unknown agrees in some characters but not in others, the user may not know how to proceed. In such an event it is always wise to check the observations on the unknown and then to start again at the beginning. If still in doubt, one should follow both paths and compare against the unknown the full descriptions of the two suggested identifications.

A final step, often forgotten, is to compare the unknown in detail with the description of the taxon that has been diagnosed, paying special attention to characters that were not used in the key. Very frequently discrepancies will be found that suggest one has made a misidentification, and which require one to reexamine the unknown.

Keys are only successful if one can discover characters that are practically constant, i.e. invarient within taxa, and different between taxa. They are thus well suited to artificial monothetic classifications, and many artificial monothetic classifications have been proposed because they led to easy keys, even though the identifications thus given were of very little scientific value. Keys are useful for higher taxonomic ranks, where very constant characters can usually be found. In bacteria they are less useful because the groups tend to be more markedly polythetic, and sufficiently constant characters are harder to find: they are apt to lead to a significant number of misidentifications or failures to identify.

D. Polyclaves

The word *polyclave* means a multiple key. The characteristic of a polyclave is that one can start with any character one chooses and take further characters in any order. Identification is made by excluding taxa in turn until only one possibility is left. Polyclaves can be stored in a computer and questioned from a keyboard, but they are very easily adapted to punched cards. This is their usual form, when they are often called 'Peek-a-boo systems'.

The principle is to allocate a fixed position on each of a set of punched cards to every taxon. A separate card is used for every character state, with a hole punched in a position corresponding to a taxon if the taxon in question *can* possess that character state. For two-state characters this means two cards: for example 'Catalase positive' and 'Catalase negative'; if taxon 15 is always positive, then its position, 15, is punched in the first card but not in the second; if it is always negative, then position 15 is punched only in the second card; if it is variable for catalase, then position 15 is punched in *both* cards.

Computer cards are conveniently used, and punched manually or by a computer program. The title of the card, e.g. 'Catalase Positive', is punched at one end (so that the punch holes corresponding to the title

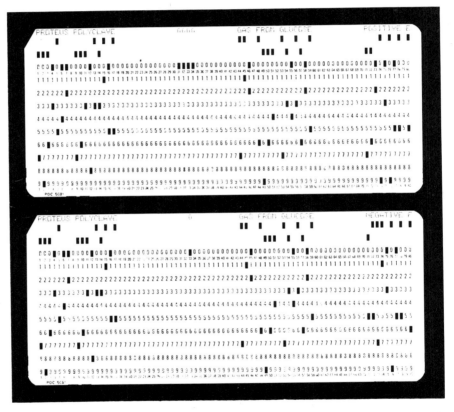

Figure 1. Two punched cards from a polyclave for species of *Proteus*. The area for the different taxa, each with a different position (here there are only four taxa) is near the middle. On the left is the title of the polyclave, on the right is the name and state of the character. The letters (E, F) at the extreme top right of each card are part of an alphabetic sequence that allows the cards to be quickly sorted back into the original order

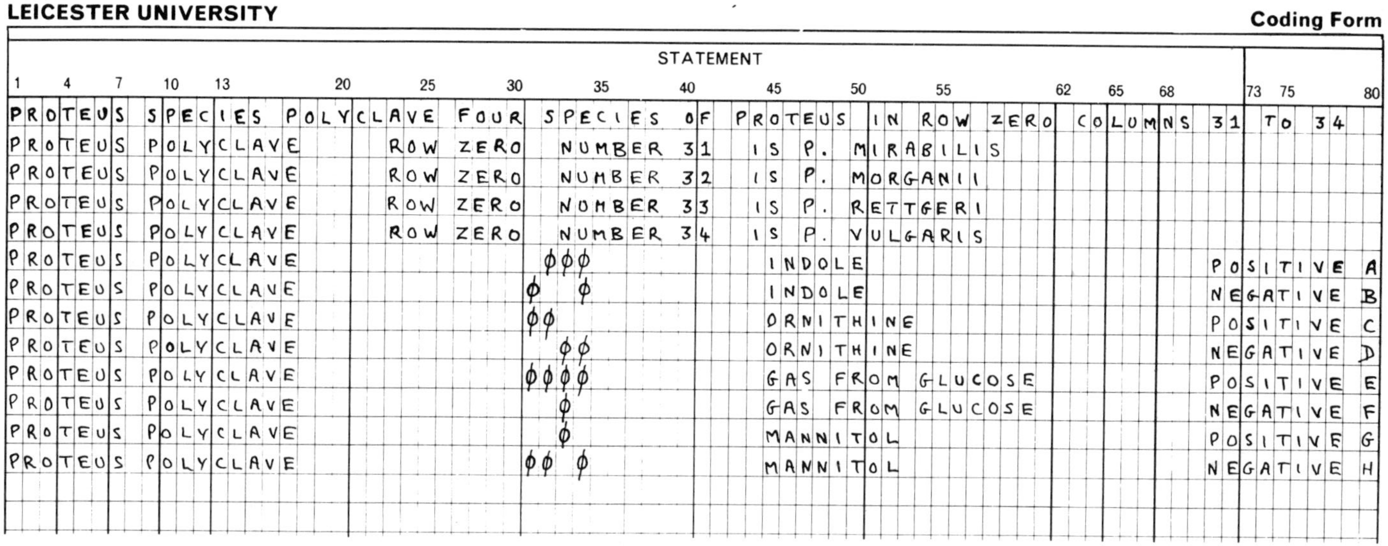

STATEMENT

1	4	7	10	13	20	25	30	35	40	45	50	55	62	65	68	73	75	80
PROTEUS		SPECIES	POLYCLAVE			ROW	FOUR ZERO		SPECIES OF PROTEUS	IN ROW ZERO COLUMNS						31 TO 34		
PROTEUS			POLYCLAVE			ROW	ZERO		NUMBER 31	IS P.	MIRABILIS							
PROTEUS			POLYCLAVE			ROW	ZERO		NUMBER 32	IS P.	MORGANII							
PROTEUS			POLYCLAVE			ROW	ZERO		NUMBER 33	IS P.	RETTGERI							
PROTEUS			POLYCLAVE			ROW	ZERO		NUMBER 34	IS P.	VULGARIS							
PROTEUS			POLYCLAVE					φφφφ		INDOLE						POSITIVE		A
PROTEUS			POLYCLAVE					φ		INDOLE						NEGATIVE		B
PROTEUS			POLYCLAVE					φφφφ		ORNITHINE						POSITIVE		C
PROTEUS			POLYCLAVE					φφφφ		ORNITHINE						NEGATIVE		D
PROTEUS			POLYCLAVE					φφφ		GAS FROM	GLUCOSE					POSITIVE		E
PROTEUS			POLYCLAVE					φφ		GAS FROM	GLUCOSE					NEGATIVE		F
PROTEUS			POLYCLAVE					φφφ		MANNITOL						POSITIVE		G
PROTEUS			POLYCLAVE					φφ		MANNITOL						NEGATIVE		H

Figure 2. The coding form for the punched cards of the *Proteus* polyclave. When punched on 80-column cards, this will produce the polyclave. The symbol φ is used to indicate zero.

are all at one side). The rest of the card can contain the positions for the taxa, and as there will normally be about 50 columns available, one can incorporate data on over 500 taxa on to one card. There are also accessory cards, e.g. one that shows the block of punching that is occupied by the taxa and, if required, a card for each taxon giving its reference number (though this is more convenient as a printed sheet). Cards are conveniently numbered or lettered so they can be quickly sorted back into the right order after use.

The system is used by picking up in any order the character cards, according to the properties of the unknown to be identified. The characters must have been divided into mutually exclusive alternatives by suitable coding. One takes cards until only one punch hole remains through the deck. Sometimes one card blocking a hole will suggest an aberrant test result that might otherwise go unnoticed. A polyclave is easy to carry about and use, but it may be very slow if there are many characters to consider. A moderate number of exceptional entries (variable character states) can be accommodated, but if there are too many exceptions then, like diagnostic keys, the system may become unworkable. Since the holes must either be punched or not punched, a polyclave must express unambiguous yes—no responses. Therefore one must choose suitable cut-off points on the scale of frequency of positive responses to score as Positive, Variable, and Negative. This can be made stringent (e.g. 100 per cent = Positive, 0 per cent = Negative, 1—99 per cent = Variable) or less stringent (e.g. 90—100 per cent = Positive, 0—10 per cent = Negative, 11—89 per cent = Variable). Very stringent conditions give greater certainty of those identifications that are made, but a greater chance that the unknown will not identify.

Figure 1 shows part of a polyclave for species of *Proteus*, constructed from the diagnostic table given earlier. Figure 2 shows the listing of the punched cards which form this polyclave. Only four characters are needed.

E. Taxon-radius Models

This identification system is only practicable with a computer. The term *taxon-radius model* is used for a type of identification system in which the taxa are treated as hyperspheres in a space of many dimensions, one dimension for each character. Each taxon is conceived as a number of OTU's (usually strains) which form a roughly hyperspherical cluster, separated from the clusters representing other taxa. If a hypersphere is drawn about the centre of gravity (*centroid*) of a cluster, with an appropriate *critical radius r* that includes a high percentage of strains of that taxon, the taxon can be treated in a different way. It can now be represented by a *hypersphere of a given position and radius.*

These hyperspheres are readily derived from a suitable reference library, and the numbers that represent them can be stored in a computer to give a taxon-radius identification system, together with a computer program for comparing an unknown with the reference data.

Taxon-radius systems are particularly easy to construct from tables of per cent positive tests. The centre of a taxon is represented by the mean of its position on each character axis in turn, and for presence—absence characters these means are, of course, just the percentages of positive characters. Under certain assumptions about the clusters it is also possible to calculate from these percentages the critical radius, r_J, of a taxon J, because the mean and standard deviation of the distances of OTU's of taxon J from the centroid of J can be estimated. Alternatively, one can plot a histogram of distances of actual strains and choose the critical radius by using this (Figure 3).

The position in the multidimensional space of an unknown, u, can be determined from its characters. Identification can then be made by calculating two things: (a) the distances of u to the centres of all the taxa, and (b) whether u lies within the spherical envelope of a taxon (ideally only one taxon, unless the taxa overlap). This system is shown diagrammatically in Figure 3.

The interpretation of the results of (a) and (b) is broadly as follows:

(1) If the unknown u is closest to a taxon J, then this is its most probable identity, but to achieve a definite identification it should lie well within the envelope of J.

(2) If u lies just outside the envelope of J it is probably an atypical strain of taxon J.

(3) If u is about equidistant between taxa J and K and only a little outside their envelopes, it is an intermediate strain.

(4) If u does not lie within or close to any sphere, it is unidentified.

In effect we are asking how far (in taxonomic space) the unknown is from certain reference points. We can think of this as being its dissimilarity from such points, and we can use the ordinary numerical taxonomic Similarity Coefficients to measure distance, because we can subtract them from 1.0 to get a dissimilarity. Thus, similarity $S = 90$ per cent or 0.90 equals dissimilarity of 10 per cent or 0.10. The actual distance is the square root of this, about 0.32. Various different distance measures can be used and there are also certain alternative ways of describing the centre of a cluster and its radius, but these are details that are not important to understanding the underlying principles. One commonly used method utilizes Bayesian probabilities

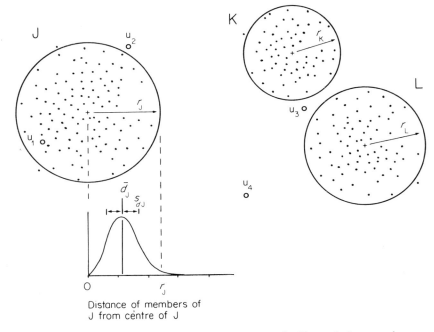

Figure 3. A taxon-radius system. Three taxa, J, K, and L, are shown diagrammatically as clusters of OTU's with envelopes defined by critical radii that take in almost all members of the taxa. The way the radius of taxon J is obtained is shown by the frequency distribution of the members of J from the centre of J at the foot of the diagram. The distribution curve allows estimation of the mean distance from the centre $\bar{d}_J$ and the standard deviation of the distances, s_{dJ}. The critical radius is set by choosing a level that is k standard deviations greater than $\bar{d}$. If k is 2.33 (as here), then in theory 99 per cent of members will lie within the envelope.

The figure also shows the positions of four unknowns, and the conclusions about their identity are as follows (see text for details): u_1 is a member of J; u_2 is an atypical strain of J; u_3 is an intermediate between K and L; u_4 is unidentified

that the unknown could show the test results that it does. This is expressed as relative likelihoods, and some criterion like 99 per cent corresponds to the complement of a taxon radius, and the likelihoods behave like the complements of a distance in a logarithmic form. The Bayesian method is illustrated in some of the figures and has the advantage of introducing differential weighting of characters. But to illustrate the principle, the simple taxonomic distance, d_{uJ} between u and the centroid of J is shown below.

$$d_{uJ} = \sqrt{\left[\frac{1}{m} \overset{m}{\Sigma} (X_{iu} - P_{iJ})^2 \right]}$$

where m is the number of characters, X_{iu} is the score of u on character i (either 1 for positive or 0 for negative), and P_{iJ} is the proportion of positives given by strains of taxon J on character i.

An example of such a distance calculation can be shown using a table of percentages for *Proteus* species. First it should be noted that these percentages are not set to 0 or 100 per cent, for two reasons: (1) because deliberate provision is made for rare exceptions, and (2) because with some statistics multiplications resulting from 0 or 100 per cent can give results of zero which are not permissible. Therefore 0 per cent is conventionally set to 1 per cent, and 100 per cent is conventionally set to 99 per cent, as in the table below.

					Taxa	
Tests		P. mirabilis	P. morganii	P. rettgeri	P. vulgaris	Unknown u
Gelatin hydrolysis 5 day		95	1	1	95	1
Simmon's citrate		50	1	95	5	0
Indole		1	99	99	90	0
H_2S in TSI		99	1	1	90	1
Ornithine decarboxylase		99	99	1	1	0
Gas from glucose		99	99	15	99	1
Mannitol acid		1	1	99	1	0
Sucrose acid		20	10	50	99	1
Distances from	mean	0.244	0.137	0.236	0.192	
centre:	s.d.	0.063	0.035	0.061	0.050	
Critical radius, r_J		0.391	0.219	0.376	0.309	

The unknown is a strain of *P. vulgaris* but it is atypical in its indole reaction. Its distance from the centre of the *P. vulgaris* hypersphere is obtained as follows. First, the percentages are divided by 100 to give proportions. Then the differences are calculated between the score in the unknown and the proportion in each test. Here they are respectively 0.05, -0.05, -0.90, 0.10, -0.01, 0.01, -0.01 and 0.01. These are the values of $X_{iu} - P_{iJ}$. They are now squared, and the squares summed, i.e. $0.05^2 + 0.05^2 + 0.90^2 \ldots$ etc., which comes to 0.8254. This is divided by $m = 8$ and the square root taken, giving 0.321. This is the distance d_{uJ} where J is *Proteus vulgaris*. The distances of u to the four taxon centres are:

Taxon	d_{uJ}
P. mirabilis	0.487
P. morganii	0.769
P. rettgeri	0.845
P. vulgaris	0.321

Clearly u is closest to *P. vulgaris,* and for this example the radius of *P. vulgaris* is 0.309, so u is only just outside, lying at a distance of 0.321. It is far outside the other hyperspheres. Therefore the conclusion that it is an atypical strain of *Proteus vulgaris* is upheld, and with one atypical test out of eight this would seem reasonable.

The way in which a taxon-radius system is used on a computer is illustrated by Figures 4 to 6. Several reference laboratories are now providing identification services of this kind for certain groups of bacteria. Although preliminary results can be obtained from the sender's test results, it is best to send the culture as well, so that checks or additional tests can be done with the reference laboratory's own standardized methods.

Taxon-radius models are the most powerful of the identification systems that are commonly used in microbiology, and they will often work well with fewer tests — perhaps half — than are needed for keys and diagnostic tables. So they offer potential savings in laboratory costs. They will work when the unknown has several exceptional test

```
PHSPROT    DEMONSTRATION: PROTEUS SPECIES    1975
0001 PROTEUS MIRABILIS
0002 PROTEUS MORGANII
0003 PROTEUS RETTGERI
0004 PROTEUS VULGARIS
000000000000000000000000000
GELATIN HYDROL 5 DAY
SIMMONS CITRATE
INDOLE
HYDROG SULPHIDE TSI
ORNITHINE DECARBOX
GAS FROM GLUCOSE
MANNITOL ACID
SUCROSE ACID
000000000000000000000000000
9550019999990120
0101990199990110
0195990101159950
9505909001990199
```

Figure 4. Identification matrix for species of *Proteus* in the form stored in the computer. The first line is the title, then come the four taxon names, a spacing row of zeros and the eight character names. Then comes another spacing row of zeros, followed by the percentage of positives for each character, one row for each taxon. The spaces between the percentages are omitted to save storage space; they are replaced in the correct position when the program is run. The percentages will be found to correspond to those in the identification matrix in the text in section IV.E. Percentages below 10 have a leading zero (e.g. 1 per cent is 01)

IDENTIFICATION PROGRAM MJSFDR

CODING SHEET FOR USE WITH MATRIX PHS PROT

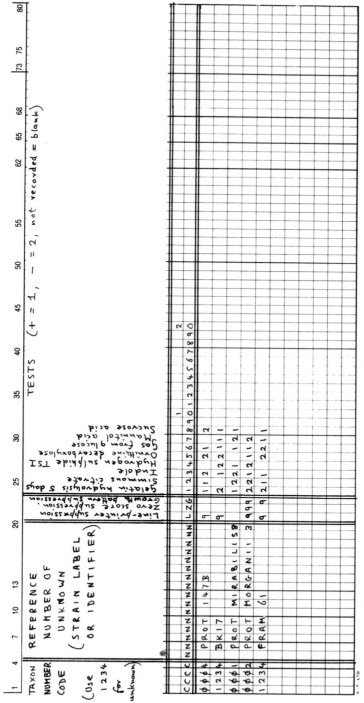

Figure 5. A coding sheet to be used with the matrix shown in Figure 4. The sheet is part of that for producing 80-column punched cards, which can then be prepared. Alternatively, the information can be entered at a keyboard terminal connected directly to the computer (either a teleprinter or a television screen).

The coding sheet contains the matrix name, the computer program against which it is to be run, and the symbols to indicate 'positive', 'negative', and 'not recorded'. Then follow, under the letters CCCC the believed identity of the strain, under the letters NNN ... the user's label for the unknown, then under the letters LZG three control symbols, followed by the numbers of the characters in correct order (the names are shown above each). These symbols appear at the terminal as a template under which the data for an unknown is entered. The test results for five strains have been entered on the sheet

```
FRIEDMAN IDENTIFICATION PROGRAM

PHSPROT    DEMONSTRATION: PROTEUS SPECIES    1975

ENTER UNKNOWN
    NUMBER CODE UNDER COLS CCCC
    NAME UNDER COLS NNNNNNNNNNNNNNNN
    LINE PRINTER SUPPRESSION OPTION UNDER COL. L (NUMBER = SUPPRESS)
    ZERO SCORE SUPPRESSION OPTION UNDER COL. Z (NUMBER = SUPPRESS)
    GROWTH PATTERN SUPPRESSION OPTION UNDER COL. G (NUMBER = SUPPRESS)
    THEN ENTER TEST RESULTS (1=POS., 2=NEG., SPACE=NO RESULT)
        UNDER NUMBERED COLS

   CCCCNNNNNNNNNNNNNNNNNNNNLZG
? 0004 PROT 147B       9
  OK?    TYPE YES OR NO
? YES
                1          2          3          4          5          6          7
   1234567890123456789012345678901234567890123456789012345678901234567890
? 112 21 2
  OK?    TYPE YES OR NO
? YES

ORGANISM  PROT 147B


  GROWTH PATTERN

GELATIN HYDROL 5 DAY + * SIMMONS CITRATE       + * INDOLE                 - *
HYDROG SULPHIDE TSI     * ORNITHINE DECARBOX    - * GAS FROM GLUCOSE       + *
MANNITOL ACID           * SUCROSE ACID          - *


RELATIVE LIKELIHOOD

   PROTEUS MIRABILIS     .794
   PROTEUS VULGARIS      .005
   PROTEUS RETTGERI      .001
   PROTEUS MORGANII      .000

  MOST LIKELY:  PROTEUS MIRABILIS      .794

  LABORATORY DIAGNOSIS  PROTEUS VULGARIS

  ENTER NUMBER OF TAXA FOR WHICH ADDITIONAL TESTS ARE REQUIRED.  MUST BE
  A SINGLE DIGIT NUMBER
? 2

  TEST RESULTS THAT ARE AGAINST THE MOST LIKELY IDENTIFICATION

ORNITHINE DECARBOX    99


  ADDITIONAL TESTS THAT WILL AID IN DIFFERENTIATING  PROTEUS MIRABILIS
  FROM  PROTEUS VULGARIS     (PRINTS DIFFERENCES)

     (NONE)


  ADDITIONAL TESTS THAT WILL AID IN DIFFERENTIATING  PROTEUS MIRABILIS
  FROM  PROTEUS RETTGERI     (PRINTS DIFFERENCES)

HYDROG SULPHIDE TSI  98    MANNITOL ACID          98
```

Figure 6. An example of a computer run with the *Proteus* matrix.

The user calls up the identification program and then the matrix he wishes to use (here the *Proteus* matrix with the label PHSPROT). The computer then prints at the terminal the significance of the symbols of the template, and then the first part results which would prevent success with keys or polyclaves. Further, they can readily take into account numerous characters, far more than one can remember, because the reference library can be as large as one desires.

F. Discriminant Analysis

Discriminant analysis is a powerful but very specialized statistical method that gives the most reliable identifications with groups that partly overlap in their properties (usually but not necessarily quantitative characters). It is best suited to use with a few taxa; commonly it is to distinguish between just two of them. The mathematics is rather elaborate (details can be found in sources in the Further Reading), but the principle is easy to show diagrammatically (Figure 7). It has seldom been used in practical microbiology, and is included only for completeness. It can be seen from the figure that the process of making the clusters as hyperspherical as possible will only be successful if the original clusters are rather similar in size, shape, and orientation. Whether the benefit in microbiology would repay the effort of combining this method with the taxon-radius system is still uncertain: most clusters in bacteria seem to be approximately hyperspherical already.

---

of the template. Under CCCC the user enters the reference number of his own diagnosis (here 0004 = *P. vulgaris*). He then enters the strain number or label under NNN . . . , and appropriate control symbols under LZG. The computer then asks the user to check the entry is correct. If it is, it prints the template for the test results, which the user enters under the appropriate test numbers. After the check for correctness the computer prints out the diagnostic information shown.

(a) the strain label
(b) the test results entered, in plain language
(c) the likelihood of identity. The statistic used here is due to Friedman, and .794 indicates a nominal likelihood of 79.4 per cent that the unknown is *P. mirabilis*, and only 0.5 per cent it is *P. vulgaris*. The other two are even less likely diagnoses.
(d) After entering the number 2 in response to the request from the computer, the program lists the test results against the most likely identification. Here it is atypical for *P. mirabilis* in the ornithine decarboxylase test.
(e) Then the computer prints any additional tests that could serve to differentiate the most likely identity from the next two possibilities. Hence if they were done and the strain run again, one might clinch a doubtful diagnosis. No extra tests are available that would separate *P. mirabilis* from *P. vulgaris*, but there are two that would separate it from *P. rettgeri*

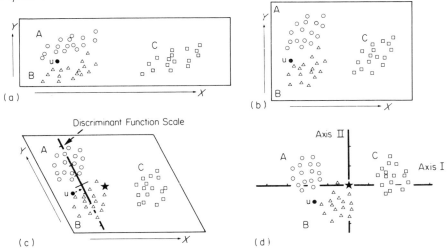

Figure 7. Discriminant analysis. The steps in discriminant analysis are shown diagrammatically for three taxa A, B, and C with two quantitative characters, X and Y, and an unknown, u.

(a) The original character states for the OTU's, which show overlap between A and B on both characters.

(b) The characters are scaled so that the average within-taxon scatter on each character is equal. The scale of X has been shrunk and that of Y stretched.

(c) The angle between the character axes has been changed so as to make the shape of the taxon clusters as nearly spherical as possible. A Discriminant Function scale for discriminating between members of A and B is shown by the heavy line, and u lies below the crossbar midway between A and B; it is therefore identified as a member of B. The taxa do not overlap in this transformed character space.

(d) The major axes of variation (Principal Axes I and II) may also be found by rotation of (c) around the grand centre of the system shown by a star. Distances in this transformed space (Mahalanobis D-space) can be measured between u and the centre of each taxon: this also shows that u is a member of B

V. Construction of an Identification System

A. Preparation of an Identification Matrix

The details of the several identification systems described earlier will allow the reader to have some idea of the kind and volume of data needed for each system. Some general points require mention here. Commonly the results of numerical taxonomic analyses can produce diagnostic data in a very convenient form. This can be obtained by feeding back into the computer the original data on the OTU's together with the lists of group members for those groups (taxa, phenons), that it has been decided to recognize as a result of the taxonomic analysis.

The choice of taxa and of tests should be carefully considered. The taxa must be as comprehensive as possible within the set of limits of the system, e.g. all species of a genus should if possible be included for a generic system. The type of habitat from which the microorganisms will come is also important; one would omit for example, bacteria that are obligate pathogens of plants when making a key for human pathogens. The choice of characters is equally important. The selection index described earlier can help to screen out poor diagnostic characters, and those that are particularly difficult to determine accurately should be discarded. Note too, should be taken of the standard selections of tests used in previously published systems, and widely used tests should be included; it can be frustrating to discover that a system cannot be checked against published sources because some standard properties like catalase or motility have been omitted. Careful standardization of the tests is part of the identification system although the full technical details do not appear in the matrix itself. These details should therefore be recorded in the associated documentation.

The construction of identification systems based on serology or bacteriophage reactions is a specialized subject. It is only necessary to note that the purpose of these systems is generally different from those that identify to the species level, because most serological and bacteriophage systems are designed to characterize a large number of serological or phage variants that can be used for epidemiology. Consequently, the battery of antisera or phages is built up in such a way that the majority of strains encountered will fall into one of the numerous variant patterns. Efforts are thus made to reduce the proportion of strains that remain ungroupable by preparing further antisera or finding further phages. Sera or phages that give clear-cut reactions are continuously sought, and new antigenic or phage-resistance schemes may repeatedly supersede old ones.

B. Evaluation of Identification Matrices

When a matrix has been constructed it is wise to evaluate its quality. Much can be done by testing it against authentic strains of the species represented, but large-scale testing is tedious. Efforts should be made to fill in any gaps in the data. Separation indices of the characters can be calculated and valueless characters removed. Also, the ability of the matrix to distinguish all pairs of taxa, the extent of overlap of taxa, and the highest scores with typical strains, can be examined by special computer programs now being developed. In all systems there will be difficulty in getting perfection, and the major factors that must be weighed against each other are the reliability of identification and the proportion of unknowns that are identifiable. Improvement in one of these may lead to worsening in the other.

C. The Effect of Errors on Identification

Errors will obviously cause mistakes in identification. These errors can be of two kinds: (a) mistakes in the descriptions of the taxa which are used to set up the identification system (e.g. describing a species as lactose variable when it is always lactose positive); and (b) mistakes in the description of the unknown u that one is trying to identify (e.g. scoring an unknown as acid-fast when it is not acid-fast). Type (a) error is often partly removed when setting up keys or diagnostic tables, but is important with the more sophisticated systems. Type (b) errors is clearly always important.

In general, small amounts of error will first spoil the identification of strains of the tightest clusters, i.e. tight clusters are most sensitive to error. Again, with some systems (e.g. keys) the error may lead to many incorrect identifications, whereas with others (e.g. taxon-radius models) it more often leads to failure to identify at all rather than incorrect identifications. The effect of error in a diagram such as Figure 3 is to make the clusters appear more diffuse and to perturb erratically the position of an unknown.

Another kind of uncertainty springs from inadequate numbers of OTU's of a taxon. This is most easily envisaged in a taxon-radius model, in which we wish to estimate as accurately as possible the position of the centre and its radius. Clearly a sample of one strain tells us nothing about its radius, and is only a poor guide to the true centre. Statistical sampling theory shows that the average discrepancy between the true centre and the observed one is about $1/\sqrt{t_J}$ where one has a sample of t_J strains of taxon J. To get the centre accurate to about 1/5 of the radius, we need about $t_J = 25$ strains. A similar relation holds for the radius. It is recommended that a minimum of 10 strains is used. The effects of a small sample are similar for other identification systems: a key based on observations of only one or two strains would carry little conviction.

VI. Sources of Assistance

A. Published Identification Systems

The microbiologist will often have to seek assistance with identifications in difficult groups. Such assistance is obtained in two ways, from specialist publications in diagnostic microbiology (books, monographs, etc.) and from reference laboratories.

In the section on Further Reading are a number of important published works. This list is not comprehensive, but two sources contain lists of reference books for microorganisms.

B. Reference Laboratories

Reference laboratories are usually supported by national institutions. There appears to be no published list of these that covers all groups of microorganisms, but there are several that deal with substantial areas. The Public Health Laboratory Service (Central Public Health Laboratory, Colindale Avenue, London N.W.9) has reference laboratories for human and animal pathogens, or has arrangements for referral to specialists on these. Plant pathogens and most microfungi are covered by the Commonwealth Mycological Institute (Ferry Lane, Kew, Surrey). Algae are studied principally at the Freshwater Biological Association, (Ferry House, Far Sawry, Cumbria). Information officers at the Royal Botanic Gardens, Kew, and at the British Museum of Natural History, South Kensington, London, may be able to provide names of experts for the more obscure groups.

It should be noted that although most of the reference laboratories and experts are very willing to assist on identification of microorganisms, their time and resources are often insufficient for them to undertake much work. This is frequently available only as a special favour if it is requested from sources outside their own organization. They will not be able to do routine identifications of common microorganisms, although they can provide information on how these should be carried out. A number of commercial organizations are also able to offer certain services and products.

VII. Further Reading

The following references refer mainly to identification systems.

Gyllenberg, H. G. (1965). A model for computer identification of microorganisms. *J. Gen. Microbiol.*, 39, 401–405.

Hope, K. (1968). *Methods of Multivariate Analysis*. London: University of London Press.
This is one of the simplest introductory texts to discriminant analysis, and is very clearly written.

Lapage, S. P., Bascomb, S., Willcox, W. R., and Curtis, M. A. (1973). Identification of bacteria by computer. I. General aspects and perspectives. *J. Gen. Microbiol.*, 77, 273–290.
A general account of a computer-based method. Two companion papers (pp. 291–315, 317–330) report on practical applications and results with bacteria.

Pankhurst, R. J. (Ed.) (1975). *Biological Identification with Computers*. London: Academic Press.
A collection of papers in the expanding field of numerical identification.

Sneath, P. H. A. (1972). Computer taxonomy. In *Methods in Microbiology* Vol. 7A (Norris, J. R., and Ribbons, D. W., Eds.) pp. 29–98. London: Academic Press.

Sneath, P. H. A. (1974). Test reproducibility in relation to identification. *Int. J. Syst. Bacteriol.*, 24, 508–523.

Sneath, P. H. A. and Sokal, R. R. (1973). *Numerical taxonomy.* San Francisco: W. H. Freeman.
This includes a general review of identification systems.

The following contain lists of key works for microorganisms.

Ainsworth, G. C. and Sneath, P. H. A. (Eds.) (1962). Some key works on the taxonomy of microorganisms. *Symposia of the Society for General Microbiology,* 12, 467–476.
Kerrick, G. J., Meikle, R. D., and Tebble, N. (1967). *Bibliography of Key Works for the Identification of the British Fauna and Flora.* 3rd ed. London: Systematics Association.

The following are on particular groups of microorganisms.

Ainsworth, G. C. (1971). *Ainsworth and Bisby's Dictionary of Genera of the Fungi.* Commonwealth Mycological Institute, Kew, Surrey.
This contains useful keys to the major fungal groups, and lists of special works.
Barnett, J. A. and Pankhurst, R. J. (1974). Amsterdam: *A New Key to the Yeasts.* North-Holland.
Buchanan, R. E. and Gibbons, N. E. (Eds.) (1974). *Bergey's Manual of Determinative Bacteria.* Baltimore: Williams & Wilkins.
Cowan, S. T. and Steel, K. J. (1974). *Manual for the Identification of Medical Bacteria,* 2nd ed. Cambridge: Cambridge University Press.
Gibbs, B. M. and Skinner, F. A. (Eds.) (1966). *Identification Methods for Microbiologists, Part A.* London: Academic Press.
Gibbs, B. M. and Shapton, D. A. (Eds.) (1968). *Identification Methods for Microbiologists, Part B.* London: Academic Press.
Skerman, V. B. D. (1967). *A Guide to the Identification of the Genera of Bacteria.* 2nd ed. Baltimore: Williams & Wilkins.
Wilson, G. S. and Miles, A. A. (1975). *Topley and Wilson's Principles of Bacteriology, Virology and Immunity.* 6th ed. 2 vols. London: Arnold.

Recent developments in methodology are described in the following:

Hedén, C. -G. and Illéni, T. (Eds.) (1975). *New Approaches to the Identification of Microorganisms.* New York: John Wiley.

The Genetic Organization of Bacteria and its Expression

M. H. RICHMOND

*Department of Bacteriology, University of Bristol, University Walk,
Bristol, BS8 1TD England*

I. Introduction

Bacterial cells multiply by binary fission, and this means that every atom and molecule in the parental cell has to be duplicated and inserted into its correct place in the two daughter cells by the time the process is complete. This enormous effort of synthesis may take place in a very short time in some microbes. For example, many common bacteria achieve this doubling in as little as an hour, and even though some duplicate themselves much more slowly, others may complete the process in as little as 20 minutes.

When considering the multiplication of bacteria, as is the case with all living organisms, the analogy of the construction of a building may be helpful. First, there must be plans to provide the necessary information for the builder to undertake his job; then one must have a supply of materials — wood, bricks, and so forth — with which to manufacture the parts that go to make up the building, and, finally, there is the construction and organization of the component parts that ultimately give rise to the complete structure. As with the building industry, energy is needed, and its provision at the right place and time is a critical part of the construction process.

Each of these stages in the construction of a building has its analogy in the duplication of a bacterial cell. Before any bacterial cell can be fashioned, instructions are needed. This is provided by the genetic information that is carried in the DNA of the parental cell. Secondly, a bacterial cell, like a building, is not fashioned by the assembly of entirely prefashioned parts. Some on-site construction from the basic building materials is necessary, and in the bacterial context, this process is the metabolic interconversion of small organic molecules to make those compounds that themselves become polymerized to become the bacterial macromolecules. These precursors of the bacterial macro- molecules — the 'building blocks', as they are often called — next have to be converted to the macromolecules which form both the structural architecture of the cell and also the catalytic machinery that keeps the cell functioning. Finally, the whole process is driven forward by a supply of energy.

Of course, the building analogy must not be pressed too far. Unlike a building, a bacterial cell is alive, that is, it carries within itself the necessary information to catalyse its own duplication. Were buildings self-reproductive in this fashion, the 'concrete jungle' would have engulfed us all by now! Nevertheless, in its need for building blocks with which to make the macromolecules that are so essential to living systems and in its requirement for energy to carry out the construction process, the analogy is useful.

A. DNA and Information

The information needed to determine the structure of those macromolecules which, in turn, go to make up the bacterial cell is carried by the cell's DNA. This molecule can be imagined as a very long molecular thread, itself composed of two strands. Each strand consists of a long polydeoxyribophosphate polymer and the two strands are wound round one another as shown in Figure 1. More detailed examination of the chemical structure of the polydeoxyribophosphate chain (Figure 2) shows that it has 'direction'. Each sugar residue is substituted with phosphates on both the 3′ and the 5′ oxygen atom, and each strand can, therefore, be thought of as running in a 'direction' (the 3′ to 5′ direction is shown by the arrow in Figure 2). In DNA, the two strands that go to make up the molecular thread are arranged to

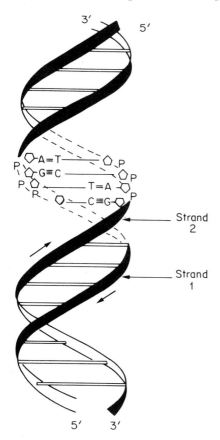

Figure 1. Diagrammatic outline of the structure of DNA. Note that the two strands run in an anti-parallel manner

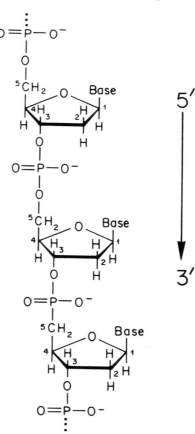

Figure 2. Structure of a section of one strand of DNA to show the 'direction' of the strand

run in the opposite direction to one another. Therefore, DNA is said to be an 'anti-parallel double-stranded molecule'.

DNA does not consist solely of deoxyribose and phosphate: each sugar residue also carries a purine or pyrimidine base. In practice, DNA contains four such bases (adenine, thymine, guanine, and cytosine), but the overall structure of the DNA molecule is such that an adenine is always placed opposite a thymine, and a guanine opposite a cytosine. This so-called 'base-pairing' (Figure 3) is the intramolecular arrangement that maintains the two polydeoxyribophosphate chains in their correct juxtaposition: that is, the base pairing maintains the conformation of the antiparallel double helix.

At the beginning of this section, it was stressed that bacterial DNA was a very long molecule but just how long is often not appreciated. In many prokaryotes, for example, the DNA molecule may be up to 1 mm long. This means that the DNA in the bacterial cell, if stretched out to the full would be more than 200 times longer than the cell in which it is located. In order to accommodate such a quantity of DNA in the bacterial cell, therefore, the molecule is extensively supercoiled.

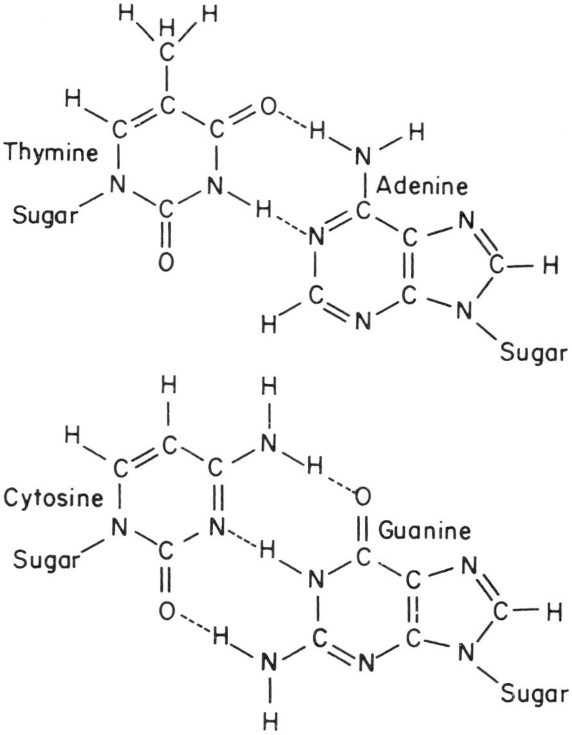

Figure 3. The structure of the adenine:thymine and guanine:cytosine base-pairs that hold the two strands of DNA in their correct double-helical conformation

Although this allows more DNA to be accommodated within the cell, it nevertheless may make the problem of separating daughter molecules at cell division more difficult.

Since most of the DNA in bacterial cells is present as a single molecule another way of thinking of the structure is to consider its molecular weight. In most normal bacteria, the single DNA molecule that comprises the bacterial chromosome has a molecular weight of about 2×10^9, and this means that the molecule contains approximately 3×10^6 nucleotide pairs.

Since the information content of DNA lies in the order of base pairs along the molecule, the total number of base pairs in turn determines the total amount of information carried. This is because the DNA base sequence determines the order of insertion of amino acids into proteins. As a consequence, the total chromosomal DNA of a bacterial cell contains enough information to code for the insertion of about one million amino acids into individual types of protein.

B. DNA:RNA:Protein

Proteins, in many ways, play the central role in the structure and metabolism of bacterial cells as they do in all living systems. Not only do they form important structural components of the cells themselves, they also give rise to enzymes, molecules which catalyse all the metabolic steps that make up the cell's economy. Even the other macromolecules (carbohydrates, lipids, etc.) that play an important part in the bacterial cell are synthesized by proteins. It follows, therefore, that the informational content of a cell's DNA must be expressed in the cell as protein if it is ultimately to be used to the full.

In practice the information that is contained in the base sequence of the DNA is translated into protein structure by means of a ribonucleic acid intermediate. This type of RNA — the so-called 'messenger RNA or "m-RNA"' — is a linear single stranded nucleic acid molecule. Unlike DNA, which is basically a poly*deoxyribo*phosphate chain with purine and pyrimidine attachments, RNA is a poly*ribo*phosphate chain substituted by purine and pyrimidine bases. A further difference between RNA and DNA lies in the nature of the purine and pyrimidine bases involved. Whereas DNA contains adenine, thymine, guanine, and cytosine, RNA contains uracil in place of thymine.

Although RNA and DNA have these clear structural differences between them, nevertheless the important point, at least as far as information transfer is concerned, is that the base sequence along the polyribophosphate chain of RNA can conserve the information present in the base sequence of the DNA. This is achieved by an extension of the base-pairing principle already discussed in relation to DNA structure; thus, cytosine and guanine recognize one another whichever type of nucleic acid they occur in, whereas the adenine/thymine recognition in DNA is replaced by adenine/uracil recognition in the synthesis of RNA. As a result of these patterns of recognition, therefore, one base pair in the DNA molecule corresponds to a single base in RNA, and the order of the purine and pyrimidine bases along the RNA strand therefore corresponds to the order of bases along one of the two DNA strands. In this way, therefore, the informational content of DNA is *transcribed* into the form of single-stranded messenger RNA.

Once the informational content of a region of the DNA has been transcribed to an RNA form, it may be 'translated' by the cell into the amino acid language of proteins. This step is achieved by the process we know as protein synthesis; the mechanisms involved will be described in detail later. At this point, it is merely necessary to stress that the order of the purine and pyrimidine bases along the RNA molecules provides the necessary information to ensure the insertion of the correct amino acid into its appropriate place in the polypeptide chains, which in turn go to make up the fully fashioned protein.

Bacterial proteins contain 19 amino acids in varying proportions and each must be specified unambiguously by the messenger RNA if mistakes in protein synthesis are not to arise. However, the informational content of the RNA is based on only four different purine and pyrimidine bases. Clearly, therefore, a sequence of more than one base must be used to specify each amino acid. In practice, a sequence of three adjacent bases on the RNA strand is used for this purpose. Thus one has a system of informational transfer in which three *base pairs* in DNA specify three equivalent *single bases* in the messenger RNA, and each set of these three bases in turn specifies the insertion of a single amino acid. Thus in bacteria, as in all living systems, there is colinearity of DNA, m-RNA, and the polypeptide chains that go to make fully fashioned proteins.

Each triplet of RNA bases that specifies the insertion of an amino acid is called a codon, and, since there are four different bases in RNA, there will be 64 distinct codons if each one is made up of three bases. But, as proteins contain only 19 amino acids, certain of them are specified by more than one codon, and the code used to translate the purine/pyrimidine language of RNA into the amino acid language of proteins is therefore said to be 'redundant'. The actual number of codons used for each amino acid and their sequences is shown in Table 1.

C. Gene:'Message' Polypeptide

Although the DNA that forms the bacterial chromosome is a single molecule of molecular weight about 2,000,000,000, the individual

Table 1. The RNA version of the genetic code

First base	Second base				Third base
	U	C	A	G	
U	Phe	Ser	Tyr	Cys	U
	Phe	Ser	Tyr	Cys	C
	Leu	Ser	C.T.	C.T.	A
	Leu	Ser	C.T.	Trp	G
C	Leu	Pro	His	Arg	U
	Leu	Pro	His	Arg	C
	Leu	Pro	GluNH$_2$	Arg	A
	Leu	Pro	GluNH$_2$	Arg	G
A	Ile	Thr	AspNH$_2$	Ser	U
	Ile	Thr	AspNH$_2$	Ser	C
	Ile	Thr	Lys	Arg	A
	Met*	Thr	Lys	Arg	G
G	Val	Ala	Asp	Gly	U
	Val	Ala	Asp	Gly	C
	Val	Ala	Glu	Gly	A
	Val	Ala	Glu	Gly	G

Abbreviations: C. T., Chain termination triplets; otherwise standard abbreviations for amino acids and purines and pyrimidines are used.* The triplet AUG codes for the insertion of N-formyl methionine when it occurs as the first codon in a cistron. Reproduced by permission of Blackwell Scientific Publications Ltd.

proteins formed by using the information carried in this molecule rarely have molecular weights more than 100,000 each. Hence, bacteria must have some way in which the continuous informational thread of the DNA is read in a discontinuous fashion.

The fragmentation of the information from the DNA to form the individual proteins occurs at two stages. First, the information is organized into functional sections in such a way that it is transcribed into a series of independent messenger RNA molecules. The molecular weight of such 'messages' varies widely depending on the part of the DNA they arise from, but commonly their molecular weights lie between 2×10^5 and 5×10^5. Thus, the full DNA content of a bacterial cell may give rise to between 2000 to 3000 individual messenger RNA molecules.

Some of these RNA molecules are translated into single polypeptide chains which then fold up to become single functional proteins. On the other hand, many messenger molecules are capable of leading to the synthesis of more than one polypeptide, and these can then either fold up individually or interact to form single proteins that contain more than one type of subunit. Alternatively, they may give rise to a number of independent proteins. Thus, at the stage of translation, the information implicit in the base sequence of the messenger RNA may be further fragmented to give rise to more than one polypeptide chain and, as a consequence, more than one protein.

The fact that a given protein can ultimately be shown to reflect the base sequence of the DNA over a restricted region allows one to identify stretches of the DNA responsible for the synthesis of given polypeptide chains and proteins. These regions are known as cistrons and structural genes, respectively. Where a single messenger RNA gives rise only to a single polypeptide, the molecule is known merely as a simple messenger, but where the messenger gives rise to a number of distinct polypeptides, it is said to be *polycistronic*. Figure 4 summarizes this arrangement, one that forms the basis of the 'colinearity of gene and protein', and which is fundamental to biological systems.

In practice, not all the DNA is transcribed into RNA, and even less is finally expressed in the form of polypeptide. This is because certain regions of the DNA have a regulatory function, and as such do not

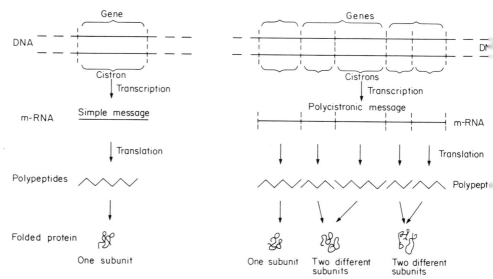

Figure 4. Colinearity of gene, RNA, and protein. In the left hand section of the diagram single gene codes for a single messenger RNA molecule, and this, in turn, gives rise to protein made up of a single folded polypeptide chain. In this case, the gene has a sing cistron. In the right hand section of the figure, an operon of three genes and five cistrons transcribed to give a single polycistronic messenger RNA. This message is then translated give rise to three proteins, one containing a single polypeptide chain and the others two, ea containing two separate subunits of different structure

necessarily give rise to a macromolecular product. For example, regulatory regions that are responsible for controlling the synthesis of messenger RNA molecules may not themselves usually produce an RNA copy. Hence, for example, the binding site for RNA polymerase – an enzyme crucial for m-RNA synthesis – is part of the bacterial DNA itself, and this region cannot code for a messenger molecule and still remain functional. Similarly, certain regions of the messenger RNA are not translated into polypeptide. Among these are binding sites for components associated with the translation of message into polypeptide while others are concerned with the termination of synthesis of one polypeptide chain or the commencement of another.

The fact that the continuous information loop of the DNA can be fragmented into sections in this way, implies that there must be some signals within the formation that cause these fragmentations to take place. As has been implied already, some of these are regions of DNA that act as binding sites for proteins. Others, in the messenger RNA, are triplets which do not lead to the insertion of amino acids into growing polypeptide chains but rather act either as stop or start signals. Such triplets are known as 'chain initiator' or 'chain terminator' triplets. The latter are designated as 'CT' in Table 1. We will return to 'stop' and 'start' signals of various kinds later.

II. Bacterial DNA

A. DNA Organization in the Bacterial Cell

The majority of the DNA in the bacterial cell forms a single circular molecule known as the chromosome. As well as carrying all the information needed to synthesize the cell proteins and to carry out the organism's metabolism, this piece of DNA also contains the necessary information to duplicate itself in step with the growth of the cell so that each daughter bacterium is equipped with one copy of the parental chromosome at division.

Even though the chromosomal DNA comprises the great majority of the bacterial DNA, there are commonly other DNA molecules in bacteria. Some of these are the so-called extrachromosomal elements or plasmids. These elements are also circular DNA molecules, but they are usually much smaller than the chromosome. Whereas the former – as we have seen – has a molecular weight of about 2×10^9 daltons, bacterial plasmids are usually in the range of 1×10^6 to 10^8. Thus, more than 90 per cent of the bacterial DNA is usually present as 'the chromosome'.

Plasmids replicate independently of the bacterial chromosome, and also carry the necessary information to balance their rate of doubling so that the daughter cells get the correct complement of plasmid DNA at division. Thus, in a sense, bacterial plasmids can be thought of as small supernumerary chromosomes.

If one tries to distinguish the concept of chromosome from that of plasmid, the only character that seems to apply unfailingly is size. Thus, one can say that the chromosome is merely the largest plasmid in the cell. On occasion, however, various writers have tried to differentiate plasmids from chromosomes on the basis of the type of information they carry. Hence, plasmids often seem to carry genes that are adventitious in the sense that it is relatively easy to find growth conditions in which the plasmid can be lost from the cell without the event being lethal. But any statement that those plasmids which can be lost are 'non-essential' is unsatisfactory since the characters in question are often needed for growth under *some* conditions and, consequently, cannot be said to be completely non-essential. A case in point is the antibiotic resistance genes which are often plasmid-mediated in bacteria. Under some circumstances, these genes can be lost without lethal effect to the bacteria concerned, but this is only true when no relevant antibiotics are in the environment. Is growth in the presence of antibiotics to be regarded as an essential character? Certainly for bacteria trying to grow in hospitals it will be. Ultimately, therefore, it is better to distinguish plasmids from the chromosome by avoiding value judgements and rely solely on size.

Pieces of DNA which can replicate themselves in a balanced way so as to provide the number of new copies needed for the daughter cells at division are said to be independently replicating units — or *replicons*. In practice, whether or not a piece of DNA is a replicon is extremely important because this fact determines whether it can survive in bacterial cells independently of chromosomal replication.

One of the most important experimental advances over the last few years has been to learn how to separate bacterial plasmid DNA from the chromosomal material, since an ability to isolate and to handle pure plasmid DNA lies at the heart of the present techniques for '*in vitro* recombination' — or 'genetic engineering', as it is more topically known. Only after the plasmid has been obtained free from the cell and from contaminating DNA can the necessary manipulations be carried out to join it in the test tube to DNA from other sources.

The separation of plasmid from chromosomal DNA is carried out in caesium chloride density gradients in an ultracentrifuge. The total bacterial DNA is first liberated from the bacterial cell by lysis under appropriate conditions. This process fragments the chromosomal DNA into relatively small linear pieces, but leaves the majority of the plasmid DNA intact in its circular conformation. This difference in behaviour has nothing to do with the base sequence of the two types of DNA: rather it occurs because the extremely long chromosomal molecule is much more sensitive to the mechanical shearing forces that occur during the isolation procedure than are the much smaller plasmid molecules.

In order to separate plasmid from chromosomal DNA, one takes advantage of this fact that the preparative procedure produces circular DNA from plasmids but linear DNA from the chromosome. Ethidium bromide is a drug that binds to DNA, but it binds to different extents to, and has different effects on the centrifugation properties of, circular and linear DNA. Hence, centrifugation of a mixed plasmid and chromosomal DNA in a CsCl density gradient containing ethidium bromide allows plasmid DNA to be separated cleanly from the chromosomal material.

Before leaving the topic of the organization of DNA in bacterial cells, one must point out that many bacteria carry virus particles (bacteriophages) and many of these contain DNA. In practice, phage DNA is carried in bacteria in one of two forms: either integrated into another replicon — commonly the chromosome — or free as additional extrachromosomal elements. In the latter state, it can be regarded as a special class of plasmid that happens to carry the necessary genes to specify the formation of those protein characters — coat proteins, tail fibres, enzymes, etc. (see Chapter 4) that go to make up a typical vegetative phage. In general, mature phage — in which the DNA is enclosed in the phage coat — is only found in bacteria in which either carried phage have recently been induced to produce mature particles, or where a lytic cycle of infection is under way. In both cases, the process is likely to prove fatal to the bacterial cell in which it is taking place.

B. DNA Replication

All DNA in the bacterial cell is duplicated by essentially the same process known as semiconservative replication. By this method, new strands are synthesized using the base sequence on each 'old' strand to generate a part of complementary 'new' strands. This process is summarized in Figure 5. In this way, the structure of the original molecule is conserved; two double-stranded DNA molecules emerge where there was one before.

The actual biochemical steps involved in this process are relatively straightforward — at least up to the point at which the new nucleotide is coupled into the growing DNA strand. Since the structure of the polydeoxyribophosphate backbone of each strand is consistent along its length, the insertion of each new nucleotide obeys the following common pattern:

$$(DNA)-3'-P + dNTP \longrightarrow (DNA)-3'-P-dN-3'-P + PP_i$$

where dNTP denotes a triphosphate of one of the four deoxyribonucleotides involved in DNA. This process is shown in greater detail in Figure 6.

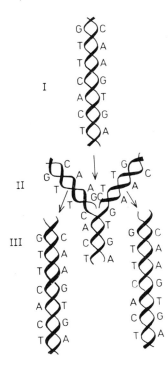

Figure 5. The semiconservative replication of DNA: I, The parental molecule, II, Replication taking place, III, The process complete

Although DNA synthesis looked at at this level of resolution seems straightforward, problems arise when the process is examined in terms of its ability to duplicate a very long molecule such as a bacterial chromosome. The difficulty is that the replication of each DNA strand proceeds in the direction $3' \rightarrow 5'$ along each polynucleotide strand, yet the experimental evidence seems to show without doubt that chromosome replication starts at one point on the replicon (the initiation site) and moves along the DNA, implicating the DNA at approximately the same point *on both strands* at a given time. Since such a continuous motion can only be in the direction $3' \rightarrow 5'$ on one of the two strands, it must be continuous on one and discontinuous on the other. This situation is summarized in Figure 7.

It is still not absolutely clear how the bacterial cell solves this problem. Certainly, if the replication of DNA is followed experimentally, there is evidence for continuous synthesis on one strand and discontinuous on the other. But how the fragments generated by the discontinuous process are then joined together is still far from clear.

The need to replicate DNA in one direction on both strands at the same time, is not the end of the difficulties. Next, there is the question as to whether replication starts from the initiator site and proceeds round the replicon unidirectionally, or whether bidirectional replication occurs. The overall patterns involved in these two systems are

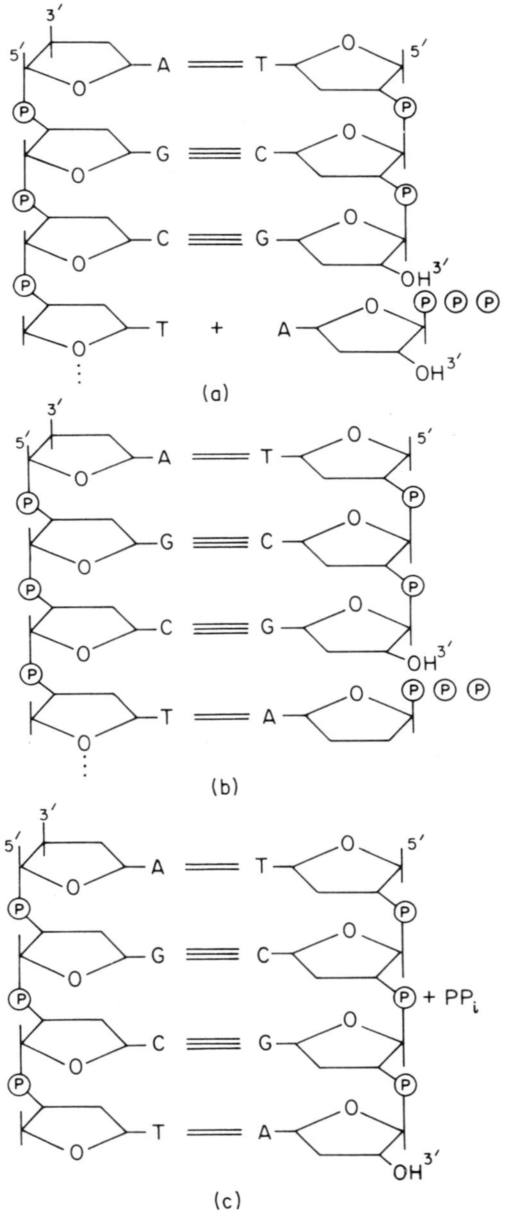

Figure 6. The biosynthesis of DNA: (a) Deoxyadenosine triphosphate is drawn to its correct position by forming a base pair with the thymine residue, (b) hydrogen bonds form, (c) the formation of the phosphate bridge takes place and pyrophosphate (PP$_i$) is expelled

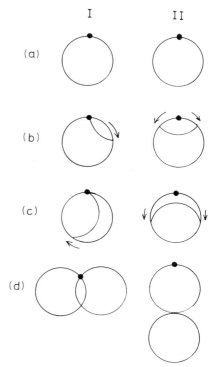

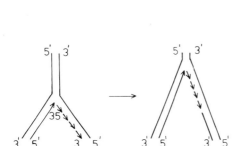

Figure 7. DNA synthesis. Since the direction of DNA synthesis only occurs in the direction $5 \rightarrow 3$, semi-conservative replication requires discontinuous synthesis on one strand. This figure illustrates two stages in the process. The products of the discontinuous synthesis are joined up to form a continuous molecule as a separate step in the replicative process

Figure 8. Four stages (a–d) in the unidirectional (I) or bidirectional (II) replication of a circular DNA molecule. In the first, termination occurs beside the initiation point (●). In the second, termination is opposite to the initiation point

summarized in Figure 8. In the one, the end point of replication (the terminator) is beside the initiator, while in the second, the terminator lies opposite the initiator.

Experimental evidence suggests strongly that chromosomal replication in bacteria is bidirectional, but the evidence for plasmids and phages is much less clear cut. On balance, it seems that examples of both uni- and bidirectional replication are to be found among the smaller bacterial replicons.

Before leaving the question of DNA replication in bacteria, it is important to mention another form of DNA replication, proposed by Gilbert and Dressler, which is known as the 'Rolling-Circle' model. The details of this model are shown in Figure 9. Whether this type of replication is ever used in bacteria for the synthesis of DNAs other than viral is uncertain. The main feature of this type of replication is that it

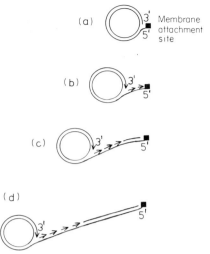

Figure 9. Four stages (a–d) during the replication of a circular double-stranded DNA molecule according to the 'rolling-circle' hypothesis. The method of replication is capable, in principle, of stopping once a single copy of the DNA has been produced. Alternatively, it can produce multiple copies of the parental structure

leads to the formation of multiple copies of the parental molecule, and correspondingly this model obviously has its attraction as a basis of vegetative phage multiplication where multiple copies of a parental molecule need to be formed rapidly.

The final problem in relation to DNA replication is the method whereby the daughter replicons are separated and distributed accurately to the daughter cells, before the growing cross-wall separates the two new bacteria at division. Basically, there would seem to be two ways in which this might be achieved. Either, on the one hand, a specific mechanism draws one copy of each replicon at some point in the division cycle, or on the other, the two copies are partitioned on a random basis. For the first to be effective, it is generally agreed that some physical contact must exist between the DNA and the structural framework of the dividing cell. In this way, movement apart of two points on the cell membrane – or some similar structure – can draw the two copies of the DNA apart. For the second to be effective, statistical considerations show that the replicons concerned would have to be present in the bacteria with an abundance of greater than two per dividing cell, if a high incidence of failure to transmit the replicon were not to be observed. Certainly this latter process is unlikely to be used by the chromosome, since experimental evidence shows clearly that there is not a high incidence of 'still-births' at bacterial division — something that should be bound to occur were the chromosomal copies to be partitioned on a random basis.

Among plasmids the situation is not so clear cut. While many plasmids are present in the cell at a level of one/two copies per chromosome, others survive as multiple copies — sometimes as many as 40–50 a cell. Although no clear decision is yet possible it seems that the first group of replicons may be distributed via a membrane attachment mechanism, while the multicopy elements may rely on dilution.

III. Bacterial RNA

A. RNA Structure

RNA is present in three main forms in the cell: messenger (or m-RNA), ribosomal (or r-RNA), and transfer (or t-RNA). In each type of RNA, however, the essential structure of the molecule is the same.

Like DNA, RNA has a backbone of alternating sugar and phosphate residues, but unlike DNA, the sugar concerned is ribose and the molecule has only one strand. In addition, the thymine which is a common component of DNA is replaced by uracil; thus RNA contains adenine, guanine, cytosine, and uracil, while DNA had adenine, guanine, cytosine, and thymine.

As far as is known at the moment, m-RNA is a linear molecule in which modification of the essential structure of the RNA occurs only on a small scale, it if occurs at all. Regions of internal double-stranding in which the molecule turns back on itself like a hairpin, are rare. With t-RNA, or r-RNA on the other hand, the RNA molecules are not only chemically modified by the addition of substitutions on some of the purine and pyrimidine bases (and even sometimes on the sugar residues as well), but also contain substantial amounts of internal base pairing. The chemical modification that produces the substitutions occurs after the basic RNA molecule has been synthesized.

In t-RNA the internal double-stranded regions of the molecule produce a much more compact structure — at least when it is considered in three dimensions — than is the case with m-RNA. Figure 10 illustrates one such example.

B. Transfer RNA Structure

As mentioned earlier, the size of messenger RNA depends on the structure of the operon on which it is formed, and may vary widely. Ribosomal RNAs are commonly about the size of the largest messenger molecules, although a small r-RNA (the so-called 5S RNA) is present in

Figure 10. A folded outline of a generalized t-RNA molecule

many living systems. The t-RNAs, on the other hand, are much more uniform in size. Commonly, they contain about 75 nucleotides, and the distribution of these bases along the RNA strand of such molecules gives them all a very similar extent and distribution of internal double-stranding.

Transfer RNA molecules are the units that are responsible for converting the purine and pyrimidine 'language' of cell DNA and RNA into the amino acid 'language' of proteins and, consequently, they occupy a pivotal position in the expression of the bacterial genome as protein. To achieve this role the t-RNA molecules are functionally differentiated in three main ways. At the 3'-end of all t-RNA molecules one finds the sequence −C.C.A-OH. This region in the molecule is single stranded and is responsible for binding the amino acid for which the t-RNA is specific in the form of a mixed acid anhydride. The second part of the structure is a region of three sequential nucleotide bases which form a single-stranded loop in the middle of the molecule. This sequence of three bases is known as the 'anticodon' and it is capable of recognizing the appropriate 'codon' of three ribonucleotide bases wherever it occurs in an m-RNA molecule. The third region on the t-RNA − one less well defined in structure − is the part of the structure which binds the enzyme molecule which catalyses the loading of a specific amino acid on to the CCA terminus of the t-RNA. This enzyme therefore catalyses two functions: first the recognition of a given amino acid, and secondly the recognition of a specific t-RNA molecule so that the 'correct' amino acid is attached to the 'correct' t-RNA and thus introduced into the 'correct' position in the protein as it is synthesized.

It is interesting in this context that evolution has not apparently been able to give rise to a single molecule that can read both the 'purine/pyrimidine language' of nucleic acids and the 'amino acid language' of proteins unaided. Evolution has, however, produced the next best thing: the catalytic protein that loads the amino acid on to

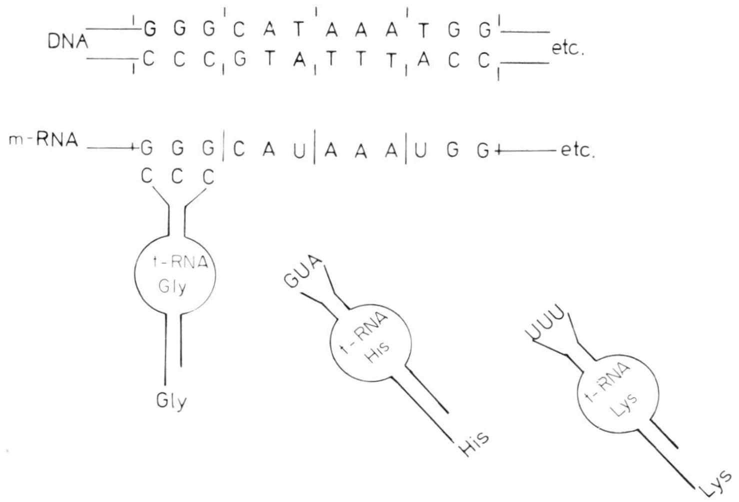
DNA
G G G C A T A A A T G G
C C C G T A T T T A C C
etc.
m-RNA
G G G C A U A A A U G G
etc.
C C C
t-RNA
Gly
Gly
GUA
t-RNA
His
His
UUU
t-RNA
Lys
Lys

Figure 11. The recognition of codons on the messenger RNA by the specific anticodons of t-RNAs. In the diagram t-RNA_{gly} is in place on the message, while t-RNA_{his} and t-RNA_{lys} are awaiting use. The next amino acid to be inserted would be tryptophan

the specific t-RNA is also responsible for recognising a molecule – the t-RNA – that is capable of reading the purine and pyrimidine language by codon/anticodon interactions.

C. RNA Synthesis

Since RNA has such a similar structure to one of the strands of DNA, it is not surprising that its synthesis is basically achieved by the same mechanism as that used for DNA save only that one strand of the DNA alone is used as the template on which to form a complementary strand of RNA. The basic biochemical reaction involved in forming the polynucleotide chain is the same as that shown in Figure 6. Since the molecule to be synthesized is a polyribophosphate, however, the ribonucleotide triphosphates take the place of deoxyribonucleotide triphosphates (see Figure 6), and uracil takes the place of thymine. Since thymine is 5-methyluracil, the change in the nature of the base does not confuse the complementary base-pairing interaction and adenine in the DNA codes for uracil in RNA.

The enzyme that carries out the process of RNA synthesis is RNA polymerase. This enzyme is bound to the DNA template at a region known as the promoter, and it then catalyses the synthesis of an RNA strand of complementary structure to one of the DNA strands starting from a point close to the RNA binding site. For this purpose the DNA is read by the polymerase in the direction $3' \rightarrow 5'$ and this means, in turn, that the RNA is synthesized in the direction $5' \rightarrow 3'$.

IV. Bacterial Proteins

A. Protein Structure

Of the macromolecules involved in the sequence DNA—RNA—protein, the proteins have in many ways the most complex structure. Not only do they contain a much larger number of different types of component unit — amino acids — but they are folded in a highly specific way. Perhaps the one type of nucleic acid that has the same sort of structural features as proteins are the t-RNAs, but in these the variability of the molecule is based only on four types of nucleotide whereas proteins contain 19 different amino acids. Moreover, the total length of t-RNA chains is about 75 nucleotides, whereas proteins commonly contain about 250 amino acids in their polypeptide chains, and many have several distinct polypeptides.

Even though proteins have this complex structure, the great majority of the chemical linkages between the amino acid units are the same — namely peptide bonds:

$$....NH.CH.CO.NH.CH.CO...$$

This structure means that proteins are basically chains of amino acids and that one end of the protein will therefore have a free $-NH_2$ group, while the other has a free $-COOH$. Occasionally, bridges are formed in the structure between two sulphur atoms:

$$...S-S....$$

but, since the cysteine residues that contain the sulphydryl groups necessary to form this type of bond are relatively very rare in proteins, the bonds are correspondingly uncommon.

It is not possible to describe the minutiae of protein structure in the space available here, and for a more detailed account the reader must consult a textbook of biochemistry. The polypeptide backbone of this structure is, of course, the means of holding the amino acids in the right order but, perhaps just as important for the catalytic function, is the way in which the individual amino acids interact with other amino acids that need not be adjacent to them in the basic polypeptide chain. In this sort of way, 'domains' on the surface of the protein may develop special properties, and it is indeed by the exercise of just such accumulated properties at restricted regions on the surface that proteins interact specifically with small organic molecules. Often these molecules are substrates; but, equally, they can, on occasion, be prosthetic groups or cofactors.

B. Protein Synthesis

Proteins are synthesized stepwise as linear polypeptides starting at the free amino terminus of the molecule, and moving steadily through the structure until the free carboxyl group is reached. Thus protein biosynthesis is essentially a linear reaction, just as is the synthesis of RNA and DNA, even though in this case the linear molecule, once formed, rapidly folds up in a highly specific way. Indeed, the evidence is strong that the protein begins to fold up immediately the first few residues have been inserted into the growing chain and long before the synthesis of the chain is complete. The order of synthesis therefore is likely to have an important effect on the way in which the folding occurs.

In view of the structure of proteins in bacteria, there are really two distinct problems to be overcome during their synthesis: first, how is the correct amino acid chosen for insertion into the growing polypeptide chain in the correct position, and secondly, how is polypeptide synthesis begun and ended?

C. Synthesis of Polypeptide Chains

As mentioned above, apart from the rare —S.S— bridges, all the bonds between the amino acids in proteins are peptide bonds. This then imposes some uniformity on the enzymic interactions that are needed for the synthesis of polypeptides — at least as they occur in proteins. In fact, there is only one: the activation of the carboxyl group of the amino acid and its use to form an amide link with the amino residue of the next amino acid. ATP ultimately forms the source of activation energy for this process, and therefore all such interactions can be summarized by the one equation:

$$\text{Amino acid} + \text{ATP} \longrightarrow \text{Activated amino acid} + \text{AMP} + \text{PP}$$

Amino acid activation consists in the formation of a mixed acid anhydride between the amino acid carboxyl group and the terminal phosphate of the particular t-RNA that loads the amino acid in question. The ATP is involved only as an intermediate, and the overall reaction can be written in two steps:

$$\text{Amino acid} + \text{ATP} \longrightarrow \text{Amino acid adenylate} + \text{PP} \tag{1}$$

$$\text{Amino acid adenylate} + \text{t-RNA} \longrightarrow \text{Amino acid–t-RNA} + \text{AMP} \tag{2}$$

As mentioned earlier, the —CCA terminus, which is common to all t-RNA molecules provides no specificity by itself. Neither does the ATP. The specificity lies with the enzymes — t-RNA synthetases — that catalyse reactions (1) and (2) above. Even though neither the ATP nor the —CCA terminus of the t-RNA provide specificity, the remainder of the t-RNA molecule and the amino acid to be loaded do provide specific information, and consequently it is the synthetase enzyme that ultimately ensures that the right amino acid is loaded to the correct t-RNA. In the process, the amino acid adenylate is not liberated from the surface of the synthetase enzyme although the pyrophosphate, and ultimately the AMP, are.

In this way, therefore, the amino acids that are needed for the synthesis of the polypeptide chains are attached to their specific t-RNAs by means of a bond which can also provide the necessary energy to drive the peptide bond formation forward. Notice in passing an interesting evolutionary development: activation with ATP — a very common means of activating molecules for subsequent chemical condensations in biological systems — is not dispensed with. However, it is used to drive a two stage activation in which AMP, derived from ATP, is attached as a terminal residue to the key intermediate.

D. Peptide Bond Formation

The bacterial cell carries a pool of t-RNA molecules, each specific for a given type of amino acid, and since appropriate synthetase enzymes are also available, the cell can build up a collection of all the amino acids necessary for protein synthesis, each amino acid being attached in activated form to its appropriate t-RNA. For polypeptide synthesis to take place it just remains to introduce the amino acids into the growing polypeptide chain in the correct order and to lock them in place by formation of peptide bonds.

The key molecule which determines the order of insertion of the amino acids is, of course, the messenger RNA. As we have seen the structure of this molecule is such that it reflects the structure of the DNA from which it was synthesized. It also, however, codes for the insertion of individual activated amino acids, the identity of the amino acid to be inserted being determined by the order of three bases — the codon — on the message. Let us take an example. The specific insertion is made by a t-RNA which bears an anticodon appropriate for recognizing the codon in the message. This recognition occurs by antiparallel complementary bases pairing between codon and anti-codon, as shown in Figure 11. Since it is normally only those t-RNAs that carry the appropriate amino acid that will recognize the codon, it is only that particular amino acid that can be selected for insertion. Thus, as shown in Figure 11, the next amino acid to be inserted will be histidine (codon: CAU), the one following will be lysine (codon: AAA), the next tryptophan (codon: UGG), and so on.

In practice the synthesis of the polypeptide occurs step-wise, that is the process occurs sequentially in a direction from the amino terminus to the carboxyl terminus of the polypeptide, the new amino acid only being attracted to the site of synthesis after the peptide bond that joins the previous amino acid to the growing chain has been formed.

As implied above, the energy needed for forming the peptide bond between the newly arrived amino acid and the previous one in the

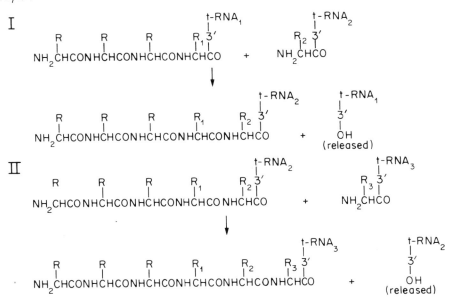

Figure 12. Stages in chain elongation during polypeptide biosynthesis. I, t-RNA₁ is already attached to the growing polypeptide The t-RNA₂ carrying its appropriate amino acid approaches, and forms a peptide bond. In the process t-RNA₁ is liberated and t-RNA₂ becomes bound to the growing polypeptide. II, the process is repeated. The loaded t-RNA₃ approaches, and t-RNA₂ is expelled. At each step the polypeptide grows by one amino acid unit

chain, comes by rupture of the bond between the amino acid and the terminal adenylate on the t-RNA. Thus, the process can be summarized as shown in Figure 12.

If one looks at the genetic code that determines which amino acid is allocated to each codon (see Table 1), one sees that there may be more than one triplet that codes for a given amino acid in certain cases. This means that there will be more than one type of t-RNA specific for certain types of amino acid — if only because they will have to have different anticodons to recognize the range of codons. Almost certainly, however, this is not the end to the variability. It seems that most codons attract more than one type of t-RNA, and each of these may well have a specific synthetase enzyme. For 64 codons in m-RNA, there are probably upwards of 100 different t-RNA molecules, but why there should be this apparent plethora of amino acid insertion systems is still unclear.

E. Start and Stop to Polypeptide Synthesis

The polypeptide chains that go to make up protein molecules do not only have a very specific order of amino acids along the chains, but the length of the chains themselves is also precisely controlled. This is achieved by the use of specific codons that signify the beginning and the end of the process.

In some ways, the start of a polypeptide chain seems to pose the greater problem, if only because there is no preceding activated amino acid to which the first amino acid of the chain can be attached. This problem is overcome by always using a special amino acid as though it were in the first position of the polypeptide chain. This amino acid is *N*-formyl methionine (Figure 13), and it is always either removed or deformylated after biosynthesis of the polypeptide is complete to liberate either methionine at the N-terminus or the second amino acid in the synthetic sequence.

Apart from the abnormal structure of the amino acid used for the purpose, the insertion of the first amino acid has many parallels to the addition of any other residue during polypeptide synthesis. The amino acid is activated and loaded on to its appropriate t-RNA using a specific synthetase for the purpose. This amino acid–t-RNA complex is then drawn to the 'start' position (codon: AUG) on the message by codon/anticodon interaction, just as though it were any one of the amino acids normally found in completely synthesized proteins.

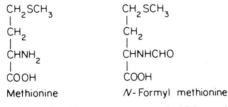

Figure 13. The structure of *N*-formyl methionine

At the other end of the polypeptide chain, the 'stop' signal is given by one of three RNA codons (UAA, UAG, and UGA). When one of these codons appear in the message, no amino acid is inserted and the chain then comes to an end. Exactly how this process occurs at molecular level is still uncertain, but it is clear that the operation of these so-called 'nonsense triplets' is different from all other codons in the message. No 'final' amino acid is added only to be removed subsequently.

In summary, then, polypeptide synthesis starts with the insertion of an *N*-formyl methionine wherever there is a UGA codon following a previous 'stop' signal. Thereafter, amino acids are inserted, as determined by the codon sequence of the message, until a nonsense (chain termination) codon is reached. At that point, no amino acid is inserted and the finished polypeptide chain is liberated from the synthetic site.

V. The Role of the Ribosome

A. Ribosome Structure

The biochemical reactions which lead to polypeptide chain elongation do not occur free in solution in bacteria; they occur on the surface of a cell organelle that catalyses the process — the ribosome.

Ribosomes are ribonucleoprotein particles which sediment in the centrifuge to form a uniform peak at 70S. They separate relatively easily into subunits of unequal size — the so-called 50S and 30S subunits. In bacteria, the ribosomal particles of 70S contrast with mammalian systems where the equivalent structures sediment at 80S.

The structural complexity of the ribosomes is illustrated by the fact that they contain three molecular species of RNA — a 16S molecule in the 30S subunit, and 5S and 23S molecules in the 50S subunit — and upwards of 50 distinct proteins. Currently, it is thought that about 20 of these proteins are in the 30S subunit and the remainder in the 50S unit. Some of these proteins are almost certainly structural elements in the ribosome, but it is equally clear that others have specific enzyme functions associated with the ribosome's role as a catalyst of protein synthesis.

B. Ribosomes and Polypeptide Synthesis — Initiation

The synthesis of the polypeptide chain — the process of chain initiation occurs when the m-RNA binds to the smaller ribosomal subunit. The binding occurs near the 5' end of the m-RNA, but whether the 30S subunit recognizes the initiator codon (AUG) itself, or a sequence of nucleotide bases that lies on the 5' side of the initiator codon is not clear. Protein initiation factors are also involved in this binding. The next step is that a t-RNA molecule already loaded with N-formyl methionine is attracted to the initiator codon (AUG), and this step is rapidly followed by the attachment of the 50S ribosomal subunit. Thus, the binding of the N-formyl methionine t-RNA occurs with the ribosome split into its component subunits. This step is, therefore, clearly different from the binding of the other loaded t-RNAs which is always to a ribosome to which the 50S unit is already attached. The process of attaching the N-formyl methionyl-t-RNA requires protein factors and also seems to need energy input in the form of GTP.

C. Ribosomes and polypeptide synthesis — chain elongation

The 50S ribosomal unit carries two sites on its surface. The first is the site to which the growing polypeptide chain, still attached to its t-RNA is bound. This is the so-called P (or peptidyl donor) site.

Adjacent to it on the 50S unit is the site to which the amino acid that is to be added to the chain is bound. This is the 'Amino Acid' or 'Acceptor' site, and is known as the A site for short.

When the 50S subunit joins the 30S unit as a part of the initiation process, the conjunction has the effect of putting the N-formyl methionyl-t-RNA into the P site on the 50S subunit. The appropriate t-RNA, already loaded specifically with its amino acid, is then drawn to the A site by the specificity of the next codon on the m-RNA, reading in the direction $5' \rightarrow 3'$. A bond is now formed between the carboxyl group of the N-formyl methionine at the P site and the new amino acid at the A site, the energy for the process coming from the rupture of the energy-rich bond joining the N-formyl methionine to its specific t-RNA.

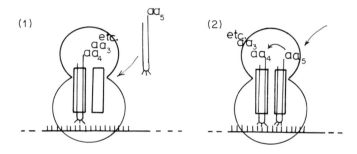

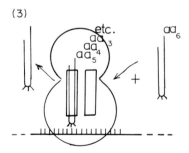

Figure 14. Three stages in the extension of a polypeptide chain by one amino acid residue (see also Figure 12). 1, The growing polypeptide chain attached to the t-RNA responsible for inserting the previous residue is held at the 'P site' on the ribosome. The next amino acid (aa5) attached to its specific t-RNA approaches. 2, The new t-RNA complex attaches to the 'A site'. It then moves into the 'P site'. 3, In the process of moving from 'A' to 'P' a polypeptide chain is formed between aa5 and the previous amino acid in the chain (aa4). In the process t-RNA$_4$ is released. The next amino acid (aa6) is already attached to its t-RNA molecule and awaits the next cycle of chain extension

The t-RNA is liberated from the surface of the ribosome during this process and leaves the immediate vicinity to be used again in the initiation of a new polypeptide chain. This sequence is shown diagrammatically in Figure 14.

Once the peptide bond is formed in this way, the process known as translocation takes place. The ribosome moves — whether as a single unit or in two steps, each involving one of the subunits, is not clear — so that the growing peptide chain (which now contains two amino acids attached through its terminal carboxyl group to a t-RNA molecule) moves from the A site to the P site. A new amino acid-t-RNA molecule is then attracted to the A site, once again using the coding properties of the messenger RNA as a guide. Peptide bond formation occurs once again, the used t-RNA is expelled, and translocation of the tripeptide to the P site takes place. By repeating this process, the polypeptide chain is built up by the step-wise insertion of amino acids, using the codon/anticodon interactions between the m-RNA and the loaded t-RNAs to achieve the desired amino acid sequence.

After the synthesis of the polypeptide chain has led to the insertion of a number of amino acids — the number is probably about 25 — a new initiation can occur with a new ribosome at the initiator region of the messenger. Thus, any messenger RNA, which commonly may contain as many as 1000 nucleotide bases, may have 20 or more copies of its information being run off at any one time, each copy being made by a single ribosomal particle. When a single message is being multiply translated in this way, it is said to be in the form of a poly-ribosome/messenger complex, and such structures can be clearly seen under appropriate conditions in the electron microscope.

D. Ribosomes and Polypeptide Synthesis — Chain Termination

When a terminator codon — UAA, UAG or UGA — is reached in the m-RNA, no amino acid is inserted at the A site and it seems that translocation leads to the release of the complete polypeptide chain from the ribosome.

As mentioned previously, some messages are polycistronic, that is they code for the synthesis of more than one polypeptide chain. Such messages have chain terminator triplets at appropriate points in the middle of polycistronic messages, and at that point it seems probable that the ribosome loses contact with the RNA. This then allows another 30S subunit to bind to the next initiator codon, and synthesis of the next polypeptide chain to start in its turn. Only when the 3' end of the messenger is reached will the ribosomes detach finally.

From this pattern of events, one sees that the ribosome is a highly specialized organelle on which the synthesis of the polypeptides is catalysed. As the process continues, the ribosomes move along the

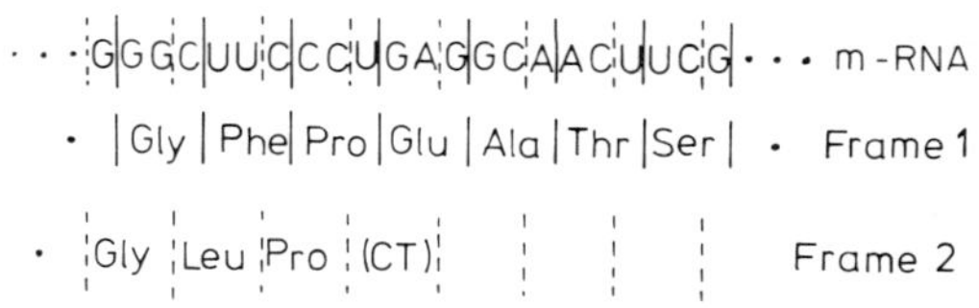

Figure 15. The relationship between polypeptide synthesis and the frame of reading of the message. A change in frame may lead to premature chain termination

message proceeding in steps of three bases at a time. The fact that the code used for the insertion of amino acids is triplet means that the informational content of the message depends critically on the precise point in the RNA sequence at which translation starts. Any shift in reading frame to right or to left will lead to a different set of bases being read as each triplet, and the outcome will be a polypeptide chain of very different composition to the one intended. This is summarized in Figure 15.

VI. Regulation of Genome Expression

Even though each protein that goes to make up a bacterial cell is coded for by a gene and uses an m-RNA as a biosynthetic intermediate, it does not follow that the cell will need equal amounts of all gene products. For example, much more leucine, isoleucine, and valine will be needed than pantothenic acid. Whereas the first three molecules are abundant on bacterial proteins, the last named is a growth factor and is needed in relatively small amounts. Yet the biosynthetic pathway to all four molecules has its early parts in common. How then is the output of the various products controlled?

A. Gene Copies

The overall process of genome expression is summarized in the sequence gene:messenger:protein, and the amount of protein produced may be controlled at a number of points. First — in some cases, at least — the number of copies of the gene itself may be adjusted. Normally, such alterations only affect genes that are on replicons other than the chromosome. Thus, the amount of a protein product — such as an antibiotic inactivating enzyme — may sometimes be greatly influenced by the number of plasmid copies present in the cell.

Clearly, one way to influence the amount of a gene product would be to control the activity of RNA polymerase since this would affect the number of messenger RNA molecules that were reproduced. This sort of transcriptional control is widespread in bacteria. Moreover, the basic mechanism is found in more than one form.

In the simplest version, the affinity of the RNA polymerase for its binding site at the commencement of the operon in question is influenced — probably by altering the precise nucleotide sequence of the polymerase binding site. By this means, therefore, the affinity of the RNA polymerase for its binding site — the *promoter region*, as it is called — is adjusted so that the number of messenger RNA molecules synthesized is balanced to meet the amount of product needed. Since operon has its own promoter gene, each one can therefore be tuned to meet the cell's requirement.

Sometimes mutations affect the base sequence of the promoter region and this, in turn, may influence the amount of transcription of the relevant operon that occurs. Such mutations are said to be 'promoter' or 'pace-setter' mutations; and they can either increase or decrease the level of gene expression.

The other way in which the synthesis of m-RNA can be controlled is by specific mechanisms which set the rate of transcription in the light of the concentration of small effector molecules. Specific increases that occur in the presence of such effector molecules are said to be *inducible systems*, while those in which gene transcription is reduced by low-molecular weight effectors are said to be *repressible*.

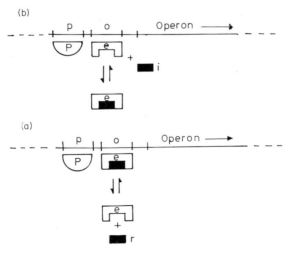

Figure 16. Regulation of protein synthesis. RNA polymerase (P) interacts with the promoter region on the chromosome (p) but cannot transcribe the operon because of the interaction of the protein effector molecule (e) with the operator (o). Figure 16a shows the arrangement for a repressible system in the presence of repressor (r), while Figure 16b shows an inducible system in the absence of inducer (i). See also Figure 17

In both inducible and repressible systems, the ability of the RNA polymerase to pass along the gene and to produce the m-RNA copy is influenced by a protein molecule that may bind directly to the DNA at a point between the promoter region and the part of the DNA that codes for the first cistron of the operon (Figure 16). The effect of this binding is, therefore, negative in the sense that it stops the expression of the gene as m-RNA even though RNA polymerase can bind to the promoter.

Both inducible and repressible systems are based on this type of mechanism, but in the inducible case the binding of the protein to the DNA occurs only when the low molecular weight effector molecule is absent, whereas in repressible systems, it only occurs when the effector is present. These situations are both summarized in Figure 16.

In practice, the low molecular weight effector influences the binding of the regulatory protein to the DNA by entering into contact with it as though it were a substrate interacting with an enzyme. Whereas

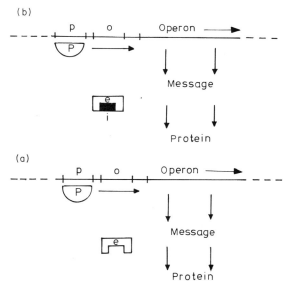

Figure 17. Regulation of protein synthesis. RNA polymerase (P) interacts with the promoter region on the chromosome (p) and then passes along the DNA in order to transcribe the operon. This is possible because in repressible systems (Figure 17a) no repressor is available to bind to the protein effector molecule (e) (compare Figure 16a). In inducible systems (Figure 17b) transcription of the operon occurs only when the effector (e) is complexed with an inducer (i) (compare Figure 16b)

Figure 16 summarizes the situation in which transcription of the gene was switched off in inducible and repressible systems, respectively, so Figure 17 summarizes the corresponding situation where expression is switched on.

The importance of inducible and repressible transcription in bacteria is enormous since it allows the cell to adjust the level of expression of an operon in response to the availability of an effector molecule. Typical examples are the depression of degradative enzymes when external concentrations of food materials are low, and the repression of endogenous biosynthesis when the product of the biosynthetic pathway is already available in adequate amounts in the immediate environment of the cell. By operating in this way, the bacteria can adjust their enzymic content most closely to meet the physiological needs of the moment.

B. Translational Control

The third point at which the level of gene expression can be controlled is by influencing the net rate of message translation. At one time, it was thought that this was achieved primarily by controlling the longevity of particular messenger RNAs. However, translational control almost certainly operates by influencing the incidence of initiation. Earlier in this chapter, the manner in which ribosomes bind at the beginning of a message before the previous ribosome has completed its passage along the molecule was described. Any alteration in the rate of initiation influences the number of polypeptide products that are run off before the message — which certainly has a finite life — is destroyed.

One form of translational control can be detected when the translation of a polycistronic message is examined. In general, the amount of polypeptide product made from such a message decreases with each successive cistron. Thus, some degree of control is manifested by the balance of termination at the end of a cistron and the initiation at the beginning of the next. This is summarized in Figure 18. Such a gradient of expression along a polycistronic message is known as polarity, and its degree is certainly due to a balance of the rates of initiation, termination, and message stability.

Modified degrees of polarity may be introduced by mutation, and such mutations are known as polar mutations. The most clear cut types of polar mutations are caused when a change in the DNA sequence of a gene produces a message that now has a chain terminating triplet in the middle of a structural gene. When this occurs, no active polypeptide product is formed from the gene in which the mutation has occurred, but the mutation also has polar effects since the polypeptides whose genes lie later in the operon are expressed at lower than normal levels.

The biochemical basis of such polar effects is still not fully

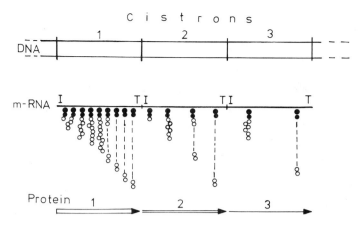

Figure 18. Diagrammatic representation of the 'polarity' in an operon of three cistrons (1, 2, and 3). Ribosomes bind to the initiator site on the m-RNA (I) and leave the message at the terminator (T). Reinitiation may then occur at the beginning of the second cistron, and subsequently again at the third. The number of ribosomes that reinitiate at each successive I site are fewer than at the one before. One, therefore, sees a gradient of expression in the operon

understood. Probably the chain termination triplet in the message make the ribosome leave the RNA and subsequent reinitiation at the beginning of the next cistron is less efficient. As a result, all subsequent cistrons are likely to be poorly expressed, since it is known that *de novo* initiation in the middle of a polycistronic message is never fully effective.

Of the three basic methods of regulating gene expression, only the first is sophisticated enough for fine and varying control to be exerted. In contrast, the other two methods have the effect of fixing the functional level of the steps at which they work, and once set, this level tends to remain fixed unless modified by mutation.

VII. Further Reading

Carlile, M. and Skehel, J. J. (1974). *Evolution in the Microbial World*. 24th Symposium of the Society for General Microbiology. Cambridge: Cambridge University Press.

Davis, B., Dulbecco, R., Eisen, H. R., Ginsberg, H. S., and Wood, W. B. (1973). *Microbiology*. Hagerston: Harper Row.

Hawker, L. and Linton, A. H. (1971). *Micro-organisms: Function, Form and Environment*. London: Edward Arnold.

Hayes, W. (1968) *The Genetics of Bacteria and their Viruses*. Oxford: Blackwell Scientific Publications.

Haynes, R. H. and Hanawalt, P. C. (1968). *Readings from Scientific American: The Molecular Basis of Life*. San Francisco: W. H. Freeman.

Lewin, B. (1974) *Gene Expression*, Vols. 1 and 2. London and New York: John Wiley.

Mandelstam, J. and McQuillan, K. (1973). *Biochemistry of Bacterial Growth*, 2nd edition. Oxford: Blackwell Scientific Publications.

Meyrell, G. G. (1972). *Bacterial Plasmids*. London and Basingstoke: Macmillan & Co.

Smith, A. L. (1973). *Principles of Microbiology*, 7th edition. St. Louis: C. V. Mosby & Co.

Srb, A. M., Owen, R. D., and Edgar, R. S. (1969). *Readings from Scientific American: Facets of Genetics*. San Francisco: W. H. Freeman.

Stanier, R. Y., Douderoff, M., and Adelberg, E. A. (1971). *General Microbiology*. London and Basingstoke: Macmillan & Co.

Stent, G. (1971) *Molecular Genetics: An Introductory Narrative*. San Francisco: W. H. Freeman.

In general, the annual Symposia of the Society for General Microbiology provide excellent specialist essays on a wide range of microbiological topics.

Interactions between Phage and Bacteria

D. KAY

Sir William Dunn School of Pathology,
University of Oxford, England

The bacterial viruses, also known as bacteriophages or 'phage' for short, parasitize bacteria. Some kill their hosts immediately, causing dissolution of the cells within 20 minutes or so, while others set up a relationship in which a few of the cells become infected and release phage while the majority of them continue to multiply, unaffected by the presence of the phage. The filamentous phages can infect all the cells of a culture which continue to multiply and divide while releasing progeny phage in a continuous stream. A particularly subtle relationship between host and virus is to be found amongst the lysogenic phages, which integrate their genetic material into that of the host and only show their presence when, very rarely, a cell produces phage.

All phages are submicroscopic particles but they vary enormously in size and structure and in the magnitude of their genomes which may be either DNA or RNA. In the simplest terms, phages consist of genetic material carrying most but not all of the information needed for their reproduction, packaged in a protective coat of protein which carries a means of recognizing its specific host cell and injecting the genetic material into it.

I. A Little History

The first published account of the bacteriophage phenomenon was made in 1915 by F. W. Twort who noticed that some colonies of the bacterium *Micrococcus* had become changed from the characteristic creamy opaque appearance to a glass-like clarity. When he transferred a little of the material from a glassy colony to a normal one, that too developed the glassy appearance and the change could be propagated by transfer indefinitely. Twort recognized that he had discovered a submicroscopic agent that killed bacteria and multiplied in the process. In 1917, F. d'Herelle independently discovered a transmissible agent which killed dysentery bacilli and multiplied at the same time. He considered that he had discovered a virus of bacteria and named it 'bacteriophage' or bacterium eater. In those days bacterial diseases were rife and medical science was much concerned with typhoid, dysentery, cholera, blood poisoning, and many other maladies of bacterial origin. The prospect of using phage as a therapeutic agent presented a challenge which was taken up by many researchers. The main sources of phage were the faeces of patients recovering from dysentery, etc. Preparations of phage were used to treat patients but no clearly beneficial effect could be demonstrated. In time the reasons became clear. The phage preparations were inactivated in the stomach and if some did reach the infected area the bacteria always contained a proportion of resistant cells which multiplied in the presence of the phage. Interest in the medical applications of phage soon declined.

However phage was interesting in its own right. It was a virus which could be easily cultivated unlike the animal viruses, which were particularly difficult to study at that time or plant viruses which had a very low infectivity and a long generation time. Phage therefore became attractive for the study of a virus which, it was hoped, would be a model for other viruses. In the 1940's a group of workers in the USA decided to concentrate their efforts on a set of seven coliphages which had been collected and named the T phages. Other workers followed suit and as a result knowledge of all aspects of these phages rapidly increased. The electron microscope enabled phage particles to be observed and measured. The 'tadpole-like' tailed structure of many of them came as a great surprise, as did the observation that they attached to bacteria by the tip of the tail. Radioactively labelled chemicals enabled biochemical experiments to be carried out, and in 1952 Hershey and Chase provided evidence that the genetic information of phage, and by inference all other forms of life, was contained in DNA. The fact had already been proved by Avery, McLeod, and McCarty in 1944 with their work on the transforming principle of the pneumococcus, but the scientific world had not comprehended the significance of it.

Phage is an advantageous tool for the study of microbial genetics because of the ease with which mutants can be isolated, crosses made, and the numerous progeny analysed. The discovery of the triplet nature of the genetic code stemmed from the work of Crick and colleagues with T4 phage in 1961. The work on the genetics, protein structure and nucleotide sequence of coliphage ϕX174 has culminated in the determination by Sanger and colleagues in 1977 of the first total nucleotide sequence of a DNA virus.

II. The Sources of Phage

Phages have been found for a wide range of bacteria. A feature of all viruses is their host specificity. Hence phages can only attack bacteria and have no effect on animal or plant cells. The specificity is usually very marked and a phage may be able to attack only a particular strain of its host species. The basis of this specificity is the possession, by the bacteria, of specific receptor substances on their surfaces or appendages which adsorb the phage particles. The possession of a specific receptor does not however ensure that a bacterium can support the growth of a phage but it is an essential prerequisite.

In order to find phages active on a particular bacterium, it is necessary to search in the environment where that organism is found. Phages for *Escherichia coli*, a common intestinal bacterium, can easily be found in sewage. Similarly, phages for the staphylococcus can be found in the pus from abscesses caused by that organism, *Bacillus subtilis* phages can be found in soil, and so on.

A. Assay Methods

The total number of virus particles can be counted in the electron microscope (see III.A, below) and the number of active particles may be determined by the plaque assay method. Phage is mixed with molten agar to which a few drops of sensitive bacteria have been added and poured onto the surface of a nutrient agar plate where it sets as a layer about 0.5 mm thick. The phage particles adsorb to the bacteria, multiply therein, lyse them and infect the surrounding bacteria. This process is repeated many times and stops only when the bacterial culture ceases to multiply. The result is that wherever an active phage particle lay in the agar layer a zone of lysis or plaque appears. By counting the number of plaques the concentration of the phage can be determined in terms of plaque forming units (p.f.u.). A high proportion of phage particles can give rise to plaques (over 90 per cent) and only one phage particle is needed to initiate the formation of a plaque. The size of the plaque and its appearance depend on many factors such as the diffusibility of the phage particles in the agar layer, the length of time taken to lyse the bacteria, and the release of diffusible bacteriolytic enzymes. Lysogenic phages give turbid plaques (see Lysogeny, Section V). Not only does the plaque assay give an adequately accurate determination of the number of particles, but it also enables different phage phenotypes to be determined (in so far as these affect plaque type) and it enables phage to be cloned because each plaque originates from one particle and contains only the progeny of that particle.

III. The Phage Particle

A. Electron Microscopy of Phage

Phage particles prepared by negative staining with uranyl compounds or phosphotungstate can be observed in great detail in the electron microscope.

The T-even phages were found to have a tadpole-like shape with a 'head' and a 'tail'. Figure 1(c) shows a T4 particle with its elongated head of angular outline and its tail with the characteristic cross-banded pattern. The tip of the tail carries an end plate and some fine fibres. The tail consists of a central tube, invisible in Figure 1(c), surrounded by a contractile sheath which is the banded structure seen in the micrograph. The function of the tail is shown diagrammatically in Figure 4. The T5 phage particle also shows an angular head (Figure 1(e))

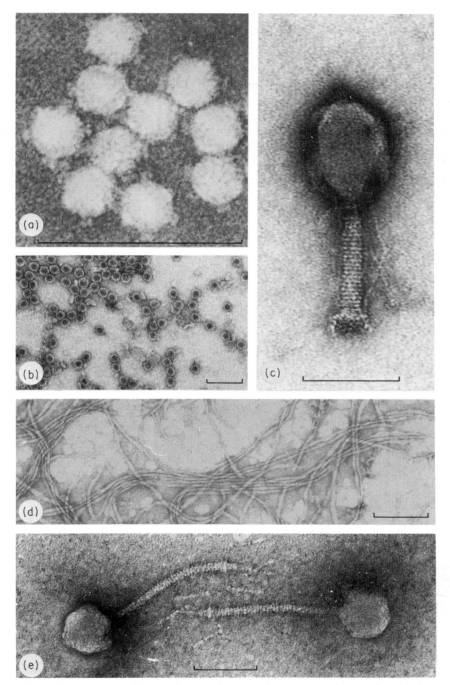

Figure 1. Electron micrographs of phages. (a) ϕX174; (b) R17; (c) T4; (d) ZJ/2, similar to M13; (e) T5. All are negatively stained with uranyl acetate. Bar equals 100 nm

but unlike T4 it is a regular polygon in outline. The tail is rather narrower than T4 and does not have a contractile sheath. It also carries some fine fibres attached near to the tip. There are other phages in which the tail is short as in coliphage T7 (not illustrated) and *B. subtilis* phage ϕ29 shown diagrammatically in Figure 2(b). This particle is remarkably complex. Its head carries a 'crown' of spikes and the collar at the junction between the head and tail also carries a number of small projections. Coliphage ϕX174 (Figure 1(a)) shows a six-sided polygonal outline with projections at the corners. A diagrammatical representation of this phage is given in Figure 2(c). The RNA-containing coliphage

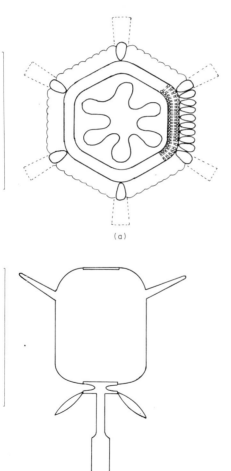

(a)

(b)

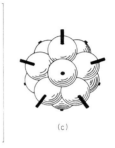

(c)

Figure 2. Diagrammatic representation of phages PM2 (a), ϕ29 (b), and ϕX174 (c), to the same scale bar = 50 nm. PM2 shows the arrangement of the lipid bilayer beneath the outer protein layer. ϕ29 shows the arrangement of the 'crown of spikes' on the head and the projections on the 'collar'. ψX174 shows the arrangement of the capsomeres on the fivefold axis of symmetry (a) and (c) based on Harrison *et al.* (1971), *Nature*, **229**, 197, and Szekly (1977), *Nature*, **265**, 685, and reproduced by permission of Macmillan (Journals) Ltd (b) based on Anderson *et al.* (1966), *J. Bacteriology*, **91**, 2082 and reproduced by permission of the American Society for Microbiology

R17 appears as an isometric particle, angular in outline but without any projections, Figure 1(b).

The lipid-containing phage of *Pseudomonas* PM2 (Figure 2(a)) is also icosahedral in shape but its outer protein coat overlies a lipid bilayer which in turn surrounds an inner protein structure enclosing the phage DNA. The filamentous phage ZJ/2 (Figure 1(d)) is a flexuous rod nearly 1 μm long and 5.5 nm wide.

The protein shell or capsid of the small virus particles is constructed of a set of identical morphological subunits called capsomeres. These consist of individual protein molecules or groups of molecules which when assembled together form a hollow shell which in its simplest form takes the shape of an icosahedron. This is a structure consisting of 20 identical equilateral triangular faces and has 12 vertices. In phage R17 and ϕX174 the capsomeres can sometimes be distinguished but in the larger phages this is not so. Presumably the capsomeres are closely fitted together and leave no interstices into which the stain can penetrate.

The tails of the T-even and other long-tailed phages show a cross-banded structure. This is due to the edge view of the stacked annuli of which the tails are constructed. These are composed of a set of identical protein subunits, probably six in number, arranged in a ring leaving a central hole for the passage of nucleic acid from the head into the host cell. The fibres at the tail tip are involved in the specific attachment of the phage particles to their bacterial host (see Section IV.C) and consist of a group of specialized protein molecules (see Section IV.E). In the filamentous phages X-ray crystallographic evidence indicates that the protein molecules are arranged in a helical array to form a tube which encloses the nucleic acid.

B. Composition of Phage Particles

Purified phages all contain protein and nucleic acid, and some contain a little lipid (Table 1). All are resistant to ribonuclease and deoxyribonuclease but when the nucleic acid is extracted from the phage particles it is sensitive to the appropriate enzyme. This indicates that the nucleic acid in the virion is protected and suggests that it is enclosed by the protein components of the particles.

The methods used to free the nucleic acid from the phage protein involve the use of phenol or detergents which loosen the bonds between the protein molecules and allow the nucleic acid to go into solution. Most phages contain DNA but some contain RNA. The phage DNA can be a linear duplex as in T4, a linear duplex with cohesive ends as in

Table 1.

Phage	Host	Dimensions (nm) Head	Dimensions (nm) Tail[a]	Nucleic acid Type[b]	MW × 10^{-6}	Genes Total[c]	Genes Structural[d]
T4	coli	115 × 85	95 × 18	2-DNA LIN	120	≲ 150	36[e]
T5	coli	75 × 75	180 × 15	2-DNA LIN	76	≲ 75	15
T7	coli	62 × 62	20	2-DNA LIN	26	30	11
λ	coli	54 × 54	150 × 15	2-DNA LIN	36	≲ 30	12[f]
φ29	subtilis	42 × 32	33	2-DNA LIN	11	≲ 19	7
φX174	coli	25 × 25	–	1-DNA CIRC	3.27	9	5
PM2	pseudomonas	61 × 61	–	2-DNA CIRC	6.05	9	4[g]
φ6	pseudomonas	65 × 65	–	2-RNA LIN	11	≲ 12	10[h]
R17	coli	25 × 25	–	1-RNA LIN	1.3	3	2
M13	coli	870 × 6	–	1-DNA CIRC	1.9	8	2[i]

[a] – no tail. [b] 2 – Duplex, 1 – Single-stranded, LIN – linear, CIRC – circular. [c] Estimated from length of NA or the number of complementation groups. [d] By Polyacrylamide gel electrophoresis of virions. [e] Contains HMC. [f] Lysogenic. DNA has cohesive ends. [g] Contains 14% lipid. [h] Contains 25% lipid. Genome is in three pieces. [i] Filamentous.

phage lambda, a closed circular duplex as in PM2, or a closed single-stranded loop as in ϕX174. Phage RNA has been found as a linear single strand in R17 or as three separate double-stranded pieces as in ϕ6. Phage nucleic acids consist of a deoxyribose (or ribose)-phosphate backbone with the bases adenine, guanine, cytosine, and thymine (or uracil in RNA) in the majority of cases but some peculiarities have been found. The T-even phages contain hydroxymethyl cytosine (HMC) instead of cytosine and some *B. subtilis* phages contain hydroxymethyl uracil in place of thymine. In the T-even phages the HMC nucleotides are glucosylated and in many phages some of the adenine bases are methylated.

The proteins of phage can be brought into solution by treating the particles with alkali, urea or detergent but if the solubilizing agent is removed, the proteins aggregate and come out of solution. This is a reflection of one of the principal properties of phage proteins, the ability to form insoluble aggregates held together by hydrophobic bonds.

If the phage proteins are dissolved in the presence of sodium dodecyl sulphate they are denatured and can be separated according to molecular weight by electrophoresis on polyacrylamide gels. The small RNA phage, R17 and the filamentous phage, fd, have only two structural proteins whereas the T-even phage has at least 36 with MWs ranging from 10,000 to 145,000. The lipid-containing phage PM2 contains four proteins and about 10 per cent of lipid while ϕ6 possesses 10 proteins and 25 per cent lipid.

IV. The Phage Multiplication Cycle

The multiplication of virulent phage can be divided into the following phases, adsorption of the virion to the host cell, penetration of the nucleic acid into the host, the eclipse phase during which the phage nucleic acid is replicated and phage proteins are synthesized and maturation of infective progeny followed by release from the cell.

A. Adsorption

In all bacteria the cytoplasm is surrounded by a membrane consisting of a phospholipid bilayer with protein attached and interspersed within it. Surrounding the membrane is the rigid cell wall consisting of a giant peptidoglycan molecule. There is an important difference between the walls of Gram-positive (G^+) and Gram-negative (G^-) cells. Both have peptidoglycan but whereas this is clearly distinguishable and separable from the cytoplasmic membrane in G^+ cells this is not so in G^- cells where the two are more closely interlinked. Moreover the G^+ cells have a layer of teichoic acid which is not found in G^- cells. Various surface proteins are attached to the teichoic acid layer. In G^- cells the peptidoglycan layer carries a layer of complex lipopolysaccharide (LPS). In addition to the outer layers of LPS some bacteria carry a polysaccharide capsule. Some are motile and carry one or more flagella, often several times longer than the cells.

Some bacteria possess long straight filaments, thinner than flagella, called pili. There are two main groups of pili, the common pili and the conjugation or sex-pili. The common pili are often very numerous and appear as a fringe of fine hairs around the cell. They are about 5 nm wide and up to 2 μm long. Not all bacteria have them. The conjugation pili are produced only by cells carrying the F or fertility factor or the colicinogenic factor Col-I. Both these factors enable the cells to produce pili, either F-pili or I-pili, which are slightly wider than common pili, very straight and up to 10 μm or more long. The significance of the F pilus in phage infection will be explained in the section on filamentous and RNA-containing phages.

All bacteria are sensitive to lysozyme. The effect of the enzyme varies from rapid and complete lysis as in *Micrococcus lysodeikticus* to the liberation of a few sugar residues in the most resistant G^- organisms. Treatment of bacteria with lysozyme and a chelating agent such as ethylenediaminetetraacetate (EDTA) results in the lysis of most strains, whether G^+ or G^-. If, however, the treatment is carried out in medium of high osmotic pressure the cells do not lyse, they swell into round objects called spheroplasts. These contain all the cytoplasmic components of the cell and an intact genome, but are devoid of a cell wall. They can carry out all the bacterial metabolic activities and under some conditions can grow and divide.

When phage is mixed with sensitive bacteria the particles attach to the cells. If phage are mixed with spheroplasts no adsorption takes place and the spheroplasts do not become infected. However if cells are infected with phage and then converted to spheroplasts the phage multiplies in the spheroplasts. This shows that the cell wall or some part of it is necessary for the adsorption of phage on to bacteria.

Cell walls can be isolated from bacteria after breaking the cells open and washing away the cytoplasmic contents. The walls can still adsorb phage and, moreover, extracts of the walls can also inactivate phage. The extracts (from G^- bacteria) contain LPS which can adsorb to the phage particles, inactivate them, and may lead to the release of the phage nucleic acid. If a lawn of bacteria on a nutrient agar plate is

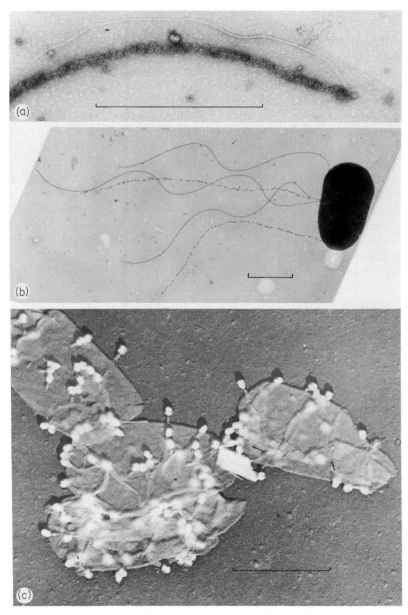

Figure 3. Adsorption of phage to bacteria. Two filamentous phage particles, one of normal length and one double length, adsorbed to the tip of an F pilus. The pilus is labelled with R17 phage particles (a). RNA phage R17 adsorbed to the F pili on a male strain of *E coli*. The other filaments are flagella (b). Phage T4 adsorbed to cell walls of *E. coli* (c). The bar equals 1 μm

spread with excess phage the great majority of the bacteria are killed, but if the plates are incubated longer a few colonies appear. These are resistant to the phage. The cause of the resistance is that the bacteria no longer adsorb the phage and although LPS can be extracted from the cells this no longer inactivates the phage.

The lipopolysaccharides of G^- bacteria compose the somatic or 'O' antigens and these serve as phage receptor substances. Examples of phages which attach to the somatic antigens of their hosts are the T set of coliphages. Those with long tails, T1, 2, 4, 5, 6 adsorb to the surface of the cells by the tips of their tails (Figure 3(c)). The number of adsorption sites on the bacterial surface is quite large and as many as 200 particles of T4 can adsorb to one cell. Several different sorts of phage can be adsorbed to one cell at the same time but, unless the phages are very closely related only one of them can multiply. The others are excluded even though they have adsorbed.

Phages have been found which attack only motile strains of the host bacteria. This has been traced to the use of the flagella as the specific phage receptors. Others attach only to capsular strains.

An extremely specialized use of a bacterial surface appendage is shown by the RNA phage R17 and the filamentous phage M13. The RNA phage adsorbs only to the sides of the F-pili (Figure 3(b)), while the filamentous phage adsorbs only to the tip of the same pili (Figure 3(a)).

The initial contact between phage and bacterium comes about as a result of the random thermal movements of the particles, particularly those of the phage particles. The primary bonds formed are probably electrostatic because they are pH dependent. The concentration of salts in the medium, particularly mono- and divalent cations, affects the rate of adsorption. Some phages, e.g. T5, show a requirement for Ca ions for adsorption and a requirement for L-tryptophan is shown by T4 coliphage. It appears that in T4 the tail fibres which make the initial attachment to the bacterial receptors are normally folded against the tail sheath but are caused to unfurl into the active spread-out postion in the presence of a small concentration of L-tryptophan (1 μg/ml).

B. Mutations Affecting Adsorption

Bacterial cultures always contain a few phage-resistant cells. These arise by spontaneous mutation and not by an adaptation to resistance brought about by the presence of the phage.

Evidence for this is provided by the fluctuation test. If a number of separate cultures of bacteria are set up in liquid media with the same small number of cells in the inoculum and allowed to grow for a few hours, the number of phage resistant organisms present can be determined by plating out a sample from each culture together with an excess of phage. In one experiment samples from 12 independent cultures were taken and found to contain highly variable numbers of resistants. These varied as follows; 0 (five samples), 1 (two samples), 3, 4, 7, 48 and 303. The average per sample was 30 and the statistical variance 6620. On the other hand if a number of samples were taken from a single culture and subjected to the same treatment with phage, the number of resistants in each was very similar with an average of 51.4 but a variance of 27. The resistants could not have arisen due to the effect of the phage on the bacteria otherwise the number in each sample from the 12 independent cultures would have been very similar. They must have arisen by spontaneous mutation which might have occurred early in the growth of the culture, thus giving the 303 resistants, or late giving the low numbers or not at all, thus giving no resistants. The phage merely acts as a selective agent which kills the sensitive cells and makes it possible to count the resistant ones.

C. Penetration

In all phage infections the nucleic acid enters the cell while the bulk, but not all the phage protein remains outside. At one time it was thought that the whole phage particle entered the cell but the famous experiment of Hershey and Chase (1952) showed that this is not so. Two stocks of phage T2 were prepared, one grown in medium containing ^{35}S as sulphate and the other in medium containing ^{32}P as phosphate. The ^{35}S was incorporated only into the phage protein, while the ^{32}P was taken up exclusively into the phage DNA. Each phage was allowed to adsorb to its host and the cultures were then subjected to vigorous stirring to shear off the adsorbed particles at the surface of the cell. The ^{32}P, and hence the phage DNA, could not be sheared off, but the majority of the ^{35}S, and hence the phage protein, could be removed. The cells from which the phage protein had been removed were still able to produce phage. The DNA of the T2 phage was therefore necessary to infect the cells while the protein of the phage played no obvious part other than enabling the phage to adsorb to the cells. It is now known that a small amount of phage protein does in fact enter the cell with the nucleic acid and plays a vital role in the infective process.

The T-even coliphages have a complex tail structure and a correspondingly involved method of penetration. The structure of the

T-even phages have been described in Section III.A and are illustrated in Figures 1 and 4. Adsorption to the bacterial receptors takes place by the distal ends of the six tail fibres and is followed by a contraction of the tail sheath by a rearrangement of the subunits of the sheath from 24 discs of six subunits each to 12 discs of 12 subunits (this is an oversimplification of the rearrangement). The retraction of the sheath reveals the hollow tail tube within it and the energy produced by the retraction appears to be responsible for driving the tube through the cell wall. The tube does not pass through the membrane but reaches to its outer surface. As stated earlier (Section IV.A) spheroplasts cannot be infected by phage but it has now been found that if particles are artificially caused to contract their sheaths, by treatment with urea for example, they can adsorb to spheroplasts and infect them. It has also been show that these contracted phages can be caused to release their DNA by treatment with phosphatidyl glycerol, a component of the phospholipid cell membrane. It is now considered that the penetration of T-even phage DNA is carried out by the following steps (Figure 4):

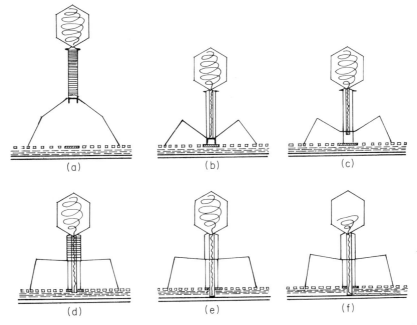

Figure 4. Adsorption of phage T4 to the bacterial surface and injection of phage DNA. Details in the text. Based on Benz and Goldberg (1973) *Virology*, 53, 225–235 and reproduced by permission of Academic Press Inc.

(1) Phage particle adsorbs to the LPS receptors on the bacterial surface by means of its tail fibres (a).
(2) The pins on the baseplate contact further receptors of unknown constitution on the cell wall and make firm unions with them (b).
(3) The base plate detaches from the end of the tail tube and contraction of the tail sheath occurs (c,d).
(4) The tail tube passes through the cell wall (e) and reaches the cell membrane where the action of a membrane component, possibly the phosphatidyl glycerol, triggers off the unplugging of the tube and the release of the phage DNA (f).
(5) Phage protein injected with the DNA may facilitate its passage across the cell membrane.

The tail-less phage ϕX174 carries spikes at each of its 12 vertices (see Figure 1(a)). On adsorption to its host cell the virion becomes partly buried in the surface. Lipopolysaccharide taken from the cell surface can attach to the virions and block their activity by preventing adsorption. Cell wall preparations not only adsorb the phage but cause the release of the phage DNA. Two of the structural proteins found in ϕX174, the products of genes G and H are found in the spikes. One of these, the gene H protein, is probably the one concerned with adsorption to the host and it is the only phage protein which enters the cell on infection.

The small phage MS2 contains two structural proteins and a genome composed of a single-strand of RNA. Adsorption only takes place to bacteria exhibiting F-pili (Figure 3(b)). If RNAase is added to the nutrient agar plates used to assay the phage, infection is inhibited and no plaques appear. Apparently the RNA of the phage passes through a stage where it is sensitive to the depolymerase. The virions themselves are resistant and so are the infected cells. It is believed that adsorption of the particle to the pilus takes place *via* the single molecule of the 'A' protein present in the virion capsid. This molecule acts as a form of keystone which holds all the subunits of the capsid together. It is not clear whether the RNA passes down the lumen of the pilus, or whether it passes down the outside of the pilus towards the cell surface and through the cell membrane.

Adsorption of the filamentous phages M13, fd, and fl, involves an end to end contact between the virion and the tip of the pilus (Figure 3(a)). The virion is a polar structure having at one end one or two molecules of the 'A' protein, the product of gene III which is responsible for the adsorption of the particle to the pilus. Like the gene H protein of ϕX174 the 'A' protein of filamentous phage enters the cell with the DNA and carries out the vital function of linking the phage DNA to the cellular replication system on the inner cell membrane of the host.

D. Replication of Phage Nucleic Acid

In order that the phage shall multiply, the following functions must be carried out in the infected cell:

(1) The parental DNA (or RNA) must avoid breakdown by the host depolymerase enzymes.
(2) The parental DNA (or RNA) must be replicated many times to make exact copies of the original infectious strand or strands.
(3) The progeny nucleic acid must avoid breakdown.
(4) Messenger RNA's must be transcribed and translated into the various proteins required for the synthesis of the progeny particles.
(5) Protein synthesis must be controlled in such a way that the requisite amounts of the proteins are made. Each gene is present only once in the phage genome but the number of molecules of each protein species can vary from a few to many thousands.
(6) The progeny nucleic acid must be condensed and packaged into the capsids of the progeny particles so that each contains a complete genome.
(7) All the appendages of the phage capsid, tails, fibres, spikes, etc., must be synthesized and assembled into active phage particles.
(8) Finally the host cell must release the progeny phage.

1. Phage T4

This phage is a very efficient parasite. It utilises the host RNA polymerase and its transcription systems, and it uses the host ribosomes to synthesize its proteins but it first carries out a programme of destruction or inhibition of those host functions which might limit or oppose its own multiplication. A key difference between the phage DNA and that of the host is that in the former all the cytosine residues have been replaced by hydroxymethyl cytosine and that these are glucosylated. This difference renders the phage DNA insensitive to the host depolymerases. The whole phage genome is not transcribed immediately but in an orderly sequence involving early and late genes. Amongst the early transcribed genes are some for DNAases which specifically recognize base sequences containing cytosine clusters. These

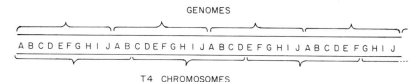

GENOMES

A B C D E F G H I J A B C D E F G H I J A B C D E F G H I J A B C D E F G H I J

T4 CHROMOSOMES

Figure 5. The circular permutation of the genes of Phage T4. Letters A—J represent the genes. The DNA is replicated as a concatenate which is cut into chromosomes slightly longer than the genome during encapsidation. Each chromosome has identical terminal sequences

enzymes therefore break down the host cell DNA, which contains cytosine, while the phage DNA, which has none, is unaffected. This has a double advantage to the phage because not only does the host DNA cease to function, but its component nucleotides become available for incorporation into progeny phage nucleic acid. Amongst the early phage genes are some for new transfer RNAs which favour the codons in phage messenger RNA rather than those of the host.

During the replication of the phage DNA there would be competition between cytidine triphosphate and hydroxymethylcytidine triphosphate if these two nucleotides were present at the same time, and the former would be incorporated into the growing phage DNA chain. This eventuality is avoided in T4 infected cells by the rapid synthesis of a phage-coded enzyme which dephosphorylates dCTP and dCDP and eliminates them from the nucleotide pool.

The molecular weight of T4 DNA is 1.3×10^8 and it contains about 165,000 nucleotide pairs. It is a linear duplex about 56 μm long.

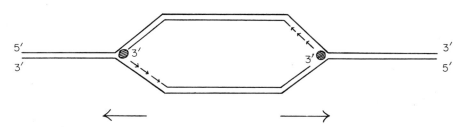

Figure 6. Replication of T4 DNA to produce 'eye' forms. The diagram shows the bidirectional replication taking place from a single initiation site. Large arrows show the direction of elongation Small arrows show the Okazaki fragments laid down in the opposite direction. These are subsequently covalently linked by a ligase. Hatched circles indicate the positions of the gene 32 protein which unwinds the duplex at the point of replication

However, when the phage is analysed genetically, the map of its genes appears to be circular. This apparent anomaly is explained by the finding that the T4 genome is circularly permuted. It also contains a little more than a whole phage genome and it begins and ends with the same nucleotide sequences. The situation is shown diagrammatically in Figure 5. The phage chromosomes are said to have a terminal redundancy and are circular permutations of each other.

Replication of T4 DNA begins with its attachment to a specific site in the host cell membrane. Some early genes must be transcribed and their products synthesized. These include gene products (gp) 32, 41, 43, 44, 45, and 62. Gp32 acts by causing localized unwinding of the duplex DNA in advance of the replicating fork (Figure 6). Replication proceeds in both directions so that 'eye' structures are formed. These have been observed in the electron microscope. The leading strand is elongated by addition of nucleotides to the $3'$ end while the lagging strand is elongated by the addition of Okazaki fragments and their subsequent joining by ligase. At the end of the lagging strands the last Okazaki fragment cannot be laid down with the result that the replicated DNA has single stranded terminations. The 'eye' structures elongate and join allowing the replicated strands to separate. Further initiation takes place and DNA replication continues. A complication now sets in due to the single-stranded ends which link together and are covalently joined by ligase to give multi-genome length molecules called concatemers. These are subject to recombination between themselves. Gp's 46 and 47 are involved in concatemer formation and in recombination which are essential for the replication of T4 DNA.

When a large pool of concatemeric DNA has been formed pieces of the concatemers are cut off during the formation of phage heads and encapsidated (see Section IV.E). These pieces are slightly more than a complete genome in length and are consequently circular permutations of each other as was the parental duplex.

2. Phage φX174

Genetic analysis of φX174 has revealed nine cistrons and it is assumed that the phage has nine genes. This is confirmed by the identification of nine gene products and by the work on the sequence of the nucleotides of the phage DNA (see Section IV.D.3). Table 2 lists the genes A to J and gives the functions and molecular weights of their products.

After injection of the genome it becomes attached to the cell membrane, probably by the spike protein gpH. This strand, referred to

Table 2. The genes of phage φX174 and their products. From Sanger *et al.* (1977) *Nature*, **265**, 687–695

Gene	Molecular weight		Function
	SDS gel	From nucleotides	
A	55–67,000	56,000	Single-strand break in RF DNA
B	19–25,000	13,845	Synthesis and packaging of progeny single-strand DNA
C	7000	—	As for B
D	14,500	16,811	As for B
E	10–17,500	9940	Lysis
F	48,000	46,400	Main capsid protein
G	19,000	19,053	Spike protein
H	37,000	35,800	Spike protein, adsorption and injection of viral DNA
J	5000	4907	Internal protein

as the plus (+) strand is immediately converted to a duplex closed loop replicative form (RF) by the action of cellular polymerases which add a complementary or (−) strand.

Several copies of the RF duplex are made by the cellular polymerases and viral messenger RNA is transcribed from them. It is not possible for copies of the (+) strand to be made on the RF DNA as it stands because there are no free polynucleotide ends. The next step is the cleavage of the (+) strand by a nuclease coded for by gene A. This produces another species of RF from which copies of the (+) strand can be prepared by the rolling circle method (Figure 7). Gene products gpB, gpC, and gpD are involved in the synthesis and/or packaging of the progeny viral strands into the capsids but their exact functions are as yet unknown. Several hundred complete virions can accumulate in the infected cell which remains intact, though of course non-viable, until some 30 minutes after infection when the cells suddenly lyse. A viral coded protein, gpE, is necessary for lysis.

3. The Nucleotide Sequence of φX174 DNA

The molecular weights of the various gene products of φX174 given in Table 2 and the total coding power of the DNA can be calculated from its molecular weight. However it turned out that a greater total molecular weight of protein is produced than could be accounted for by

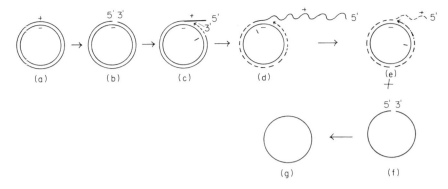

Figure 7. Replication of φX174 DNA by the rolling circle method. The viral parental strand (+) is converted to a closed circular duplex and replicated by the cellular enzymes to give the replicative form (a). A single stranded break is formed in the (+) strand by a viral coded enzyme (b). The circle rolls about a fixed point while a new (+) strand is laid down (pecked line) (c) and a strand of viral DNA is released (d). A second round of replication begins (pecked line with dots) so releasing another strand of viral DNA (e). The progeny DNA (f) is cyclized by a cellular ligase to give a covalently linked circle (g) which is then packaged into the phage capsid

the molecular weight of the genome. This discrepancy was resolved by the work of Sanger and coworkers, who have determined the total nucleotide sequence of φX174 DNA which consists of 5375 nucleotides. Previous work had shown that the order of the genes, was A, B, C, D, E, J, F, G, H. A combination of many techniques has enabled the position of the genes on the chemical map to be determined. Data from various sources has been combined into the circular genetic map in Figure 8. The features to note are that the gene B lies completely within gene A together with its polypeptide initiation and termination sequences and its mRNA initiation site. The termination and initiation signals of genes A and C appear to overlap as do those of genes C and D. Gene E lies within gene D and the terminator of gene D overlaps the initiation site of gene J. This is shown in detail in Figure 9 where the whole gene J ribosome binding-site sequence is given. It can be seen that all three possible reading frames are used so as to provide the initiation signal for the gene J polypeptide, and the termination signals for the polypeptides of genes D and E within the mRNA sequence which serve as the gene J ribosome binding site. The observation that nucleotide sequences can be formed in such a way as to allow for overlapping information explains the unexpectedly high coding power of the φX174 genome.

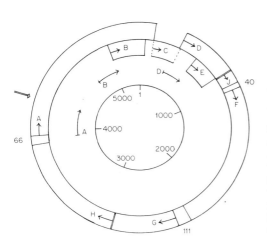

Figure 8. The genetic map of φX174. The genome is a closed loop of single stranded DNA of which the sequence of the 5375 nucleotides has been determined. The inner circle indicates the numbering of the nucleotides. The outer circle shows the sequence of the genes A—J, their lengths, and the positions of their initiation and termination signals. Gene B lies within Gene A, and gene E within gene D. Gene C overlaps genes A and D. Curved arrows in the intermediate circle show the initiation sites for the mRNA of genes A, B and D. Externally an arrow shows the site of initiation of replication and numbers give the number of untranslated nucleotides in the intergenic spaces. Based on data of Sanger *et al.* (1977) *Nature*, **265**, 687—695

Figure 9. The ribosome binding site nucleotide sequence for the gene J mRNA of φX174. 'Init' shows initiation and 'term' the termination codons used in genes E, D, and J. Based on data of Sanger *et al.* (1977), *Nature*, **265**, 687—695

4. Filamentous Phages

The DNA of these viruses is a closed single-stranded loop of molecular weight 2.4×10^6. Eight complementation groups have been found and it is assumed that the genome contains eight genes. Two of the gene products are structural, gpVIII is the bulk coat protein of MW 5196 and is present in about 2000 copies per virion, while gpIII is the minor coat (or 'A') protein of molecular weight 56,000 present in only one or two copies per virus particle. Two of the remaining genes, II and V, are concerned with viral DNA synthesis while the function of the other four is unknown. The progeny virions leave the cells by a

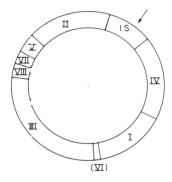

Figure 10. The genetic map of filamentous phage fl. showing the relative positions and sizes of the eight genes I–VIII. Initiation of complementary strand synthesis begins in the intergenic space I.S. Simplified from Horiuchi and Zinder (1976), *Proc. Nat Acad. Sci. (Wash.)*, 73, 2341

process resembling extrusion, not cellular lysis. It should be said that the host—virus relationship is not entirely stable and cell death does occur with sufficient frequency to permit the phage to be assayed by the standard plaque assay methods.

The genetic map of phage fl is shown in Figure 10. All the genes are closely adjacent to one another except for II and IV which are separated by an intergenic space, a length of DNA which does not become transcribed into messenger RNA. The phage DNA is replicated by a process which is similar to that used by the small icosahedral phage ϕX174. The gpIII molecule(s) remain attached to the cell membrane and anchor the parental phage DNA to it.

The cellular DNA polymerase is unable to initiate synthesis of a complementary strand but is able to elongate an existing complementary polynucleotide already attached to a viral strand. This difficulty is overcome by the action of a host RNA polymerase which lays down a short length of complementary RNA on the viral strand. It does this at a particular point within the intergenic space which contains the origin of DNA replication. After a short length of RNA has been laid down, the DNA polymerase takes over and the RNA segment is eliminated. This results in the formation of a closed circular duplex (RFI) still attached to the cell membrane.

RFI cannot be replicated to give viral strands directly; it has to be nicked by the formation of a single strand break. This step is performed by gpII, an enzyme which specifically recognizes a sequence of nucleotides in the region of the origin of replication. This nicked circular structure is known as RFII and is replicated by the rolling circle method as described for the DNA of phage ϕX174 (Figure 7). The newly synthesized viral strands might now either be incorporated into new phage or they might be converted to replicative forms and indeed some are converted to replicative form DNA. However there is another peculiarity of filamentous phage. There is never a pool of completed intracellular phage as there is with other phages but there is a pool of viral DNA molecules. These would all be converted to RF form unless they were withdrawn in some way and protected ready for coating with gpVIII at the appropriate time. This is achieved by the action of gpV, a small protein of molecular weight 9688, which specifically attaches to single-stranded DNA, covers it completely and renders it safe from conversion to RF form. It is in this form, complexed with gpV, that the pool of phage DNA is accumulated.

The coat proteins, gpVIII and gpIII are synthesized at or near the cell membrane and are used immediately to coat the phage DNA. This is performed by an exchange reaction in which the gpV is stripped off and recycled while the gpVIII is put on and the coated virion extruded at the same time. Finally, the one or two molecules of gpIII are added and the completed virion separates from the host cell.

5. RNA Phages

Phages with an RNA genome have been found for several species of bacteria but the best known are the closely similar ones f2, MS2, and R17. All are isometric particles about 25 nm across, in which there are 180 molecules of the 'B' protein and one of the 'A' protein. The genome is a piece of single-stranded linear RNA containing 3569 nucleotides in the case of MS2. The phage genome acts as its own mRNA. It codes for the A and B structural proteins and an RNA-dependent RNA polymerase subunit.

The complete sequence of the nucleotides is known, as are the amino acid sequences of the proteins it codes for. A block diagram of the genome showing the order of the genes, their initiation and termination codons, their lengths, and the lengths of the untranslated intergenic spaces is given in Figure 11.

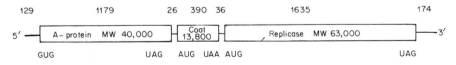

Figure 11. The genetic map of RNA phage MS2 of which the sequence of the 3559 nucleotides is known. The numbers indicate the number of nucleotides in each gene and in the untranslated segments. The initiation and termination codons are given. Based on Fiers et al. (1976), Nature, 260 500, and reproduced by permission of Macmillan (Journals) Ltd

During replication the parental strand is converted into an RNAase-resistant, double-stranded, replicative form by the synthesis of a complementary (−) strand. New viral strands are synthesized by the replicase by complementary strand formation on the (−) strand of the RF RNA and the displacement of the parental strand. It is also possible for the parental strand to be conserved while the viral strand is being made by a process which does not totally displace the parental strand. The phage genome is used directly as mRNA and is translated into the three phage proteins.

Phage R17 is a good example of the self-assembly mechanisms which can occur *in vitro*. Under suitable conditions the B protein can self-assemble into capsid-like structures even without any RNA being present. When viral RNA is present the B protein can assemble round it and form a non-infective, virus-like particle which is sensitive to RNAase, but if the A protein is present as well, active RNAase-resistant virions can assemble *in vitro*. Their properties are indistinguishable from particles made *in vivo*.

E. Morphogenesis of Phage T4

In the final stages of the intracellular development of the phage a pool of DNA and proteins is accumulated and molecules are withdrawn from this pool at random and assembled into whole phage particles. The T4 virion is assembled by a sequence of events in which protein molecules are fitted together in a well-defined order. Evidence for these morphological assembly pathways was obtained largely by the use of conditionally lethal phage mutants. If phage is treated with certain mutagens such as hydroxylamine or *p*-nitrosoguanidine it is possible to convert certain amino acid triplets into termination triplets, in particular one which is transcribed as UAG in mRNA and is known as an 'amber' mutation. These mutants cannot grow on their normal host, in which case the mutation is clearly lethal, because the protein specified by the mutated gene is truncated at the point of mutation. However, there are certain strains of host bacteria that possess a supressor of this type of mutation, which can insert an amino acid in place of the foreign termination codon. The polypeptide chain can therefore be completed and, as is often the case, the new amino acid does not affect the function of the protein. These strains of bacteria, known as permissive strains, can support the growth of the mutant phages which are clearly only conditionally lethal mutants. If a large number of amber mutants is obtained it is possible to classify them into groups and obtain an estimate of the number of genes in the phage.

Amber mutants can be used to determine the function of each mutant gene. If the wild-type virions are analysed by polyacrylamide gel electrophoresis, the number and molecular weight of all the structural and internal proteins can be found. If then the non-permissive host is infected with a mutant and the gene products caused to be radioactively labelled they can be examined on gels and it can be determined whether any of the structural proteins are missing and, if so, it is a reasonable assumption that the mutant gene is the structural gene for that protein. Other functions such as polymerases, nucleases, and lytic enzymes can be studied and linked to particular genes by the same procedure.

In another type of experiment, non-permissive cells are infected with amber mutants and after a time to allow synthesis of viral products the cells are lysed artificially. From the nature of the products found the function of the mutant gene can be inferred. For example if the products include filled phage heads but no tails then the mutant gene must be concerned with tail synthesis, and if tails but no heads are found then the gene must be involved in head formation. Sometimes the cells contain aberrant structures made of phage coded proteins. Examples of these are polyhead, a tubular structure which can have the same diameter as a phage head but many times as long, and polysheath which is made of the tail sheath protein assembled into long tubes. These structures are formed when the structural proteins are produced but the genes which control or direct their assembly are mutant.

Further insight into the morphogenetic pathways has been obtained from *in vitro* complementation experiments. If an extract of non-permissive bacteria infected with a mutant deficient in head formation but which contains fully formed tails, is mixed with one from a mutant deficient in tail synthesis but which contains perfectly formed heads it is found that the heads and tails automatically assemble together to give active phage, provided that fully formed tail fibres are present as well.

The morphological development pathway for phage T4 is given in Figure 12. In the earliest stage of head formation a 'core' composed of gp22 and $gpIP_{III}$ is formed in association with the host cell membrane. Gp23 activated by gp31, which itself is not a structural protein, and gp20 are added to give a structure called Prohead I. Gp23 is the main structural protein of the T4 capsid while gp22 determines the short diameter and gp20 the long diameter of prolate icosahedral capsid. Under the action of gp21 and possibly gp24 a polyhedral capsid, Prohead II is formed and the gp23 is cleaved to a smaller size, losing 10,000 in molecular weight. It will be recalled that the replicated phage DNA is in the form of very large molecular weight repeat molecules

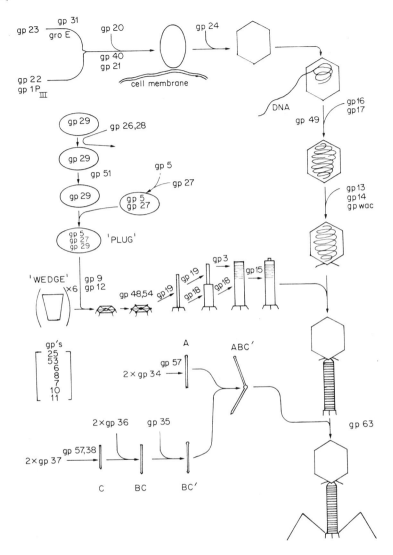

Figure 12. The morphogenetic pathway of phage T4. The gene products involved either structurally or catalytically are marked thus gp23 . . . three independent pathways produce filled heads, tails, and tail fibres. Tails combine with heads are are converted to active phage by the addition of the tail fibres. Reproduced, with permission, from the *Annual Review of Biochemistry*, Volume 44. Copyright © 1975 by Annual Reviews Inc. All rights reserved.

called concatemers (see Section IV.D.1). The proheads are developed in association with this DNA and by the action of gp49, gp16, and gp13, which control the cutting of the concatemeric DNA into 'headsize' pieces. The complete phage genomes are packaged into the developing heads to form Prohead III. The formation of the head and the cutting and packaging of the DNA are linked processes which take place simultaneously. Further enzymic cleavage of gp20, gp24, and gpIP$_{III}$ takes place and the mature head is produced by the action of gp13, gp14, and gp wac, the latter three gene products rendering one apex of the icosahedron receptive to the tail structure.

The tail is assembled separately by a process which begins with the linking of at least seven different gene products into 'wedges'. Six of these are assembled into the hexagonal base plate to which is added a central plug followed by gp9, which is the site for tail fibre attachment, and gp12, the short pins beneath the base plate. The base plate is then primed by the addition of gp48 and gp54 which act as substrates for the polymerization of the tail tube, this being composed of rings of subunits of protein coded for by gene 19. As the tail tube grows in length, so does the surrounding tail sheath as rings of its subunit protein, gp18, are laid down. Finally gp3 and gp15 are added to the completed tail to act as a connector to the head. Tails and head assemble spontaneously but are not infective because they lack the tail fibres.

The tail fibres are assembled separately from two molecules of gp34 which form the proximal half, and two molecules of gp37 which form the distal half. The dimer of gp37, known as antigen C, is lengthened by the addition of two molecules of gp36 to give antigen BC. To this is added gp35 which modifies antigen C to C' so forming antigen BC' which can self assemble with antigen A to give the complete bent fibre ABC' which under the action of gp63 attaches to the already assembled head and tail to give an active phage particle.

V. Lysogeny

There is another class of phage, unlike those discussed so far, known as temperate phage which, although it can kill its host, does not normally do so. These are also known as lysogenic phages and are typified by coliphage Lambda (λ). A culture infected with a lysogenic phage does not lyse but continues to grow and divide normally. The culture fluid, however, always contains a low concentration of the phage and this cannot be removed by washing or treatment with antiphage serum. A

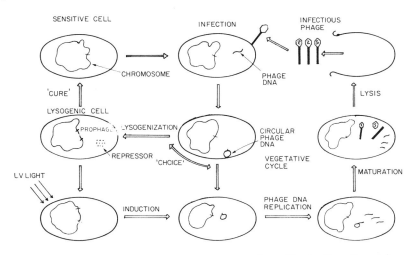

Figure 13. Lysogenization by a temperate phage illustrating the main features of the lysogenic state and the alternative vegetative phage cycle. Modified after Casjens and King (1975). *Annual Review of Biochemistry*, **44**, 555.

susceptible strain which gives plaques with a lysogenic phage is known as an 'indicator' strain.

The reason why lysogenic cultures always contain a low level of phage is because once every thousand generations or so a daughter cell suddenly lyses and liberates 100 phages or more while the remaining cells carry on dividing. The bacteria possess receptors for the phage and the particles do adsorb but the phage DNA cannot replicate. The presence of the lysogenic phage has rendered them immune to infection by that phage, though not to other phages. The immunity is specific to the phage which conferred it.

If a lysogenic culture is treated with a small dose of ultraviolet light, the whole culture is caused to lyse and to liberate many phages per cell. Ultraviolet light is an inducing agent, as are X-rays and certain substances such as mitomycin C. This shows that although occasional cells of a lysogenic culture do lyse naturally, all of the cells are capable of lysis and production of phage. The property of carrying phage in the lysogenic state is therefore an inheritable property. If lysogenic bacteria are broken open, no phage can be found inside them. The phage must therefore be carried in a non-infective form which is known as 'prophage'. It turns out that the phage genome becomes integrated into the bacterial chromosome and multiplies with it.

Lambda DNA is a linear duplex 17 μm in length, but after storage for a short time it becomes circular. The reason is that each end of the duplex carries a single-stranded length of DNA which are complementary to each other. By random collisions the ends touch each other and remain in contact so producing the circular molecules. After injection the duplex is also cyclized and the adjacent nucleotides covalently joined by a cellular ligase. At this stage the DNA can either engage in vegetative replication leading to phage production and lysis, or it can establish the lysogenic state.

A. Vegetative Cycle

The cyclic DNA becomes attached to the cell membrane at a specific site and replication takes place symmetrically. It begins at a particular point and proceeds bidirectionally until the two forks meet and the daughter molecules separate. The cyclic progeny are replicated by the rolling circle method. This is shown for single-stranded DNA in Figure 7. In the present case a complementary strand is laid down alongside the single strand as it is displaced from the (−) strand to give a duplex of great length called a concatemer rather than single length duplexes. Phage mRNA is transcribed from the cyclic DNA and is translated in an orderly sequence. The concatemeric DNA is cut into pieces containing exactly one phage genome by the action of an endonuclease which makes single-stranded nicks 12 nucleotides apart, and so regenerates the sticky ends. At the same time the genomes are packaged into the capsids. Dissolution of the host cell is caused by the product of gene S which stops cellular metabolism and gene R which lyses the cell wall and releases the progeny phage.

B. Establishment of Lysogeny

In the genetic map of phage (Figure 14) the point marked with an arrow between genes A and R is where the two ends of the linear viral duplex have been joined. The genome consists of three distinct regions, the right operon which is involved in the vegetative functions of DNA replication, head and tail synthesis and lysis, the left operon which is involved in integration and recombination, and the immunity operon whose products and their interactions with the DNA decide whether it shall become vegetative or lysogenic. In lysogenic bacteria there are no mRNAs produced for the vegetative and recombination regions but there is a small mRNA corresponding to the immunity region. The reason for this is that a powerful repressor has been syntheized and it has blocked the transcription of those regions of the genome. This

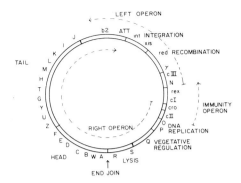

Figure 14. Genetic map of phage λ. The linear duplex DNA is cyclized at 'end join'. It attaches to the bacterial chomosome at 'att'. Some of the genes in the right, left, and immunity operons are marked with their functions. Dotted lines indicate direction of transcription. After Davis *et al.*, (1973) *Microbiology* Harper and Rowe, New York

substance, the λ repressor, has been isolated and studied in detail. It is a protein of molecular weight 30,000 which exists as a dimer and specifically binds to the phage DNA. It is the product of gene cI and it blocks the action of the left promoter which in turn prevents gene N being transcribed, and the right promoter which prevents genes, O, P, and Q from being transcribed. These last three genes are early transcribed genes and without their products, the late genes cannot be transcribed. The repressor therefore completely obstructs all the functions required by the vegetative phage. The immunity region also has a gene *cro* which codes for a protein which tends to block the continued production of the repressor. If the *cro* gene product is synthesized to a greater extent than the repressor, or if its action is enhanced by the physiological condition of the host or by the composition of the medium, then the vegetative region of the genome will be transcribed and the phage can multiply. If, on the other hand, the balance is in favour of the repressor then the vegetative functions will be blocked and the lysogenic state will prevail.

C. Integration

Once the phage genome has been cyclized it becomes attached by the *att* region to a particular site on the host chromosome. The phage DNA is then inserted into the host genome by a single reciprocal crossover which is mediated by the product of the phage gene *int* which has been transcribed before the repressor was present in sufficient amount to prevent it. The insertion is shown diagrammatically in Figure 15. The phage DNA has now become a 'prophage'. The replication of the prophage takes place under host control by the normal mechanism of

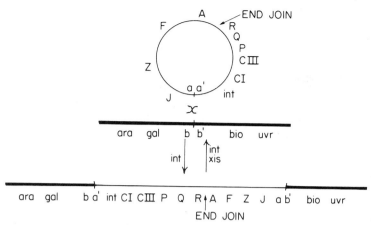

Figure 15. The Campbell model for the integration of λ DNA into the host chromosome. The attachment site is aa' and the host site is bb'. Site specific recombination between aa' and bb' requiring gene *int* causes integration while recombination between ba' and ab' under influence of genes *int* and *xis* results in the excision of the prophage. Heavy line denotes bacterial DNA. Based on Guarneros and Echols (1973) *Virology*, 52, 30–38 and reproduced by permission of Academic Press Inc.

bacterial DNA replication. The immunity operon is continually transcribed and the cellular concentration of repressor is maintained, thus blocking the production of phage.

D. Excision

In order for the prophage to become vegetative it must be freed from the bacterial chromosome. This is mediated by the products of the genes *int* and *xis* which enable a reversal of the integration step to take place. A reciprocal crossover between the attachment sites at the end of the prophage takes place (Figure 15) and the circular phage DNA is liberated into the cytoplasm. All the vegetative functions of the phage are then free to act and the cell is committed to phage production and lysis.

E. Transduction

Some times, during the excision of λ prophage a mistake is made and the host gene *bio* or *gal* is included in the progeny phage genome. The phage is inactive because, being able to accommodate in its capsid only a limited amount of DNA, some phage genes are left out. Phage particles of this type can inject their DNA into a host cell. This becomes inserted and the host now possesses a gene from the first host cell. This is called specialized transduction. Other lysogenic phages, e.g. P22 and *B. subtilis* phage PBS1, can encapsidate larger pieces of host DNA and transduce them into new hosts. All transducing phages are defective. By selective plating methods recipient strains to which any host genes have been transduced can be isolated. This is termed generalized transduction and has proved extremely useful in mapping bacterial chromosomes.

There are examples of phages carrying genes for functions which are non-essential for the phages themselves and transmitting them to their host bacteria. One example is the transfer of the genes for diphtheria toxin formation by corynebacterium phage from toxin-producing cells to non-producers. This is called phage conversion. The phenomenon has wide implications in the spread of pathogenic properties through bacterial populations.

VI. Further Reading

Davis, B. D. *et al.* (1973). *Microbiology.* New York: Harper and Row.

Duckworth, D. H. (1976). 'Who discovered Bacteriophage'. *Bact. Revs.*, **40**, 793—802.

Fiers, W. *et al.* (1976). Nucleotide sequence of MS2 RNA. *Nature*, **260**, 500—507.

Hayes, W. (1968). *The Genetics of Bacteria and their Viruses.* Oxford: Blackwell.

Sanger, F. *et al.* (1977). Nucleotide sequence of ϕX174 DNA. *Nature*, **265**, 687—695.

Stent, G. S. (1963). *The Molecular Biology of Bacterial viruses.* San Francisco and London: W. H. Freeman.

The Determinants of Microbial Pathogenicity

H. SMITH

Department of Microbiology, University of Birmingham, P.O. Box 363, Birmingham B15 2TT

I. Introduction

Pathogenicity or virulence is the capacity to produce disease and mankind suffers from the fact that some microorganisms have this capacity in ample measure. There are two reasons for investigating microbial pathogenicity. First, it might explain why so few microorganisms are pathogenic, and second it might help the design of new measures against infectious disease.

A. The Place of Pathogens in the Microbial World

Pathogenic microorganisms receive much attention but they occupy only a small part in the total microbial world. In fact they are peculiarities. The great majority of microorganisms are harmless and many are beneficial. Bacteria pathogenic for man and animals occur as a few species in genera which contain many non-pathogenic species. Furthermore, the few pathogenic species contain strains (isolates) that produce disease and others that do not. These remarks apply equally to microfungi and protozoa; the pathogenic species are few and far between. Viruses are different; they are obligate parasites and all are potentially pathogenic. However, virus species also contain many strains that do not produce disease in animals although they replicate in tissue culture cells.

Microbial pathogens have biochemical processes which set them apart from other microorganisms and determine disease production. These biochemical differences — the determinants of microbial pathogenicity — are the subject of this essay.

B. The Persistence of Infectious Disease

Although the worst effects of infectious disease have been controlled by public health measures, infectious disease remains a major problem in human and veterinary medicine with the economically important nuisance and chronic aspects gaining prominence as the fatal consequences become less frequent. Effective chemotherapy is still lacking for virus diseases. Antibiotic resistance of bacteria and drug resistance of protozoa are increasing. The present pandemic of gonorrhoea shows a formerly controlled disease can rebound with changing social conditions. Gram-negative bacteria and microfungi that do not normally attack healthy people cause fatal infections in debilitated or immunosuppressed patients. Many vaccines do not produce complete protection. Clearly infectious disease is still with us and likely to remain for many years to come. This fact, after the euphoria of the antibiotic era, explains the current interest in microbial pathogenicity. New methods are needed against infectious disease, and one approach is to recognize and to neutralize the determinants of pathogenicity.

C. The Plan of the Essay

Methods for studying mechanisms of infectious disease are outlined first. Then, present knowledge of the determinants of microbial pathogenicity is summarized and gaps emphasized. Bacteria form the main subjects since studies of them have indicated the methods and difficulties of investigating infectious disease, and have produced concepts of pathogenesis that apply to other microbes. The pathogenicity of viruses, fungi, and protozoa is then discussed in relation to the concepts used for bacterial pathogenicity. In each section there is discussion of the main aspects of pathogenicity.

(1) Entry to the host: mucous membrane interactions.
(2) Multiplication *in vivo*.
(3) Interference with host defence.
(4) Damage to the host.
(5) Tissue and host specificity.

II. The Methods of Studying Infectious Disease

There are two methods of studying infectious disease. Natural infections can be observed either in single animals (clinical observations) or within populations (epidemiology), and disease can be investigated in greater depth by deliberate introduction of pathogenic microbes into animals. This essay deals mainly with results from the second method. However, the pattern of disease in individuals is summarized as background to a discussion of the experimental study of disease.

A. General Pattern of Natural Disease in Individuals

Pathogenic microorganisms must enter the host. Occasionally this occurs through the skin, especially after wounding. Sometimes there is direct injection by insect bites as in bubonic plague, but most infections take place on or over the mucous surfaces of the respiratory, alimentary, and urogenital tracts, with the pathogen rather than an external agency providing the impetus for penetration. Sometimes, for instance in cholera, penetration does not occur and the microbes cause harm by growth on the mucous surface. In other cases, such as in dysentery, penetration is minimal involving only the surface layers of the tract. However, many pathogens pass through the mucous membranes to infect the host tissues, as in typhoid fever.

After gaining access to the tissues the pathogen faces antimicrobial substances in the body fluids and in special cells called phagocytes which ingest and kill microorganisms. Some antimicrobial mechanisms are present at entry and others are mobilized within a few hours in the so-called inflammatory response. At this primary lodgement phase the pathogen is at its weakest in relation to host defence and the success or failure of most infections is probably decided within 2—6 hours of entry. Successful pathogens may localize at the initial site, for example in dysentery, but mostly they spread through the lymph channels which run close to the mucous surfaces and then in the connecting blood stream. The microbes may be free in the body fluid or contained within white or red blood cells. Spread is helped by the efficient circulatory system but it is hindered by special phagocytes fixed in filtration systems set across lymph and blood vessels in the lymph glands, spleen, and liver. After spreading from the initial site, some infections become generalized, as in anthrax, but others localized in particular tissues such as the meninges of the brain in meningitis.

On mucous surfaces, at primary lodgement, during spread and after localization, pathogens damage the tissues. This damage can be fatal and is always unpleasant. Usually, however, the host eventually eliminates the pathogen and recovers from the disease. Recovery is often, but not always, accompanied by immunity to further attacks.

B. Experimental Study of Disease

First the causative microbial species must be identified. Then, mechanisms of disease production must be recognized and explained in biochemical terms. The second task is made easier, if strains (isolates) of differing disease producing capacity can be identified and compared biologically and chemically. Methods for identifying the causative organism and for measuring differences in the disease producing capacity of strains are described below.

1. Identifying Causative Microorganisms: Koch's Postulates

The criteria below were enunciated by Koch in 1891 as a result of his identification of *Bacillus anthracis* as the cause of anthrax, and to this day fulfillment of Koch's postulates provides the rigid proof of causation which is sought for each newly appearing disease. Unfortunately in some cases they cannot be fulfilled completely.

(1) The organism should be found in all cases of the disease in question and its distribution in the body should be in accordance with the lesions observed.

Ideally, as in anthrax, the microbe can be recognized in tissues by microscopy and its presence confirmed by isolation and testing *in vitro*. However in some diseases, for example leprosy, the microbe can be seen but cannot be cultured *in vitro*. In others, such as hepatitis, the causative microbe can neither be seen nor cultured. Fortunately the microbe can usually be detected by immunological methods not available to Koch.

(2) The organism should be cultivated outside the body of the host, in pure culture, for several generations.

This has been accomplished for most pathogens but some remain recalcitrant, such as the agents of leprosy and glandular fever.

(3) The organism so isolated should reproduce the disease when introduced into other susceptible animals.

The major snags occur here. For veterinary diseases such as anthrax the natural host can be inoculated. However ethics restrict experiments in humans. Only rarely, such as tuberculosis in the guinea pig, can a disease typical of the human form be reproduced in small laboratory animals. Although typhoid bacilli will kill mice the diseases are unlike human infections. In some cases primates suffer human-like illnesses, for example chimpanzees with gonococci. Inoculation of human volunteers has occurred for diseases (common cold and gonorrhoea) that are not dangerous. However this cannot be done for example with human cancers of suspected viral origin.

2. Quantitative Comparisons of Disease Producing Capacities: Pathogenicity, Virulence

The terms *pathogenicity* and *virulence* are nearly synonymous and mean capacity to produce disease. They should be used in a comparative sense: a microbial population is more (or less) pathogenic or virulent than another population. Pathogenicity is used mainly with respect to species, e.g. *Bacillus anthracis* is more pathogenic than *Bacillus subtilis*, and virulence with respect to strains within species, e.g. the Vollum strain of *B. anthracis* is more virulent than the Weybridge strain. Quantitative comparisons of pathogenicity and virulence are made by inoculating a suitable animal species with graded doses of the two species or strains and comparing the numbers that produce a certain disease effect; for examples, those needed to kill half a group of animals (the 50% lethal dose: LD_{50}) or to produce skin lesions of a certain size.

3. Virulence Determinants

To produce disease microorganisms must be able to:

(1) Enter the host, usually by surviving on and penetrating the mucous membranes of the respiratory, alimentary, and urogenital tracts.
(2) Multiply in the physical and chemical conditions of the host tissues.
(3) Interfere with the action of humoral (in body fluids) and cellular defence mechanisms of the host.
(4) Damage tissues thereby producing the unpleasant and possibly lethal effects.

The microbial products responsible for these steps in disease production are *virulence determinants*. Virulent strains can accomplish all four steps; avirulent or attenuated (of lowered virulence) strains fail to accomplish effectively one or more of them. Each step is complex as is described later. The cardinal point to make here is that virulent microorganisms must possess several different determinants to accomplish the whole disease process. Occasionally, for example in tetanus, the disease is dominated by one virulence determinant, a powerful toxin, but such cases are rare.

If virulence is determined by a number of factors, avirulence or attenuation can result from loss of some, but not necessarily all, of the determinants, whose effect can be interdependent or additive. In the first case, loss of one determinant results in an almost complete loss of virulence. The virulence of *B. anthracis* is determined by capsular poly-D-glutamic acid which inhibits host defence, and a toxin which kills the host. The Vollum strain produces both and is virulent, the Weybridge strain produces toxin but not capsular material and is avirulent as is the H M strain which produces capsular material but not toxin. In the second case, for example the virulence of *Staphylococcus aureus*, the loss of one determinant results in some, but not a complete, reduction in virulence. Presumably another determinant accomplishes the same essential step in disease production although less efficiently.

4. Recognition of Virulence Determinants: Comparison of Properties of Strains of Different Virulence

If the biochemical and biological properties of virulent and avirulent strains of a pathogenic species are compared the majority will prove similar since virulence results from relatively small differences in microbial genomes. A few properties *virulence markers* will be associated with virulent rather than avirulent strains. They may be virulence determinants, but are not necessarily so. For example, virulent strains of *Brucella abortus* contain more catalase than avirulent strains and metabolize glutamate less well. However, neither virulence marker seems to influence disease production. On the other hand, a virulence marker for *S. aureus* is an ability to clot fibrin in plasma (so-called coagulase activity) and this activity may inhibit host defence.

Comparisons that yield most information on virulence determinants are those in biological tests related to the four steps in disease production, entry, multiplication *in vivo*, interference with host defence, and production of damage. The variety of such tests will become clear later but they include: ability to multiply in segments of mucosal tissue isolated *in vivo*; ability to multiply in extracts of tissue known to be infected in disease; ability to resist antimicrobial substances in blood, serum, and tissue extracts; ability to resist ingestion and killing by host phagocytes, and ability to kill the host or damage its tissues.

Always it should be remembered that virulence is determined by more than one factor and hence any avirulent strain examined may possess one or more virulence determinants. Comparison with a virulent strain will reveal only the determinants not possessed by the avirulent strain, and to obtain the full picture, more than one pair of strains may have to be compared.

5. Purification and Identification of Virulence Determinants

Once biological activity related to virulence has been recognized, fractionation of the determinant concerned can proceed using the appropriate test as an assay. This might be based on a direct action of the determinant, for example in fractionating a toxin from extracts of a virulent strain. Or a test can be used indirectly by assaying the activity of an avirulent strain with or in extracts of the virulent strain. For example, if a virulent strain resists killing by serum or ingestion by phagocytes more than an avirulent strain, the determinants might be assayed in extracts of the virulent organisms by adding dilutions to the test system containing the avirulent strain, thereby reducing its killing by serum or ingestion by phagocytes.

III. Bacterial Pathogenicity

A. Difficulties of Studying Bacterial Pathogenicity

The fact that pathogenicity is due to a number of determinants causes difficulty in recognizing and fractionating any single determinant. However, the main difficulty arises because virulence can be measured only *in vivo* and is markedly influenced by changes in growth conditions due to selection of types and to phenotypic change. Virulence is reduced by subculture *in vitro* because bacteria lose the capacity to form one or more of the full complement of virulence determinants manifested in infected animals. Bacteria grown in infected animals are different chemically and biologically from those grown *in vitro*, especially in relation to virulence determinants. Hence, bacteria grown *in vitro* can be incomplete in virulence determinants. This is the essence of the difficulties encountered in studies of pathogenicity.

The determinants of pathogenicity can be produced in laboratory cultures if the correct nutritional conditions can be found as for the classical bacterial toxins. However, for problems of pathogenicity which have defied solution by conventional procedures, one approach is to study bacterial behaviour *in vivo*. Aspects of pathogenicity might be revealed which later could be reproduced *in vitro*. This essay describes examples of virulence determinants that were recognized by studying bacterial behaviour *in vivo* by the following methods. Bacteria and their products can be separated directly from the diseased host for biological and chemical examination. The behaviour of organisms growing in animals and their repercussion on the host can be examined directly. Some light can be shed on microbial behaviour *in vivo* by observations in organ culture and finally, tests *in vitro* can often be made more relevant to microbial behaviour *in vivo*.

Investigating mechanisms of pathogenicity is even more difficult for diseases caused by mixed infections for example dental caries or peridontal disease. We are only scratching the surface of the complex interactions occuring in mixed infections. A few investigations provide templates for future work, for example in heel abscess of sheep a mixture of *Corynebacterium pyogenes* and *Fusiformis necrophorus* is needed for pathogenesis. The former provides a growth factor and an anaerobic environment for the latter which produces an aggressin (see later) that prevents phagocytosis of the former as well as of itself.

B. Entry to the Host: Mucous Membrane Interactions

Most infections start on the mucous membranes of the respiratory, alimentary and urogenital tracts, membranes that are protected by moving lumen contents, surface mucus, and often by the activity of commensal (i.e. indigenous) microorganisms. Bacterial components could contribute to mucosal infection and penetration: (1) by promoting adherence to the epithelial surface, thus resisting the flushing action of moving lumen contents; (2) by resisting competition (for food and space) with the commensals including resisting antimicrobial materials (e.g. fatty acids) produced by them; (3) by resisting humoral and cellular defences in mucous secretions, not the least being a high or low pH, and (4) by promoting penetration of epithelial surfaces.

1. Adherence

Pathogens and potentially pathogenic commensals adhere to mucous surfaces selectively. Enteropathogenic *Escherichia coli* adhere to the ileum rather than to the duodenum of pigs and calves. In humans *Streptococcus mutans* and *St. sanguis* adhere to the teeth, *St. salivarius* to the tongue, and *St. mitis* to the buccal mucosa.

Some, but not all, strains of *E. coli* enteropathogenic for piglets attach to the brush border of the upper small intestine by a plasmid-controlled surface protein, the K88 antigen. Strains lacking the K88 plasmid are avirulent. Colostrum of sows immunized with K88 antigen protects piglets against challenge with a K88 positive strain which does not then attach to the brush border. Strains of *E. coli* infecting calves, sheep and humans also appear to possess 'sticking' antigens.

St. mutans, which causes dental plaque, adheres to teeth in two phases; first a weak reversible association through unknown bacterial components with salivary glycoproteins which form a pellicle to the teeth; then a stronger attachment through two sticky glucose polymers (glucans) synthesized from dietary sucrose.

Virulent strains of *St. pyogenes* adhere to the epithelium of the human throat by their cell wall M-protein. Removal of the M-protein with trypsin or pretreatment with M-protein antiserum inhibits attachment.

Gonococci adhere to epithelial cells of the urogenital tract in human gonorrhoea, in organ culture, and in cell suspensions but the surface components responsible are not clear. Pili, present on virulent but not on avirulent strains, promote adhesion of gonococci grown *in vitro* to tissue culture cells. However, both pilated and non-pilated gonococci grown *in vitro* adhere to human fallopian tube and to human endocervix. Also *in vivo* gonococci may have different surface components.

2. Competition with Commensals

Competition and antagonism between mucous membrane-commensals and pathogens were indicated by the survival of introduced pathogens (e.g. shigellae) when some indigenous bacteria had been removed by antibiotic treatment. This was confirmed by mixed culture experiments in gnotobiotic animals, and *in vitro*. Thus, competition and antagonism are proven but the mechanisms of destruction or survival of pathogens are still not clear. The microbial activity which determines the low E_H of the lower bowel may prevent establishment of some pathogens which require at least microaerophillic conditions. Other possible inhibitory mechanisms are nutrient deprivation, bacteriocin production, and production of bacteriostatic materials. Fatty acids produced by intestinal fusiform bacteria are inhibitory, in a reducing environment, to typhoid and dysentery organisms, and Döderlein's bacillus (a lactobacillus) produces lactic acid from glycogen in vaginal epithelium which prevents growth of pathogens. How these inhibitory mechanisms of commensals are overcome by small numbers of pathogenic bacteria, for example, in the initial stages of typhoid, is unknown.

3. Resistance to Host Defence Mechanisms in Mucus

Mucous surface secretions contain non-specific and immunospecific (e.g. the immunoglobulin IgA) antibacterial substances. Some prevent adherence, for example the action of parotid fluid on streptococci and that of K88 antibody of *E. coli*. Others are bacteriostatic or bactericidal, sometimes by virtue of pH being alkaline in some sites and acid in others. Also phagocytes are extruded from the mucous surfaces into the lumen and can carry bacteria ingested at the surfaces. How bacteria survive these antimicrobial mechanisms is not clear. Presumably aggressins effective against humoral and cellular defences in the blood and tissues (see below) operate on mucous surfaces.

4. Penetration

Microscopy has revealed some penetration processes, for example the typhoid bacillus goes through epithelial cells not between them.

However, the bacterial products involved are obscure except perhaps for those of *Shigella flexneri*. These bacilli multiply in intestinal epithelial cells. They rarely invade beyond the epithelium but spread laterally producing necrotic ulcers and diarrhoea. Penetration is the paramount virulence attribute and mutants lacking it are avirulent. The nature of the surface O antigen may be responsible for penetration. Hybridization of *Sh. flexneri* and *E. coli* produced *Sh. flexneri* derivatives expressing either *E. coli* O25 or O8 somatic antigens. Comparisons, in various penetration tests, indicated that a *N*-acetyl-glucosamine—rhamnose—rhamnose repeat unit in the O antigen of *Sh. flexneri* may be the determinant of penetration. Replacement of this unit by one containing D-mannose (that of *E. coli* O8) resulted in a hybrid unable to enter cells, but substitution of a unit containing rhamnose (that of *E. coli* O25) retained the ability to penetrate.

C. Multiplication *in vivo*

Avirulence can arise from inability to grow in the environment *in vivo*. Nutritionally deficient mutants of *Salmonella typhi* were avirulent unless injected with their required nutrients. However, for most bacteria the tissues probably contain sufficient nutrients for some growth. Nutritional considerations will however, affect rate of growth *in vivo*; the more rapid it is, the greater the chance of establishing the infection against the host defence mechanisms.

At present little information exists on the influence of particular host nutrients on bacterial pathogenicity: the best documented case is that of iron supply.

Iron is essential for growth and to ensure its supply bacteria synthesize chelators of iron called siderochromes. For example, *Salm. typhimurium* and *E. coli* make enterochelin, a cyclic trimer of 2,3-dihydroxybenzoic acid. Some siderochromes are excreted to form a complex with iron which then enters the bacterium. Others are surface components: lipophilic mycobactin formed by *Mycobacterium tuberculosis* in the cell wall, transports iron across the lipid layers into the cell. Serum, mucus, milk, and phagocytes contain the iron-binding compounds transferrin and lactoferrin. Alone, and sometimes in conjunction with antibody and complement, they seem to prevent the growth of many pathogens by denying them essential iron since inoculation of iron compounds with the bacteria increased pathogenicity. In the normal situation where external iron is not supplied, the possession of a powerful siderochrome to compete successfully with transferrin or lactoferrin for iron may be as essential to a virulent bacterium. Indeed, the virulence for mice of an attenuated strain of *E. coli* was enhanced by an iron chelator from a virulent strain.

D. Interference with Host Defence

Defence mechanisms can be humoral (in body fluids), cellular or a combination of both, acting non-specifically against any pathogen or specifically against a single invading species when an immune response is illicited. During the first few hours within the host, pathogens must cope with antibacterial mechanisms already present and then with phagocytes [short-lived polymorphonuclear phagocytes (PMN) and long-lived mononuclear phagocytes (MN)] mobilized by inflammation soon after the tissues are irritated. These phagocytes ingest bacteria by invagination of their outer membrane and kill them by discharge of cytoplasmic granules containing antibacterial substances into the vacuoles formed around the bacteria by invagination. In this primary lodgement period, antibacterial mechanisms are weighted against the few invading bacteria and many infections are eliminated here. If survival is achieved spread of infection is opposed by MN phagocytes fixed in the lymph nodes, spleen, and liver, and proliferation despite the activities of these phagocytes results in acute disease. This usually subsides and infection is eliminated because within a few days there is a developing immune response which not only increases the efficiency of the phagocytic defence but also provides antibodies capable of directly neutralizing virulence determinants.

Pathogens are unique among microorganisms in being able to overcome the host defence mechanisms. They produce compounds— aggressins — which inactivate host defence mechanisms, resist their action or even prevent their appearance. They can act by: (1) conferring resistance to humoral antibacterial agencies in blood, milk and other body fluids; (2) stopping mobilization of phagocytes by inflammation; (3) hindering contact with phagocytes; (4) preventing ingestion by phagocytes; (5) interfering with intracellular killing by phagocytes, and (6) inhibiting the immune response or reducing its effectiveness.

1. Interference with Humoral Defences

Resistance to humoral bactericidins is characteristic of virulent strains of many bacterial species. The aggressins responsible are known only in a few instances. The lipopolysaccharides of Gram-negative organisms interfere with humoral bactericidins. Capsular poly-D-glutamic acid of B. anthracis interferes with the bactericidins of horse serum. The acid polysaccharide, K antigens of E. coli infecting the urinary tract interfere with the lytic action of antibody and complement. A cell wall component of Br. abortus containing protein, carbohydrate, formyl residues, and about 40% of lipid interferes with the bactericidins in bovine serum. Gonococci collected directly from urethral pus and from plastic chambers implanted subcutaneously in guinea pigs were more resistant to the bactericidins of human serum than after being subcultured on laboratory media. The surface component produced *in vivo* and responsible for resistance is unknown.

2. Interference with Mobilization of Phagocytes

Virulent staphylococci suppress the inflammatory response by virtue of a cell wall mucopeptide which acts by preventing release of kinins.

3. Prevention of Contact with Phagocytes

Chemotaxis has been inhibited *in vitro* by cell wall fractions from tubercle bacilli and by a number of staphylococcal products but whether these materials act in infection is not known.

4. Interference with Ingestion by Phagocytes

Once ingested many bacteria are usually killed and digested. Resistance to ingestion is the main virulence mechanism of these bacteria and in many cases the surface and capsular products responsible are known (Table 1). Resistance of *E. coli* to phagocytosis

Table 1.   Bacterial surface components which prevent ingestion by phagocytes

Pathogen	Capsular or surface component
Pneumococci	Polysaccharide
	Possibly peptido-glycan
Meningococci	Polysaccharide
Streptococci	M protein, hyaluronic acid
Staphylococci	Mucopeptide
Gonococci	Possibly pili or lack of protein called leucocyte association factor
B. anthracis	Poly-D-glutamic acid
Y. pestis	Fraction 1: polysaccharide—protein complex
Enterobacteriaceae	Complete O antigens
Salm. typhi	Vi antigen (poly-N-acetyl-D-galactosaminouronic acid)
E. coli	Acid polysaccharide K antigens
Treponema pallidum	Possibly hyaluronic acid
Pseudomonas aeruginosa	Slime
	Possibly leucocidin

by mouse PMN phagocytes seems to depend on a complete poly-saccharide side chain in the cell wall O-antigen: a mutant lacking colitose was more susceptible to phagocytosis than the wild type. Similarly for *Salm. typhimurium*; the tetrasaccharide sequences absequosyl—mannosyl—rhamnosyl—galactose in the O antigen seem important for phagocytosis resistance. The antiphagocytic moiety of the M protein of streptococci seems to be separate from that which determines serological specificity but its chemical structure is not known.

The surface aggressins (Table 1) are non-toxic and their mode of action in inhibiting ingestion is still not clear. They may act purely mechanically, by inhibiting adsorption of serum opsonins, as seems to occur for *B. anthracis* and staphylococci, or by rendering the bacterial surface less foreign to the host as might happen for the M protein of streptococci.

In addition to surface components, extracellular products of bacteria having a toxic action on phagocytes also function as aggressins. Examples are the leucocidin of staphylococci, toxin of *B. anthracis* and streptolysin O of streptococci.

5. Prevention of Intracellulur Digestion by Phagocytes

Ability to resist the bactericidins of PMN phagocytes contributes to bacterial survival and dissemination in the early phases of disease, since surviving bacteria are liberated when the short-lived PMN phagocytes die. This process seems to occur in brucellosis, staphylococcal infections, and gonorrhoea.

Virulent strains of *Br. abortus* survived and grew in bovine blood phagocytes (predominantly PMN with some MN), whereas attenuated strains were gradually destroyed. The virulent strains produced, under growth conditions occurring *in vivo* and simulant ones *in vitro*, a cell wall substance which interfered with the bactericidal mechanisms of the phagocytes. Thus, brucellae obtained from infected bovine placental tissue, or from cultures supplemented by bovine placental extracts, survived better intracellularly than the same strain grown in unsupplemented media. Cell wall preparations of organisms from infected bovine placenta and from supplemented media inhibited the intracellular destruction of an avirulent strain of *Br. abortus*. The surface antigen responsible for resistance to the bactericidins is unknown.

Staphylococci grown in rabbits were more resistant to killing by rabbit PMN phagocytes and their extracts than were staphylococci grown *in vitro*. This resistance might have been due to a surface covering of host protein, deposited by the action of free and bound coagulase.

Although some gonococci are killed within human PMN phagocytes, others survive and grow intracellularly: this occurs in urethral pus and in phagocytosis tests *in vitro*. The aggressin concerned has yet to be identified.

Intracellular survival and growth in long-lived MN phagocytes contribute to bacterial survival for long periods within the host and produce chronic disease. This occurs for the typical intracellular pathogens, brucellae, and tubercle bacilli.

Brucellae grew in bovine MN phagocytes. These phagocytes may have been inhibited by the same cell wall material that inhibits bactericidins of the mixed PMN and MN phagocytes of bovine blood (see above).

Myc. tuberculosis (and *Myc. microti* which causes vole tuberculosis) and *Myc. lepraemurium* resist intracellular bactericidins by different mechanisms. In mouse peritoneal MN phagocytes infected with virulent *Myc. tuberculosis* and *Myc. microti*, the granules (lysosomes) did not discharge into the vacuoles (phagosomes) which contained intact bacteria. The way in which the discharge is prevented is not clear; material in electron-transparent layers surrounding the bacilli may be involved but secretion of cyclic AMP into the vacuoles has also been implicated. With *Myc. lepraemurium*, lysosomal discharge into the phagosomes occurred normally but the pathogen was resistant to this discharge, possibly due to a surface type C mycoside (a peptido-glycolipid), which was isolated from the livers and spleens of infected mice.

6. Interference with the Immune Response

Bacteria interfere with the effectiveness of the immune response by antigenic shift and by direct suppression.

In antigenic shift, surface antigens are changed so that already formed, potentially protective antibodies to original surface antigens become ineffective. The surface antigens of *Borrelia recurrentis* seem to change during relapsing fever. Similarly, antigenic variation may occur in infections of *Campylobacter fetus* in cows and with oral streptococci.

Suppression of both antibody and cell-mediated immune response occurs in bacterial infection; examples are seen in *Ps. aeruginosa* infections, tuberculosis, leprosy, and syphilis. The bacterial products involved usually are not clear although membrane fragments from Group A streptococci, the capsular polysaccharide of *Klebsiella pneumoniae* and endotoxins have been implicated in some cases.

E. Damaging the Host

When a bacterial species produces a fatal or serious disease, sometimes poisons or toxins have been produced by them *in vitro* and sometimes they have not. These situations are discussed in turn.

1. Toxins are Produced

Bacterial toxins can be divided into four categories.

a. *Toxins produced outside the host.*   The toxin of *Clostridium botulinum* and the enterotoxin of staphylococci are produced in food stuffs. When the host eats the infected food material a chemical poisoning occurs. It is not an infection but clearly the microbial toxin is responsible for disease. Botulinum toxin is a protein neurotoxin acting on the autonomic system by interfering with acetylcholine synthesis or release. The staphylococcal enterotoxin causes diarrhoea and sickness and is a protein with a known amino-acid sequence.

b. *Toxins of overriding importance in infectious disease.*   *Cl. tetani* and *Corynebacterium diphtheriae* produce powerful exotoxins that are responsible for almost the whole disease syndromes since immunization with toxoid (formalin-treated, detoxified toxin) protects against disease. Both toxins are proteins with no abnormal amino acids or toxic moieties. Tetanus toxin is a neurotoxin acting on the central nervous system possibly by binding onto a ganglioside receptor on neurones and interfering with synaptic inhibitors, the normal control agencies in the central nervous system. Diphtheria toxin interferes with protein synthesis and consists of two parts. One determines entry into a cell and the other, an enzyme, inhibits the formation of peptide bonds.

c. *Toxins which are significant but not the only factors responsible for infectious disease.*   These toxins are responsible for some pathological effects of infection. However, they are not the sole determinants for sometimes they do not produce all the pathological effects of disease and usually immunization with toxoid does not protect against infection. Examples are the α-toxin of staphylococci which affects membrane permeability of many types of cells, the exfoliating toxin of some staphylococci, the rash-forming toxin of streptococci and possibly the α-toxin (lecithinase) of *Cl. welchii*. The most important representatives of these toxins are the endotoxins, lipopolysaccharides found in the cell walls of many different Gram-negative bacteria. Their basic structure consists of a heptose polymer core with attached lipid moieties containing 2-keto-3-deoxyoctanoic acid and polysaccharide side chains. When extracted from cell walls by fairly drastic means (trichloracetic acid or warm aqueous phenol) and injected into animals they cause pyrexia, diarrhoea, prostration, and death. The toxic effects are the same, irrespective of the source, and derive mainly from damage to host cell membranes and 'triggering' complement by the alternate pathway. In some infections, for example, typhoid fever and brucellosis, endotoxins are liberated from the cell walls of invading bacteria and are responsible for pathological effects. On the other hand, in other Gram-negative infections the cell wall endotoxin is never liberated in significant quantities and the pathological effects are due to an exotoxin.

d. Toxins produced in vitro *but of unknown importance in disease.* Many toxic substances have been isolated from cultures but their relevance to infection has never been examined. Some may be laboratory artifacts having no connection with disease *in vivo.* Examples are some of many enzymic products of staphyloccocci, streptococci, and other organisms.

2. *A Toxin has not been Produced* in vitro

This situation still occurs in bacteriology, for example in pneumococcal pneumonia, and it is the general case for diseases produced by microbes other than bacteria. There are two explanations. First, a relevant toxin is formed but has defied recognition, and second, host damage is caused by processes other than direct toxicity of microbial products.

a. Hitherto unknown toxins recognized by studying bacterial behaviour in vivo *or in biological tests relevant to the disease.* The first toxin recognized in this way was that responsible for death from anthrax. It was found in the plasma of guinea pigs dying of anthrax; later it was reproduced *in vitro,* purified, and shown to consist of three synergistically acting components. Then, in the past decade, the role of toxins in acute diarrhoeal diseases of man and animals has become clear by discarding toxicity tests in mice in favour of more relevant biological and animal tests in which organisms and their products were examined for their effects in the gut.

An enterotoxin from *V. cholerae*, responsible for the gross fatal fluid loss which occurs in cholera, was recognized by using two tests. First, a ligated segment of small intestine in a living rabbit would fill with fluid following intraluminal injection of *V. cholerae* and its products. Second, *V. cholerae* and its products caused fluid accumulation and diarrhoea in suckling rabbits when introduced into the gut lumen by a gastric tube. The extracellular enterotoxin is a protein consisting of two moieties, one of which is toxic and the other promotes entry to the cell. It increases the normal secretion of the small intestine by activating acenyl cyclase. Using similar 'gut reaction' tests, enterotoxins have now been demonstrated for the following bacteria: *E. coli* (scours in young pigs and calves, and diarrhoea in babies); *V. parahaemolyticus* (a marine vibrio which causes food poisoning), *Cl. perfringens* (food poisoning) and *Sh. dysenteriae*.

A toxin of *Ps. aeruginosa*, probably responsible for death of infected burned patients, was first revealed by studies on extracts of infected rabbit tissue. Like diphtheria toxin it inhibits protein synthesis. More recently, toxins formed by the *Leptospira* species have been recognized in infected tissues, just as the anthrax toxin was originally discovered.

These discoveries of relevant toxins in important bacterial diseases warn against attributing damage in other microbial diseases to causes different from direct toxicity, until the possible formation of toxins has been investigated using realistic biological tests.

b. The role of immunopathology in bacterial disease. The immunological reactions of the host, although usually protective, can sometimes have unpleasant consequences as sufferers from hay fever know well. Work with *Myc. tuberculosis* in guinea pigs showed that hypersensitivity to bacterial products can be dangerous, even fatal, for the host. Furthermore, skin tests indicate that hypersensitive states occur in many bacterial diseases. The reactions are usually of the delayed type indicating that cellular mechanisms are involved but antibody-mediated, cytotoxic, and immune complex reactions also occur. Thus non-toxic bacterial products could cause harm by evoking hypersensitivity reactions. The pathology of tuberculosis appears to be due to cell-mediated hypersensitivity to products of *Myc. tuberculosis*, particularly the cell-wall waxes. Similarly, the cardiac, rheumatoid, and nephritic sequellae to streptococcal disease can be attributed in part to immunopathological phenomena in which the M protein and C polysaccharide cell wall antigens are involved. Endotoxins of Gram-negative bacteria may cause damage by hypersensitivity reactions. Immunopathology also appears to play a part in syphilis; the phospholipids of *Treponema pallidum* are related to the cardiolipins of host mitochondria. Hence antibody to the treponemal phospholipids may produce antibody-mediated cytotoxic activity against host cells.

F. Tissue and Host Specificity

Two of the most striking and largely unexplained phenomena in microbial pathogenicity are tissue specificity, the ability of microbes to attack some tissues in preference to others; and host specificity, the ability of microbes to attack some animal species and not others. There are many examples in bacteriology, for example gonococci are largely confined to the urogenital tract and gonorrhoea is restricted to man. The specificities are determined by variations of the host environment in relation to promotion or restriction of microbial proliferation. The most important host influences are: the location of host receptors for initial attachment; the physical and nutritional environment, and the nature and strength of humoral and cellular defence mechanisms.

1. Selective Adherence

Many bacteria adhere selectively to a mucous surface at initiation of infection. *V. cholerae* and *E. coli* attach preferentially to the epithelium of the upper rather than the lower bowel, whereas *Sh. flexneri* adheres to colonic cells rather than to those of the upper bowel. With regard to host specificity, *St. pyogenes*, which rarely infects rodents, attaches more strongly to human than to rodent buccal epithelial cells.

In rare cases the host receptors are now known. Compounds related to salivary glycoproteins and blood group substances have been implicated in the adherence of oral streptococci to teeth and buccal cells, and an epithelial glycopeptide, possessed by some piglets but not others and inherited in a simple Mendelian manner, seems to be responsible for the adherence of K88 antigen positive *E. coli* to the upper intestine.

2. Physical and Nutritional Environment

Myc. ulcerans and *Myc. marinum* enter human skin through abrasions and cause chronic skin ulcers but remain localized because their optimum growth temperature is low (30–33 °C). *Corynebacterium renale* and *Proteus mirabilis* localize and cause severe damage to the kidney of cattle and man respectively because they possess ureases which enable them to use the urea in kidney tissue for growth. A similar explanation has been offered for the localization of *Leptospira* spp. in the kidneys of domestic and pet animals. Brucellosis in many animals (e.g. humans) is a relatively mild and chronic disease; the causative organisms have no marked affinity for particular tissues. However, in pregnant cows, sheep, goats, and sows there is prolific growth of brucellae in the placenta, foetal fluids, and chorions, leading to abortion. The presence of erythritol, a growth stimulant for brucellae, only in the susceptible tissues of susceptible species explains this tissue and host specificity in brucellosis.

3. Variation in Strength of the Host Defence Mechanisms

Kidney tissue is prone to infection by many bacteria. This appears to be due to inhibition of mobilization of phagocytes and their bactericidal activity by the relatively high salt concentrations in the kidney. The susceptibility or resistance of different species of mice to infection with enterobacteriaceae is reflected either in the ability or lack of ability of macrophages to support bacterial multiplication *in vitro*, or in the relative capacities of the mouse species to mount a cell-mediated immunity.

IV. Viral Pathogenicity

Viruses differ from other microbes in their unique method of replication within the host cell and the mechanisms of replication receive much attention from virologists. But viral, like bacterial, pathogenicity is not determined solely by ability to replicate in the tissues. Virulent and attenuated strains replicate in host cells *in vitro* yet they differ fundamentally in behaviour *in vivo*, presumably — as for bacteria — due to different capacities to enter the host, to counteract host defence mechanisms, and to damage tissues. As yet, these latter facets of pathogenicity have received scant attention.

A. Difficulties of Studying Viral Pathogenicity

Difficulties arise from the facts that pathogenicity is determined by more than one factor and virus behaviour in tissue culture is often different from that *in vivo* where pathogenicity occurs. In addition, quantitative comparisons of the virulence of different strains are inaccurate. Disease effects in animals (LD_{50}; lesion size) must be related to plaque counts or egg infectious doses. The latter detect only a small proportion of the total virus particles and therefore may not measure all the particles (which could vary for different strains capable of multiplying in experimental animals). Hence only virus strains for which conventional tests have indicated large differences in virulence should be compared in order to recognize virulence markers and determinants. Comparisons of such well-tested and well-separated strains have been rare but informative.

B. Entry to the Host: Mucous Membrane Interactions

Knowledge of the factors promoting mucosal invasion of viruses is scanty. As for bacteria, the main mucosal defences against viruses are the flushing action of moving lumen contents, the competitive action of commensals, and inhibitory materials in mucus.

1. Adherence

The mechanisms whereby viruses penetrate the mucus blanket are not known but degradation of mucus by neuraminidases of myxoviruses and paramyxoviruses might contribute. Adherence is easily explained for viruses that infect epithelial cells, such as influenza virus, rhinoviruses, poliovirus, and foot-and-mouth disease virus, since attachment is the first stage of replication and viral surface components are involved. However, what happens with viruses that seem to penetrate the mucosa without establishing infection in the membrane itself, for example in rinderpest and African swine fever, is unknown.

2. Competition with Commensals

In cell culture and in animals, bacteria and fungi affect virus infection by inducing interferon. Also, mycoplasmas, notable inhabitants of mucous surfaces, inhibit the replication of herpes virus and measles virus in cell culture. If such microbial interactions occur on mucous surfaces, virulent viruses must be resistant to them but the mechanisms are not known.

3. Resistance to Host Defence Mechanisms in Mucus

The pH of mucus can have an antiviral effect (e.g. that in the stomach on acid labile rhinoviruses and influenza virus); viral inhibitors are found in homogenates of lung and intestinal mucosa; bile dissociates enveloped viruses, and viruses are destroyed by extruded phagocytes. How virulent viruses survive against these defences is unknown.

4. Penetration

Virulent viruses penetrate mucous membranes since they can be found in blood and other tissues but the mechanisms of penetration are unknown.

C. Multiplication *in vivo*

Although ability to replicate is not the only factor in viral virulence it is essential and the more rapid it is, the more likely is disease.

Investigations are complicated not only by complexity of biochemical factors required for replication, but also by the difficulty of distinguishing the influence of their absence from that of host factors (defence mechanisms) which actually destroy virus or inhibit replication. Ability to replicate depends on host-cell features that are involved in attachment and penetration of virus, uncoating, provision of energy and precursors of low molecular weight, synthesis of viral nucleic acid and proteins, assembly, and release. The host-cell characteristics that complement virus components and determine these stages of replication might be called 'replication factors'. They are counterparts in virology of the environmental factors necessary for bacterial multiplication in host tissues or fluids. Tissue culture experiments show that 'replication factors' vary from cell type to cell type and determine whether proliferation occurs and at what rate. They also show that replication is influenced by changes in environment of the cell. In animal infection, a similar variation of availability of 'replication factors' will affect viral proliferation and therefore pathogenicity.

The viral components of particular importance are those on the surface that complement components of host cell membranes and promote entry into cells by fusion or virapexis. When vesicular stomatitis virus lost a surface glycoprotein on treatment with bromelin or pronase, it became spikeless and non-infectious, but when treated with the purified glycoprotein its infectivity was restored. Virulent strains of poliovirus adhered to primate nerve receptors more strongly than avirulent strains, suggesting differences in surface components responsible for virulence, and chromatographic differences in the capsid peptides have been noted. Some viruses incorporate into their envelopes components of host cell membranes. For further infection the fusion of the enveloped viruses with the membranes of the fresh cells requires a similarity between the interacting membranes. Membrane fusion and so infection, may be prevented by too great a dissimilarity of the virus envelope formed by budding from one cell type (for example a tissue culture cell) and the plasma membrane of another cell type (for example a target cell in an animal). This may explain in some cases the lower virulence of virus produced in tissue culture compared with animal passaged virus.

D. Interference with Host Defence

There has been much work on host defence against viruses but little on the mechanisms whereby they overcome this defence.

1. Interference with Humoral Defence Mechanisms

Non-specific factors include the low pH of inflammatory exudates and inhibitors in tissues and serum present before infection or induced by it. Virulent viruses resist these inhibitors, for example virulent strains of influenza virus withstood the action of mouse serum more than avirulent strains, but the mechanisms are unknown.

2. Interference with Cellular Defence Mechanisms

Cellular defence factors include those present or induced in any cell the virus attacks, such as interferon. Then, there are those present in the phagocytes and other cells of the reticuloendothelial system.

Virus species and strains differ both in the amount of interferon they induce and in their susceptibility to it. Sometimes virulent strains induce less interferon or are more resistant to it than attenuated strains, but not always. However, a strict correlation between virulence and induction of, or resistance to, interferon would not be expected if virulence is determined by more than one factor. Interferon is produced in virus infection and a capacity to reduce its production or resist its action would be an advantage to an invading virus. How could a virus achieve these ends? Early inhibition of host cell RNA and protein synthesis would depress interferon production. Also, some viruses produce *in vitro* antagonists of interferon. Whether they are found in infection and play any role in virus invasion as do bacterial aggressins has yet to be assessed.

The role of macrophages (MN phagocytes) in defence against virus disease has been investigated far more than that of PMN phagocytes. Macrophages kill some viruses but not others. The ability to survive and possibly replicate within macrophages appears to be a main virulence mechanism of some viruses. Infected wandering macrophages can spread infection and infected fixed macrophages can start infection in organs such as the liver. The viral products which determine virus ingestion, survival or replication within macrophages are unknown. Surface components might promote viral survival by interfering with intracellular inhibitors but equally, survival may be due to overall inhibition of macrophage function by cytotoxic action. Some viruses such as myxoviruses, vaccinia virus, and measles, are cytotoxic to macrophages since they inhibit phagocytosis of bacteria.

3. Interference with the Immune Response

Viruses could reduce the protective effect of antibody by being 'bad' antigens, by antigenic shift, and by infecting and inhibiting the function of antibody-forming cells. Virus strains vary in their ability to evoke antibody and 'slow viruses' such as the scrapie agent induce none. Host-cell membrane constituents in the envelope proteins of some viruses may make virus antigens more 'host-like' and therefore 'bad' antigens, but this has not yet been proved. Antigenic shift contributes to the ability of influenza and other viruses to attack fresh hosts but has not been detected in the course of virus infection as it has in bacterial and protozoal diseases. Most virus infections depress but do not stop antibody synthesis and sometimes it is increased. Cytotoxic activity could operate in antibody forming cells. Cellular immunity is depressed in most virus infections and some viruses grow in lymphocytes and produce immunosuppression with or without cytotoxic damage.

E. Damaging the Host

Harm to the host is the culmination of damage to individual cells which may result from a passive role of the virus — a simple repercussion of replication, such as the depletion of cellular components essential for life, or mechanical harm due to production of virus. On the other hand, virus cytotoxicity and immunopathology may be important.

1. Virus Cytotoxic Activity

Pathologically important cytotoxicity can operate at two levels, biochemical damage without noticeable morphological damage (e.g. in nerve cells). and that occurring with cell fusion, lysis, or death, the usually observed cytopathic effects. Progress has been made in investigating the cytotoxicity of a few viruses but only in tissue culture. How far the findings can be extended to other viruses and to animal infections remains to be seen.

Cytopathic effects can occur in tissue culture without the presence of infectious virus. Influenza virus and Newcastle disease virus damage cells which are either incapable or poorly able to support replication. Cells are also damaged by poliovirus and vaccinia virus in the presence of chemical inhibitors of virus replication. Pathological damage can also occur in animals in the absence of new infectious virus, e.g. injection of large quantities of influenza virus and pox-viruses causes immediate toxic effects and, in mice, liver damage was caused by Coxsackie virus which replicated in the pancreas and not the liver.

Some virion components exert cytotoxic effects. The capsid penton of the adenovirus caused cell rounding and detachment from glass, and the hexon of adenovirus, although not noticeably cytopathic, inhibited macromolecular synthesis. Capsid components have been implicated in inhibition of protein synthesis by poliovirus and in the cytotoxicity of reovirus. Like live virus, inactivated Sendai virus, Herpes virus, and Newcastle disease virus cause cells to fuse into polykaryocytes or syncytia. Virus envelope components seem to be involved for cells were fused by fragments of the envelopes of all three viruses. Not all virion cytotoxins are surface components. A double-stranded RNA from a bovine enterovirus rapidly caused death of cells without formation of infectious virus and a toxic RNA has been obtained from influenza virus.

Thus, there is increasing evidence that viruses produce cytotoxins, but how do they act? Inhibition of macromolecular synthesis or other interference with the functions of the host cell could be produced directly by a virus product, just as the toxic component of diphtheria toxin interferes with protein synthesis. On the other hand, the virus-induced product might release autolytic enzymes from the cell's own lysosomes.

2. The Role of Immunopathology in Viral Disease

Immunopathology is likely to occur in viral diseases because the obligate parasitism increases the chances of occurrence of host-cell bound virus antigens. Also some viruses incorporate host cell membrane constituents into their structure. Hence antibodies and cell-mediated immune mechanisms elicited against these virus/host complexes could react with membrane constituents of both infected and normal cells, and produce cell damage. Such cell damage has been demonstrated in various *in vitro* systems. *In vivo*, it appears that immunopathology may be involved in measles, pox virus rashes, pneumonia from respiratory syncytial virus, yellow fever, mumps, and coxsackie-B virus infection. However, the evidence is mostly suggestive. But for lymphochorio-meningitis in mice and aleutian disease in mink sufficient experimental evidence has shown that immunopathology is the major cause of observed damage.

F. Tissue and Host Specificity

Tissue and host specificity occurs in virus infections, for example poliovirus infects only enteric and neural tissues of primates and variola infects only primates. As for bacterial specificities the explanation for viral specificities can be the location of susceptible cells and barriers in relation to the route of entry of the virus, the presence or absence of the necessary 'replication factors' in cells, and the variation of non-specific and immunospecific defences. Since virus susceptibilities change when cells differentiate in normal tissue cultures, the latter cannot be used to investigate the host and tissue specificities of natural infection. Short-term studies with either primary cell cultures or suspensions of relevent tissues have yielded most of our information and recently organ cultures have been used. In both cases, the parent susceptibilities were usually retained in the *in vitro* system. Using these methods *in vitro*, coupled with whole-animal experiments, the explanations for some examples of viral specificities have been elucidated.

1. Location of Susceptible Cell in Relation to Route of Entry, and Barriers to Spread of Infection

Influenza in ferrets is a good model for influenza in man. After respiratory inoculation of ferrets, infective virus was found predominantly in the nasal turbinates with some in the trachea and lung, but none in any other tissues. Yet in organ culture, ferret bladder and uterus, like human bladder and endometrium, were highly susceptible to infection with influenza virus, and after local inoculation urogenital infection occured in ferrets. Hence, localization of influenza virus in the respiratory tract is due to susceptible cells being in the path of the in-coming virus combined with barriers in the blood stream, and between it and the tissues, which prevented spread of virus to other highly susceptible tissues. This recent work with influenza virus recalls earlier work with influenza and myxoma virus in mice, where blood-borne virus was unable to infect liver parenchyma cells, whereas if the Kupfer cell barrier was circumvented by bile duct inoculation, parenchymal infection ensued.

2. The Presence or Absence of Replication Factors

In the examples below, the headings indicate the stage of replication for which the important factors were present in the susceptible cell and absent from the insusceptible cell.

a. Adsorption and penetration. Human and monkey cells susceptible to poliovirus infection adsorbed virus and produced membrane fragments with associated viruses, in contrast to non-primate cells insusceptible to poliovirus infection. Corresponding to tissue localization in infection, homogenates of human and monkey intestine, brain, and spinal cord adsorbed poliovirus whereas homogenates of other tissues did not. The importance of envelope-cell receptor interaction was underlined by showing that cells insusceptible to infection with intact virus (rodent and avian cells) translated envelope-free poliovirus RNA if this was introduced into the cell. Only one cycle of replication occurred because the released poliovirus possessed the 'wrong' capsids for further cell entry. Neither the nature of the operative group on the virus surface nor that of the host cell receptors is known.

Avian RNA tumour viruses and feline leukaemia viruses provide other examples of host specificity determined by the interaction between virus surface components and host receptor substances, and they were investigated as in the classic studies on poliovirus.

b. Uncoating. The HMV (PRI) strain of mouse hepatitis virus infected PRI but not C_3H mice. Host specificity was reflected in ability of the virus to infect and destroy liver and peritoneal macrophages from these two types of mice. Virus adsorbed to and penetrated both resistant and susceptible macrophages. However, after adsorption and penetration, eclipse and replication ensued only in the susceptible cells. Thus, susceptibility and resistance of the mice to hepatitis virus appears to be determined by the presence or absence of a system in the liver macrophages which removes the virus surface coat.

c. Component production and assembly: Production of defective interfering particles. C_3H/RV mice are resistant to infection with West Nile virus and congenic C_3H/He mice are susceptible. The differences in susceptibility in whole animals was reflected in ability of the virus to replicate in spleen cells, peritoneal exudate cells, embryo fibroblasts, and brain explants from the two types of mice. This similarity suggested that replication factors involved in virus–cell interaction were responsible for determining specificity rather than host defence mechanisms seen in whole animals. Recent experiments have suggested that cells from the resistant mice are incomplete in factors needed to produce fully infective virus, since they produced more defective interfering virus particles than the cells from the susceptible mice.

3. Variation in Host Defence Mechanisms Against Viruses

Variation of pH from tissue to tissue can contribute to specificities of infection, for example influenza virus is acid labile and does not pass the stomach; in contrast, acid-stable poliovirus infects the intestine. Differential induction of interferon has not been proved to determine tissue or host specificity but variation of the immune response explains host specificity in some cases. Strains of mice resistant to murine leukaemia virus, others resistant to ectromelia virus and others resistant to cytomegalovirus mounted a stronger immune response to infection than mouse strains that were susceptible to these viruses.

V. Fungal Pathogenicity

Fungi produce animal diseases such as thrush, dermatitis, 'farmers' lung', and mycotic abortion. Some have mycelial and yeast forms differing in virulence and thus they seem particularly appropriate for comparative studies. Despite this advantage little is known of the mechanisms of fungal pathogenicity.

A. Entry to the Host: Mucous Membrane Interactions

Products of fungi responsible for adherence to and penetration of mucous surface are unknown. Bacterial commensals seem to have an inhibitory action on fungal pathogens, since antibiotic treatment can result in spontaneous fungal infection. How their inhibitory activity is overcome in normal fungal infection is not known. Fungistatic materials are present on mucous surfaces such as those in the conjunctival excretions and saliva and the fatty acids in teat secretions of domestic animals. However, we do not know how fungi resist there inhibitory materials.

B. Multiplication *in vivo*

The morphology of the dimorphic fungi is different *in vivo* and *in vitro*. With some exceptions, notably *Candida albicans*, yeast forms occur *in vivo* and mycelial/arthrospore forms *in vitro*. The yeast forms seem to be more pathogenic and immunogenic than mycelial/ arthrospore forms and differ from them in cell wall chemistry and antigens. The host nutrients which determine the morphological form *in vivo* have not been recognized, but factors (often SH compounds) achieving conversion of saprophytic to parasitic forms *in vitro* have been recognized and might also operate *in vivo*. The particular nutrients that are important for rapid fungal growth *in vivo* are unknown for any fungal species, but *Candida* spp. at least have siderochromes which presumably ensure their supply of iron.

C. Interference with Host Defence

Serum and tissue extracts contain fungicidal materials, some of them complement dependent. Whether virulent strains of fungi resist these materials better than avirulent strains is not known, nor are any fungal components that might be involved. Phagocytes ingest and kill some fungi and PMN phagocytes appear more effective than MN phagocytes. Some fungi resist ingestion by phagocytes and for *Cryptococcus neoformans*, the aggressin is a capsular polymer containing uronic acid. Many fungi resist intracellular killing. *C. albicans*, *Coccidioides immitis*, and *Histoplasma capsulatum* survive and grow with MN phagocytes. Fungal surface components are probably responsible for this resistance to intracellular killing but none has been identified.

D. Damaging the Host

In some mycotic diseases mechanical blockage by large mycelia probably damages the host. Fungi produce in foodstuffs toxins such as aflatoxin and sporidesmin. They may be produced in mycotic infection but this has to be demonstrated. Recently fungal toxins from *in vitro* cultures have been described which if released *in vivo* could be important in disease. A nephrotoxin, possibly of cell wall origin, was extracted from the washed mycelium of *Mortierella wolfii*. The cell walls of a virulent strain of *Blastomyces dermatitidis* produced granulomatous reactions in mice similar to those seen in human disease and the mice died; in contrast, cell walls from an avirulent strain produced no granulomatous reaction although the mice died. Toxic preparations have been obtained from *C. albicans*. Extracellular peptidases, collagenase, and elastase cause damage by dermatophytes.

Hypersensitivity occurs in many mycoses and immunopathological phenomena probably explain the pathology of some skin and respiratory mycoses but products responsible are ill-defined.

E. Tissue and Host Specificity

The growth of *Aspergillus fumigatus* in placental tissue, which causes mycotic abortion in sheep and cattle, may have a nutritional basis. A material which stimulates spore germination is in extracts of bovine foetal placenta but its nature is as yet unknown. Also, as regards the influence of host defence mechanisms, the lack of sebaceous glands which excrete mycostatic fatty acids may allow growth of the dermatophyte *Trichophyton* spp. in certain areas, for example between the toes.

VI. Protozoal Pathogenicity

The microbial determinants of protozoal infections have received little attention. The major difficulty in identifying virulence determinants appears to lie in the versatility of protozoa. The different phases of their life cycles, their different morphological forms and their antigenic plasticity all have profound influences on pathogenicity. Keeping one strain in one form for repeated experiments is a major operation.

A. Entry to the Host: Mucous Membrane Interactions

Many protozoa are injected into the host by vector bite but for those that invade across mucous surfaces little is known of the mechanisms concerned except for *Entamoeba histolytica.* When entamoebae invade the lamina propria of guinea pigs, PMN phagocytes are destroyed, epithelial cells are shed and vascular damage occurs, probably as a result of PMN destruction. How far products of the pathogen are responsible for the changes is not known. In lysing tissue culture cells, entamoebae transfer lysosomes (granules containing destructive enzymes) which protrude from their surface, into the culture cells by a trigger mechanism. Lysosomal transfer may not operate *in vivo* during mucosal invasion because mesenchyma cells in direct contact with entamoebae are unharmed, and the PMN phagocytes that are destroyed are free.

B. Multiplication *in vivo*

Snippets of information are available on nutrients important for protozoal multiplication *in vivo.* Some gut bacteria seem to be an essential food for *Ent. histolytica.* Malarial attacks seem to be stimulated by p-aminobenzoic acid and methionine if they are given to the host, and this parasite seems to grow within red cells because it needs haemoglobin to form haemazoin.

C. Interference with Host Defence

Lysis by antibody and complement is the main humoral mechanism and antigenic shift is used to sidestep it. Antigenic shift has been best studied in the trypanosomes. They have a surface coat of a closely packed monolayer of glycoproteins which are expressed sequentially during a persistent infection. Glycoproteins purified from populations taken at successive times during infections in animals had the same molecular weight (about 70,000) and carbohydrate composition (about 6% by weight) but differed profoundly in amino acid composition and sequence. The number of changes that can take place from one clone of inoculated trypanosomes is not known nor is the mechanism of surface replacement.

Protozoa are also ingested and killed by PMN and MN phagocytes, especially those from immunized hosts. Some protozoa such as *Plasmodia* species need to be opsonized by antibody before ingestion and others such as *Toxoplasma gondii* required similar treatment before being killed intracellularly; in either case antigenic shift would interfere with the defence mechanism. *Leishmania* species, *Trypanosoma cruzi,* and *T. gondii,* survive and grow intracellularly in phagocytes. *T. gondii* prevents granular discharge in a manner similar to tubercle bacilli. However, no protozoal product has yet been identified as being responsible for resistance to ingestion or digestion by phagocytes.

Immunosuppression occurs in protozoal disease but which microbial components are involved is not known.

D. Damaging the Host

A microbial toxin has not yet been recognized as being unequivocally responsible for the main pathology of any protozoal disease. In malaria, increase of capillarly permeability leading to shock and brain damage, appears to be mediated by kallikrein, kinins, and adenosine, but a malarial toxin responsible for release of the host products has not been demonstrated. *T. gondii* produces a factor which promotes penetration of host cells. The lytic effects of *Entamoeba* species are known but mechanisms of cell destruction are not clear. As stated above, lysosomal enzymes may be transferred from the entamoebae to the host cells by tubules formed between the two cells. Toxic products of trypanosomes and toxoplasms are known but their importance in disease is not clear.

Hypersensitivity and autoallergic phenomena occur in protozoal infections but their responsibility for important pathological effects is hard to judge. In malaria, antibodies to host antigens changed by red blood cell parasitization may, by opsonization, promote phagocytosis and destruction of unparasitized red blood cells. Also antibody-complexed surface glycoproteins are shed from trypanosomes in chronic disease and may be deposited on kidney membranes to produce Arthus type reactions.

E. Tissue and Host Specificity

Tissue and host specificity occur in protozoal infection. The connection between growth in red cells and the requirement of the malarial parasite to form haemazoin has been mentioned. The host specificity for *Plasmodium knowlesi* infection has been correlated with merozoites adherence to different red blood cells and not to the antiprotozoal or other properties of the different sera. The nature of the red blood cell receptors of the susceptible species (human and three species of monkey) is not known.

VII. Conclusions

The conclusions of this essay are simple. The determinants of microbial pathogenicity are hard to identify; however, there is sufficient progress on bacterial pathogens to show that the determinants of viral, fungal, and protozoal pathogenicity can be elucidated, and to indicate the lines of approach.

VIII. Further Reading

Ajl, S. J., Kadis, S., and Montie, T. C. (1970). *Microbial Toxins.* London and New York: Academic Press.

Dubos, R. J. and Hirsch, J. G. (1965). *Bacterial and Mycotic Infections of Man.* 4th ed. Philadelphia: Lippincott.

Dunlop, R. H. and Moon, H. W. (1970). *Resistance to Infectious Disease.* Saskatoon: Modern Press.

Howie, J. W. and O'Hea, A. J. (1955). *Mechanisms of Microbial Pathogenicity.* Cambridge: Cambridge University Press.

Mims, C. A (1964). Aspects of the pathogenesis of virus disease. *Bacteriological Reviews,* 28, 30.

Mims, C. A. (1976). *The Pathogenesis of Infectious Disease.* London: Academic Press.

Schlesinger, J. (1975). *Microbiology 1975.* Washington: American Society for Microbiology.

Smith, H. (1968). The biochemical challenge of microbial pathogenicity. *Bacteriological Reviews,* 32, 164.

Smith, H. (1972). Mechanisms of virus pathogenicity. *Bacteriological Reviews,* 36, 291.

Smith, H. (1976). Survival of vegetative bacteria in animals. *Symposium of the Society for General Microbiology,* 26, 299.

Smith, H. (1977). Microbial surfaces in relation to pathogenicity: *Bacteriological Reviews.* 41, 475.

Smith, H. and Pearce, J. H. (1972). *Microbial Pathogenicity in Man and Animals.* Cambridge: Cambridge University Press.

Smith, H. and Taylor, J. (1964). *Microbial Behaviour* in Vivo *and* in Vitro. Cambridge: Cambridge University Press.

Smith, W. (1963). *Mechanisms of Virus Infection.* London and New York: Academic Press.

Resistant Forms

RALPH A. SLEPECKY

Biological Research Laboratories, Syracuse University, Syracuse, New York

I. Introduction

Prokaryotic microbes exhibit wide diversity in form, structure, and physiology as a result of the many evolutionary paths they have taken since Pre-Cambrian times, when it is believed they first appeared. Many form non-dividing resting stages having little or no metabolism, which allow them to remain dormant for long periods of time. Associated with this dormancy is resistance to various environmental stresses normally lethal to the progenitor vegetative cells and to non-resting stage formers. Besides possessing such resistance the dormant forms can survive long periods of nutrient deprivation while most prokaryotes unable to form dormant resting stages usually cannot. Upon restoration of nutrients to the environment the resting form is capable of returning to the vegetative state. Hence the resting cell is not a dead end but part of the organism's life cycle.

II. Known Prokaryotic Resting Cells

As can be seen in Table 1 the list of known resting forms is not large when one considers the extensive array of known prokaryotic species. The list is headed by the ubiquitous endospore formers which make up not only the largest group known, but also contain the most dormant resting forms. Endospores from this group have been known to survive

Table 1.   Prokaryote resting stages

Type of structure	Organism	Formation
Endospores	Bacillus Clostridium Sporosarcina Sporolactobacillus Desulfotomaculum Thermoactinomyces Actinobifida	After an asymmetric division the larger cell engulfs the smaller which eventually becomes a spore.
Exospores (conidia)	Actinomycetes	Small cells formed by separation of aerial hyphae eventually become spores.
Exospores	Methylotrophic bacteria	Rod shaped bacteria bud off spherical shaped cells which become spores.
	Rhodomicrobium vanneilli	Sequential formation of angular exospores at the end of filaments.
Spores (akinetes)	Some blue green algae, e.g. Anabena cylindrica	Formed within chains of vegetative cells.
Myxospores	Myxobacteria Sporocytophaga	Direct conversion of rod-shaped vegetative cells into spherical spores.
Cysts	Azotobacter	Modified vegetative cells; symmetric division precedes encystment
Bdellocysts	Bdellovibrio	Modified vegetative cell produced in prey but remain viable in absence of prey.

adverse conditions for great lengths of time, the latest evidence documents their recovery from samples estimated to be 1900 years old. The second group are the actinomycetes, filamentous bacteria abundant in soils and known for their antibiotic production (e.g. streptomycin). There are many actinomycetes and in general their reproduction is similar to that found in fungi. They produce two types of mycelia — substrate and aerial. Exospores (conidia) are formed on the latter in a sheath and result from a modified division process within the filament. In addition to conidia they produce other types of resting spores — 'zoospores' and arthrospores. 'Zoospores' are quite active metabolically but arthrospores have been little studied. A third group producing resting forms include methylotrophic bacteria and *Rhodospirillum vanneilli*. All methylotrophic bacteria (methane utilizers) oxidize methane, are found extensively in nature, and possess a complex arrangement of membrane structures. All form either exospores or cyst-like bodies but the latter have been little studied. *Rhodospirillum vanneilli* is the only representative of the non-sulphur purple bacteria (facultative phototrophs which are unable to oxidize elemental sulphur to sulphate) known to possess resting forms. The purple photosynthetic bacteria do not form resting cells. Other photosynthetic bacteria that form resting cells are included in the group cyanobacteria (blue-green algae). Another group, the myxobacteria are a complex group of Gram-negative cells that glide along the surface of solid media and require complex media for growth. The group includes *Cytophaga*, existing only as vegetative cells; *Sporocytophaga*, cells that can form myxospores without fruiting body formation; and *Myxococcus* and *Polyangium*, cells that release myxospores from fruiting bodies. *Azotobacter* are large Gram-negative obligately aerobic rods. They fix nitrogen non-symbiotically and produce cysts. The genus *Bdellovibrio* is comprised of unusually small cells (0.3 μm in diameter), widespread in soil and water, which parasitize other bacteria. They produce cysts within the host bacterium. This brief exposition gives only the key identifying features of each group. The method of formation of each type of resting cell is given in Table 1.

There is reason to suspect that other dormant forms remain to be discovered. The exospores and cysts of methylotrophic bacteria and of photosynthetic bacteria and the bdellocysts have only recently been described. Until recent years endospore formation was thought to be restricted to the Gram-positive *Bacillus* and *Clostridium* species. As a result most of our knowledge, not only of sporulation and germination events but of the natural history and ecology of endospore-forming bacteria has been derived from these two rod-shaped organisms. Seldom are we reminded that this morphogenesis occurs in cocci (*Sporosarcina*) as well as bacilli, and even less often is attention paid to the presence of this trait in a Gram-negative staining genus (*Desulfotomaculum*), or in the actinomycetes (*Thermoactinomyces* and *Actinobifida*), or amongst the lactic acid bacteria (*Sporolactobacillus*). Endospore formation is alleged to occur in several other genera (*Fusosporus, Arthromitus, Coleomitus, Bacillospira, Sporospirillum, Oscillospira, Metabacterium, Thiobacillus,* and *Corynebacterium*), but whether endospore formation actually occurs in members of these genera is uncertain because the information available about these supposed spores is incomplete or ambiguous. Whether the spore-like objects seen with the light microscope fulfill the requirements of the term endospore and whether the 'spores' were formed by the organisms to which the trait is attributed remains for future careful study. Although there are reservations about endospore formation in the genera listed above, there is no *a priori* reason why this trait (or the ability to form any type of resting cell) may not be far more widespread in the bacterial world than we now perceive. The ability of an organism to form a resting cell is, after all, nothing more than the end product of a set of physiological and biochemical events. There are numerous examples of other physiological-sets which are found in more than one genus and which are not restricted to a single Gram staining category; many examples can be found for energy sources (H_2-chemolithotrophy), metabolic end products of anaerobes (the propionic fermentation), as well as morphological traits (presence of flagella). When one recognizes that the well-characterized, endospore-forming bacteria include both aerobes and anaerobes, rods and cocci, and both Gram staining types, and represent an enormous array of the physiological spectrum such as cellulose fermentation, methane production, nitrogen fixation, nitrate and sulphate reduction, and others, there is even more reason to believe that a systematic search for endospore formation (or indeed, other resting cell formers) amongst other bacteria should be undertaken. Clearly such a search will have to take into account the diversity mentioned for there is no single growth environment that permits sporulation or encystment to take place for the now-recognized resting cell formers. As has been so persuasively argued by others, the environment exerts profound, differential effects on the ability of organisms to express their inherent, potential activities.

It is well recognized that resting-cell forming prokaryotes are present in an enormous variety of habitats and, accordingly, widely distributed geographically. Undoubtedly this is, in large part, a reflection that as a group these bacteria comprise a nutritionally versatile array, able to thrive in both anaerobic and aerobic environments, in a wide range of acidities and temperatures, by using photosynthesis, respiration, fermentation, anaerobic respiration, or a combination thereof, for energy provision. Although no single organism is known to possess the spectrum of life styles encompassed by the group as a whole, some of the organisms regarded as 'more common' in terms of distribution are those which grow relatively rapidly on rather complex laboratory media containing mixtures of sugars, nitrogenous materials, vitamins, etc., and which often have a pH near neutrality. Even if carefully prepared soil preparations are placed and incubated under both anaerobic and aerobic conditions, however, there is every reason to expect that a fraction, of unknown proportions, of the bacteria present will not be recovered and therefore not be detected.

III. Protective Layers in Resting Cells

In all but one case (endospores) the resting stage is formed by transformation of the entire cell. One or more protective layers are deposited outside the cytoplasmic membrane. That deposition may have been preceded by formation of smaller cells as in actinomycetes or by budding as in methylotrophic bacteria. Myxospores are usually produced in fruiting bodies formed by aggregation and differentiation of myxobacteria but in some cases may result directly from the vegetative myxobacteria cells without fruiting body formation. Endospore formation differs in that they are formed within the cytoplasm of another cell but similar protective layers develop. The cell harbouring the spore eventually lyses giving rise to a free spore. It is interesting to note that different species of actinomycetes produce exospores (*Streptomyces*) and endospores (*Thermoactinomyces*) suggesting close evolutionary ties with regard to spore formation.

The various resting cells are readily recognizable as different entities from the vegetative precursor cells in the light microscope, and fine-structure analyses have revealed greater differences particularly in the outer surface layers in some of them. Endospores are relatively more complex than *Azotobacter* cysts which in turn have more surrounding layers than the myxobacteria. The others have not been as extensively studied as these but it appears that in general the outer layers are thicker modified counterparts of the vegetative cell outer layers. In all cases these outer layers surround a central core (protoplast) which contains all the essential structures and constituents, nucleic acids, protein, ribosomes, etc., needed to sustain subsequent vegetative growth. These substances are, to varying degrees, in a reduced water state which is presumably responsible for the dormancy exhibited by these forms. In some instances a particular substance may be found, e.g. dipicolinic acid (DPA) in all endospores. Specifically, the protoplast of the endospore is delineated by a cytoplasmic membrane which is surrounded by the cortex, made up of peptidoglycan similar in composition to the cell wall of the vegetative cell. However, this peptidoglycan is less cross-linked than its vegetative counterpart and contains none of the accessory polymers such as teichoic acids which are usually present in vegetative cells. The cortex is enclosed on the outside by protein coat(s), the number depending on the species (Fig. 2). In some spore formers the spore may possess a loose fitting structure known as the exosporium; in others this has been recognized as a tight fitting layer. In *Azotobacter* cysts the cytoplasmic membrane is bounded by a thin peptidoglycan cell wall and then by two outer coats, an electron transparent *intine* consisting mainly of carbohydrate and lipid and a layered electron dense *exine* which contains phospholipid and carbohydrate (Fig. 7). The less complex structures myxospore has its protoplast surrounded by a wall whose structure is typical of Gram-negative bacteria enclosed within a capsule composed of galactosamine and glycine (Fig. 9).

IV. Induction of Sporulation or Encystment

It is difficult to generalize about the induction of sporulation or encystment in these diverse systems. Nutrient limitation seems to play a role in all cases while enrichment of media with some readily metabolizable, energy-rich substrate tends to suppress sporulation or encystment. These observations strongly suggest that catabolite repression may play a key role. For example, sporulation occurs in endospore-forming bacteria during stationary phase in batch culture when nutrients become limiting, or upon shift down to more minimal conditions during outgrowth (microcycle sporulation). Rarely does it occur during logarithmic phase and then only to a small degree. In myxobacteria omission of certain amino acids induces myxospore formation. On the other hand, nutrient limitation may not be the sole inducer since myxospore formation can occur as a result of the addition of glycerol to the medium. The same pattern is found in the encystment of *Azotobacter* where *n*-butanol and its product of oxidation, beta-hydroxybutyrate, are inducers; however, encystment can occur under conditions of divalent cation limitation. That conidia of actinomycetes are mainly found in older colonies where nutrients would be limited, or that exospores of *Rhodomicrobium* are found in the centre of colonies, the oldest part where nutrients are depleted, agrees with the general concepts presented. Similarly, exospores of methane-utilizing bacteria are formed in batch culture after stationary phase is initiated. In *Anabena cylindrica*, spores (akinetes) are formed in filaments next to heterocysts, sites of nitrogen fixation, during phosphate limitation as well as in other conditions. The relationship between the spores and the differentiated heterocysts is not clear but it is believed that akinete formation may be dependent on their nearness to the heterocysts.

V. Resistance of Prokaryotic Resting Cells

Where studied, the resting form is usually more resistant to various environmental stresses than its counterpart vegetative form. It is difficult to make comparisons among the various types but Table 2 presents some limited data where some comparisons can be made. Except for resistance to ultraviolet irradiation, bacterial endospores are the most resistant in all categories. Since endospore resistance has been recognized for a long time and is of major importance in food processing and medical sterilization, their resistance has been most carefully studied. Their resistance to physical and chemical agents other than those in the table is noteworthy. Vegetative cells are killed at 88,000 p.s.i. for 14 hours hydrostatic pressure while their spores have been shown to require 176,000 p.s.i. for 14 hours. Spores are about 10,000 times more resistant to hypochlorites than are vegetative cells. Two hundred ($r \times 10^3$) of X-rays were required to kill 50 per cent of treated *B. megaterium* spores, while 50 per cent of treated *E. coli* cells were killed at 5.6 ($r \times 10^3$). In every process designed for killing microorganisms the spores are more resistant than vegetative cells and, with few exceptions, are the most resistant biological entities known. One exception is *Micrococcus radiodurans,* the most radiation resistant organism yet discovered.

VI. Mechanisms of Resistance

No entirely satisfactory explanation has been made for the endospore's remarkable resistance to desiccation, heat, radiation, chemicals or enzymes. Dipicolinic acid (pyridine 2,6-dicarboxylic acid) was long thought to play a major role in heat resistance, however the recent isolation of DPA$^-$ mutants which are still heat resistant may relegate that compound, found in concentrations as high as 15 per cent of the dry weight of the spore, to other roles possibly in the maintenance of dormancy and/or the initiation of germination.

Temperature adaptation is important in any consideration of heat resistance. Spore death temperature is directly proportional to maximum growth temperature. For example, spores of *B. macquariensis*, a psychrophile isolated in Antarctica, are killed at a much

Table 2.   Resistance of various prokaryotic resting cells to different agents

	Heat[a]	Ultraviolet irradiation	Desiccation[b]	Sonication
Endospores	Cl. botulinum type E: (D_{80C} = 0.6 min) Cl. thermosaccharolyticum: ($D_{132.2C}$ = 4.4 min) (vegetative cells killed at 50 °C)	Twice dosage required to kill spores than that required for vegetative cells	Retained viability after 6 years	More resistant than vegetative cells
Myxospores	Myxococcus ruber: survive 70 °C for 30 min (vegetative cells killed at 50 °C)	Five times dosage required to kill myxospores than that required for vegetative cells	Survived six days	More resistant than vegetative cells
Azotobacter cysts	Azotobacter vinelandii: slightly more resistant than vegetative cells	Slightly more resistant	Survived 12 days	More resistant than vegetative cells
Bdellocysts	Slightly more resistant than vegetative cells	Not resistant	Survived 6 days	More resistant than vegetative cells
Methylotrophic bacteria exospores	Methylosinus sporium: survived 75 °C for 70 min (vegetative cells killed within 10 min)	Slightly more resistant than vegetative cells	Slightly more resistant than vegetative cells	More resistant than vegetative cells
Rhodomicrobium exospores	Rhodomicrobium vanneilli: survive 100 °C for 30 min (vegetative cells killed at 50 °C)	Slightly more resistant than vegetative cells	Slightly more resistant than vegetative cells	Unknown
Exospores (conidia)	Slightly more resistant than vegetative cells	Slightly more resistant than vegetative cells	Slightly more resistant than vegetative cells	Unknown
Spores (akinetes)	Not resistant	Slightly more resistant than vegetative cells	Slightly more resistant than vegetative cells	Unknown

[a] D value, treatment which destroys 90 per cent of the spores.
[b] Under comparable conditions, drying followed by storage in vacuum

lower temperature than those of *B. stearothermophilus*, a thermophilic soil isolate.

Current hypotheses on the mechanism of heat resistance of endo-spores centre on the dehydration of the protoplast (core) and the expandable cortex with its counterions. In the heat resistant form the spore coat is relatively impermeable to multivalent cations. The cortex (of high water content) contains the expanded electronegative peptido-glycan and mobile counterions exerting high osmotic pressure. The protoplast (or core), of low water content, is osmotically dehydrated by the surrounding cortex and is therefore heat resistant. In the heat sensitive form, there is a modified coat leaky to multivalent cations. The neutralized cortex, collapsed and free' of counterions, exerts low osmotic pressure. The protoplast becomes partly hydrated and therefore heat sensitive. This theory, called the osmoregulatory expanded cortex theory, fits all the known facts but has yet to be proven or disproven. Germinated spores lose their heat resistance and yet, under special conditions, can be dehydrated to become both heat resistant and dormant once again. It is noteworthy that the heat resistance of the non-sporeformer *E. coli* is increased in dehydrated cells. The expansion or contraction of the cortex is thought to account for the dehydrated state of the spore protoplast. This reduced water content may also play a role in radiation resistance. It is thought that conformational differences between DNA in spore and vegetative cells may be associated with differences in hydration levels. The high resistance of *M. radiodurans*, mentioned previously, is mainly due to the possession of efficient repair mechanisms for radiation-induced changes in DNA. While excision repair occurs in some vegetative cells of endospore formers it has not been found in the spores. Resistance to lytic agents and chemicals may be due to the spore being impermeable to these agents. To what extent these postulated mechanisms apply to other resting forms is not yet known. It is interesting to note that most resting forms share the property of desiccation resistance and it has been suggested that this property is of most significance in the evolution of resting cells.

VII. Resting Cell Formation as Models for Differentiation

The formation of a resting form having low metabolism from an actively metabolizing vegetative cell provides that cell with unique survival advantages and represents a distinct differentiation. Similarly, germination, the breaking of the resting cell's dormancy and subsequent emergence of a vegetative cell represents another type of cell differentiation. In each case the ordered formation of a new cell occurs as a distinct sequence of events involving the synthesis of new mRNA and new developmental proteins. Thus, the processes of resting cell formation and the change back to the vegetative cell present excellent models for studying differentiation with the added attendant advantages of microbial systems: ease of handling, use of large numbers of cells, fast growth, synchrony, availability of mutants, etc. One criticism of prokaryotic differentiation models has been concerned with the lack of inter-cell communication found in eukaryotic systems. Indeed, the extent of such communication in microbial cells is unknown (although chemotaxic responses to cell-secreted chemicals is highly likely); however, in the myxobacter system inter-cell relationships have been shown to give that model an extra dimension. Other prokaryotic models present other advantages. Some cyanobacteria (blue-green algae) form two types of differentiated cells – heterocysts and akinetes (spores). The heterocysts are spaced at regular intervals and influence the formation of these spores. The system allows for the study of biochemical mechanisms governing the formation of multicellular patterns.

VIII. The Endospore Model, Cycle of Germination, Outgrowth, and Sporulation

The endospore models were recognized early and therefore more knowledge has been accumulated using these systems than with the others. The cycle of germination, outgrowth, growth, and sporulation (shown schematically in Figure 1) has been studied from many different aspects. Firstly, free spores usually must be activated in some manner. Activation is a reversible process which conditions the spore for germination and increases the number of spores undergoing germination as well as the rate of germination. Spores can be activated by a variety of treatments, notably exposure to heat; other resting forms do not require activation. Germination, the breaking of the spore's highly dormant state, follows. A series of degradative reactions is triggered in an unknown manner by simple compounds such as certain amino acids and ribosides or their mixtures (no universal germinant exists), and can be monitored by the loss of spore refractility as seen in the phase contrast microscope and by decrease in optical density. No metabolic activity can be detected during the first 2 minutes of germination of spores requiring either alanine or glucose for germination. Generation of ATP or production of known metabolic products of these initiators have not been found. Mutants deficient in key glycolytic pathway enzymes can germinate, thus ruling out glycolysis in the case of spores requiring glucose for germination. The

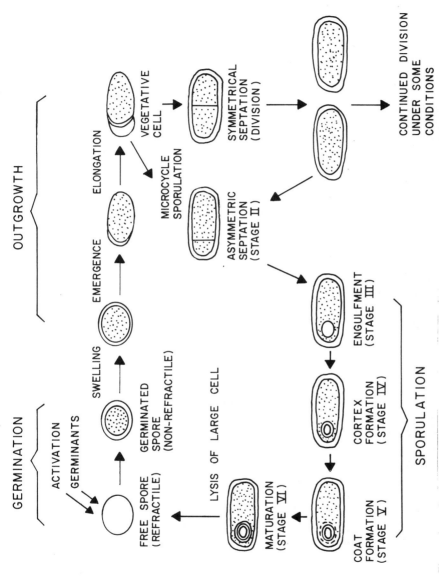

Figure 1. Cycle of germination, outgrowth, growth, and sporulation of a typical spore-forming bacterium (modification of a figure in Gould, G. W. in *Methods in Microbiology* (J. R. Norris and D. W. Ribbons, Eds.) 1971. London Academic Press

same spores can be germinated by non-metabolizable glucose analogues as well. These observations have led to a hypothesis that suggests that receptor proteins, possibly in the inner membrane, undergo conformational changes which alter permeability leading to an autocatalytic loss of heat resistance and stimulation of energy changes which initiate metabolism leading to vegetative growth. An opposing view based on the utilization of inhibitors of the electron transport system which have marked effects on germination, sees respiration and ATP creating a proton motive force and establishing a proton gradient. This early establishment of a membrane potential in spores is later followed by a repolarization of the membrane leading to germination. The proton motive force is used for transport of ions from core to cortex to neutralize other ions.

In addition, upon germination the spores lose their resistance to heat, radiation, injurious chemicals, and dilute stains. Concomitantly with 'phase' darkening, the spores swell, break out of their coats, and exude up to 30 per cent of their dry weight; about half of the exudate consists of a calcium chelate of the spore-specific substance, dipicolinic acid, and the rest of it consists of peptidoglycan fragments and amino acids. The earliest measurable events are the loss of calcium, DPA, and heat resistance.

The dormant spore lacks the ability to produce amino acids; and amino acid biosynthesis is absent early in germination. During the first minutes of germination 20 per cent of the spore's protein is degraded providing the source of amino acids for biosynthesis during outgrowth. The spore's enzymes are not degraded. Rather, three distinct spore proteins have been described. These proteins, A, B, and C, comprising 15 per cent of the protein in the spore and unique to the spore, are sensitive to proteolysis and are located in the spore coat. Their molecular weight is about 8000 daltons and although they are not histones, they bind to the spore DNA. The proteins are degraded by a unique protease which has an absolute specificity for these proteins.

Outgrowth is the period during which the spore gradually becomes a vegetative cell and requires new macromolecular synthesis. RNA synthesis begins rapidly within 2 minutes of germination and is followed within a short time by protein synthesis. DNA is replicated relatively late in outgrowth just before division. The vegetative cell is then capable of undergoing various morphological and biochemical changes which lead either to a series of symmetric cell divisions and subsequent spore formation or to the production of a spore without intermittent cell division. The latter is known as microcycle sporulation. Many fine structural analyses of cells during sporulation have revealed seven distinct stages (Figure 1, Figure 2 and Figure 3). Prior to

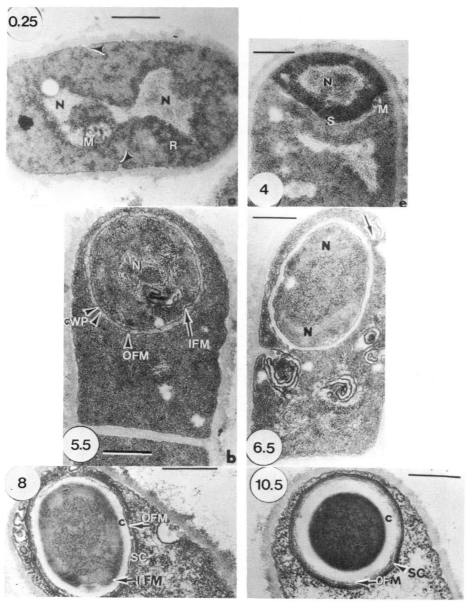

Figure 2.   A composite of electron microscope photomicrographs of thin sections of *Bacillus megaterium* showing the various cells and stages of sporulation depicted schematically in Figure 1. The circled numbers refer to hours after T_0 (the end of logarithmic phase of growth in a batch culture). 0.25 h shows a typical vegetative cell. 4 h is a Stage II cell illustrating the asymmetric septation (S). 5.5 h is a Stage III cell (engulfment), CWP indicates cell wall primordium; IFM, inner forespore membrane; OFM, outer. 6.5 h is a Stage IV cell (cortex formation).; 8 h, Stage V (coat formation); 10.5 h shows the fully mature spore with all the surface layers intact. N = nuclear area, SC = spore coats, OFM and IFM = outer and inner fore-spore membranes, C = cortex, M = mesosome, CWP = cell wall protein. The bar indicates 0.5 μm. (R.A. Greene, Ph.D. Thesis 1966, Syracuse University.)

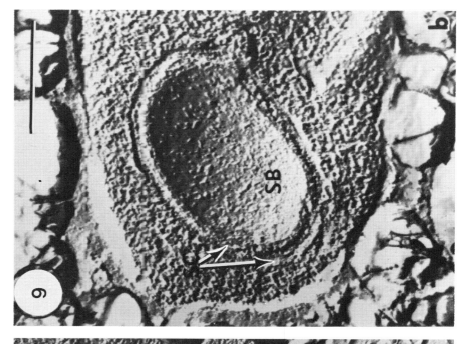

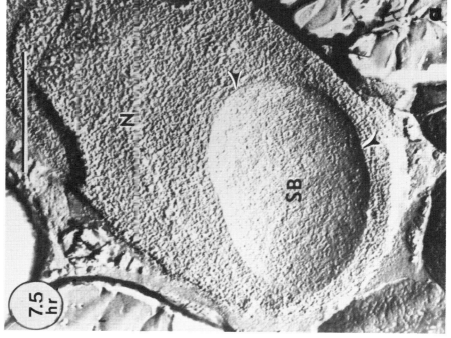

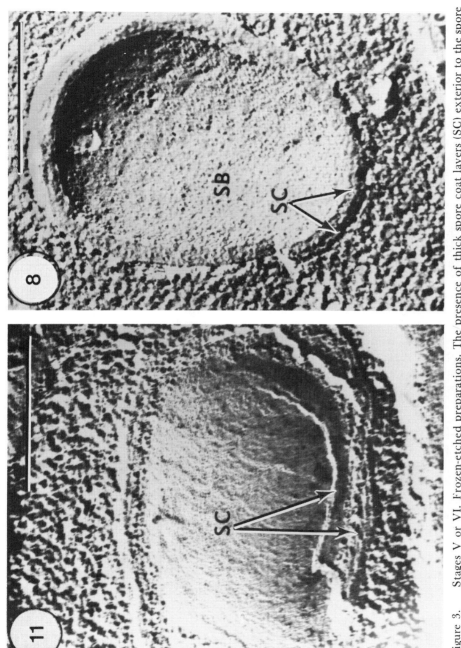

Figure 3. Stages V or VI. Frozen-etched preparations. The presence of thick spore coat layers (SC) exterior to the spore body (SB) is pronounced. The bar indicates 0.5 μm. (R. A. Greene, Ph. D. thesis 1966, Syracuse University.)

sporulation, the nuclear material is in an axially disposed filament. This is Stage I. However, since such a pattern does not appear to be unique to sporulating cells, current practice is to refer to stages prior to Stage II as being preseptation cells. Segregation of the chromatin material to the poles of the cell occurs concomitantly with the invagination of the plasma membrane and its associated mesosome in an asymmetric position on the cell. The invagination moves concentrically toward the centre of the cell and fuses to complete the spore septum (Stage II). The mode of formation of this septum is similar to the formation of the symmetric division transverse septum. However, in sporulation the division of the cell is not equal and subsequent proliferation of the larger cell's cell membrane leads to complete engulfment of the 'forespore' and liberation of the immature spore, surrounded now by a double unit membrane, into the cytoplasm of the larger cell (Stage III). This is a key step since this double membrane now has different transport properties owing to the opposing direction of the two membranes. At this time the cell is 'committed' to complete the process of sporulation. Cortex material similar to vegetative cell wall peptidoglycan is laid down between the unit membranes, and its deposition corresponds in time to the accumulation of dipicolinic acid and calcium (Stage IV). Studies with cortex-less mutants show that the cortex is needed for refractility of the spore and for accumulation of DPA. It plays a fundamental role in making spores dehydrated but the mechanism is not understood. Protein coats are synthesized around the outside of the spore (Stage V) and in some species an additional layer, the exosporium is synthesized. The distinct protein of the coats can be detected in preseptation stages by sensitive immunological methods perhaps indicating the extent to which vegetative cell functions play a role in the sporulation process.

Since the coat may play important roles in protection of the spore and its subsequent germination, it has been the subject of many investigations. Electron microscopy reveals that *B. cereus* contains an outer coat showing a cross patched pattern, an inner pitted layer, and a third layer, the undercoat, while other species show distinct differences. *B. subtilis* possesses a very thick multilayered coat with an outer striped layer and *B. thuringiensis* has a coat deficient in the outer cross patched layer. Chemical differences show up as well within the major structural polypeptides and coat associated proteases. These differences may be responsible for the variation found in germination and resistant properties of various species. After the spore has matured (Stage VI), at which time the spores become resistant to heat, lytic enzymes lyse the sporangial cell (often referred to as the 'mother' cell but probably more apt would be the 'sister' cell) and the free spore is liberated (Stage VII).

IX. Biochemistry and Genetics of Sporulation

Many other biochemical and physiological events in addition to those indicated above are linked with the morphologically identified stages. Some of these are listed in Table 3 and Figure 4. Some vegetative enzyme activities disappear but some remain while other vegetative enzymes are modified; some enzymes are sporulation specific. The total known events are far from complete since the genetic evidence would lead one to suggest that 40 to 50 operons are involved in sporulation. The ordered appearance of these cytological and biochemical changes implies a sequential reading of the genome but what induces the reading and how it is regulated are among the challenging questions. The genetic data are consistent with there being a single linear dependent sequence to Stage III with a more complex pattern of gene expression beyond that stage of development. With regard to the genetics, it is quite complex and our present understanding far from complete. Most of the definitive work has been done with *B. subtilis* which is amenable to transduction and some mapping has been accomplished. More recently, the isolation of phages which can convert certain asporogenic mutants into sporogenic cells, studies on

Table 3. Some biochemical and physiological events associated with sporulation

Stage		Event
I	Preseptation	Exo-protease Antibiotic production Protein turnover New mRNA
II	Asymmetric division	Alanine dehydrogenase RNA turnover
III	Engulfment	Formation of spore protoplast Alkaline phosphatase TCA enzymes active Radiation resistance
IV	Cortex formation	Cortex peptidoglycan formation Development of refractility Accumulation of Ca Synthesis of dipicolinic acid
V	Coat formation	Cysteine incorporation Octanol resistance
VI	Maturation	Heat resistance

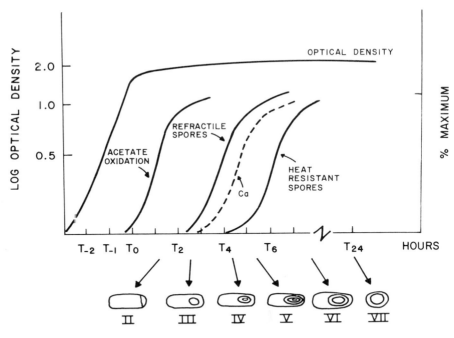

Figure 4.   Typical patterns of sporulation events in a batch culture. T_0 refers to the time at the end of logarithmic growth and when sporulation events are initiated. The diagrams at the bottom of the Figure refer to the sporulation stages (II to VII) depicted in Figure 2. For some other events and biochemical changes see Table 3

the formation of merodiploids, and the exciting possibility of cell fusion and cloning, wherein it may be possible to insert the sporulation operons into the paragon of genetic manipulation, *E. coli*, may enhance our knowledge. Two major types of sporulation mutants have been described: Sp⁻ which are completely asporogenous, and Osp (oligosporogenic) which can sporulate but at extremely low frequencies. To distinguish sporulation mutants from auxotrophic mutants which may affect growth but not sporulation, the sporulation mutants are referred to as *cacogenic* mutants with their revertants known as *protogenic*. The Osp mutants presumably are damaged in some regulatory function. In addition, it is possible to have mutants with blocks at various morphological stages during sporulation. These are referred to as Spo0, SpoI, SpoII, etc., depending on the stage blocked. The primary biochemical defect, however, is not known for most sporulation mutants and to add to the difficulty most mutants show pleiotropic effects. Current evidence suggests that the sporulation process is controlled by a large number of unlinked genes (possibly as many as 200) dispersed over the chromosome and there is some evidence for clustering of closely linked genes.

X. Initiation and Regulation of Sporulation

Sporulation, therefore, consists of a series of complex sequential biochemical changes (estimated at about 20) to produce the observed changes in morphology, all of which must be determined by distinct genetic loci. The manner of the initiation of the reading of the genome and its regulation is not known. Much work has centred on a presumed trigger of sporulation. The studies have been diverse and cover many different types of probes and investigations to take into account the various observed biochemical changes. Some investigators believe that certain polypeptide antibiotics (e.g. tryrothricin), produced at about the same time that sporulation is initiated, may repress vegetative functions. The availability of supposed antibiotic negative mutants which still sporulate appears to rule out such a role for antibiotics but perhaps the sensitivity of the antibiotic assay procedures may not detect the few molecules necessary to allow sporulation. Most endospore producing bacteria have been shown to produce antibiotics. With the exception of exospore-producing actinomycetes, antibiotic production has not yet been demonstrated in other prokaryotic resting cell formers. The antibiotics are secondary metabolites produced during late stationary phase during nutrient depletion and at about the same time as sporulation is initiated. In addition to the possibility that they may play a direct role in sporulation it is postulated that they may be involved in influencing the quality of the spores. Gramicidin-deficient mutants give rise to heat sensitive spores which can be made heat resistant by the addition of the antibiotic. Additional roles of antibiotics may be as inhibitors or activators of germination. It may be desirable to inhibit germination until conditions become favourable. The excretion of the antibiotic during germination may kill competitors for scarce nutrients or, since it has been shown that some dormant spores have protozoan predators, some spores may be less attractive due to possession of certain antibiotics.

Extracellular or intracellular proteases have been implicated in sporulation since a protease may be responsible for the protein turnover and a proposed modification of RNA polymerase discussed later. Unfortunately, several different proteases are elaborated and, as with the antibiotic story, some protease negative mutants can still sporulate.

Changes in specific tRNA's have been observed and it is suggested that they may be involved in the synthesis of modified or new proteins during sporulation. The isolation of rifampicin-resistant asporogenic mutants as well as RNA polymerase altered in one of the beta subunits, have led to the hypothesis that a modified polymerase may then transcribe in a different manner. Recently, several new proteins associated with the RNA polymerase of sporulating cells have been described. In a similar vein, modified ribosomes have been detected in cells during the time of sporulation initiation. Such modification might result in the synthesis of proteins which are altered or different from those made on vegetative cell ribosomes.

A number of spore loci described to date have been shown to be concerned with Stage 0. These Spo0 mutants have pleiotropic phenotypes and can make neither polar septa (the asymmetric division) nor Stage 0 associated products: proteases, antibiotics, and transformation-competent cells. Some Spo0 phenotypes show ribosomal alterations. It has been suggested that the products of the genes must be able to act with ribosomes through accessory proteins that select for certain messages.

It has long been conjectured that there may be a repressor of a sporulation specific mRNA but such a repressor has not yet been isolated. Many metabolites of low molecular weight which change concentration at the end of vegetative growth have been proposed as suppressor candidates, including ATP, cGMP, and adenosine polyphosphates. Cyclic AMP found in myxobacter and some eukaryotic cell differentiating systems has not been detected. Unfortunately the evidence is primarily of a correlation type and no causal relationship has been demonstrated for any compound.

Suggestions have been made that DNA may be involved in initiation. The capacity to initiate sporulation is identified with a specific stage of DNA replication. Shifting of cells either from early stationary phase in batch culture or during outgrowth (microcycle sporulation) to replacement medium results in sporulation only after DNA is initiated. Initiation of sporulation also may be a consequence of a modification of the DNA by folding in such a way that the resulting transcription is different from that of the unfolded chromosome. Recently, some studies with specific inhibitors of nucleic acid synthesis brought about the initiation of sporulation, further suggesting an involvement of DNA in initiation.

One interesting and important approach to understanding sporulation involves the separation of the two compartments after Stage II by lysozyme treatment followed by sonication giving two distinct cell

Figure 5. Freeze-etched preparation of *Bacillus megaterium* showing both the symmetrical and the asymmetrical (stage II) types of division. Bar = 0.5 μm

'populations' of 'mother' cells and forespores which can then be subjected to various analyses. Such studies reveal that forespores have some degree of autonomy once engulfment is completed. The forespore is capable of synthesizing protein and although levels of synthesis are different, protease and TCA enzymes are detected. Many new protein bands are detected by gel electrophoresis which suggests that new and distinct proteins are synthesized in the forespore compartment. The differences noted are not thought to be due to changes in the ribosomes since few differences in either the sedimentation characteristics of the ribosomes or in the protein composition of the ribosomes have been seen. The forespore can synthesize DPA but a challenging question is how the calcium, known to be accumulated in the mother cell, ends up in the spore since it must be transported through the forespore membrane which possesses reverse polarity from the mother cell. DPA is believed to play a role in transporting calcium across the membrane. These types of experiments show that while the two compartments have some autonomy both are required in concert for full development of the spore implying some inter-cell communication.

Thus, a number of alterations in molecular components of the cell have been correlated with induction of sporulation but the key change, if any, has yet to be determined. The varied observed alterations serve to remind us of the complexity of the process which at one time appeared deceptively simple.

Once the cell is induced, the amount of time necessary for the completion of the sporulation process (6 to 8 h) suggests that there is a gradual modification of the vegetative metabolic pattern rather than a relatively sudden switch to a sporulation genome. The lack of evidence for a distinct trigger may imply that the cell simply alters a major cell capability during certain conditions which leads to sporulation. Hence, the hypothesis that bacterial endospore formation may be viewed as a modified prokaryotic cell division is attractive. It has been proposed that the bacterial spore is a small cell formed by an asymmetric cell division and enveloped by the larger sister cell, both processes being morphological expressions of unbalanced growth resulting from shift down conditions.

Table 4 compares the morphological stages of sporulation with sporulation interpreted as a modified cell division. Sporulation seems to be the resultant of two factors. One is the innate ability of spore forming cells to complete a round of DNA replication and to position the replicated molecules by means of a septum as late as 2—3 h after the end of growth. In essence, this means cell division without net growth.

The second factor is the effect of change of metabolic pattern resulting from exhaustion of growth-limiting substrate on this cell division. The metabolism of the cells changes from that typical of logarithmic growth, through a shift down pattern, to a pattern of endogenous metabolism, which includes turnover and consequent reallocation of cellular materials. This results in unbalanced synthesis — there is an accumulation of new end products (secondary metabolism) and, most importantly, chemical modifications and changes in the relative timing of synthesis of components common to the vegetative and sporulating cells. The morphological changes that occur during sporulation seem to be principally the result of changes in the relative timing and quantity of membrane and cell wall materials that are synthesized. The finding of small quantities of coat protein in preseptation cells agrees with this. In essence, a precise balance of metabolites is necessary. In this way, different shapes may be made from similar materials and the membrane seems to play a particularly important role in these shape changes. The hypothesis that sporulation is a modified cell division is acceptable for the following reasons. First, the spore contains an independent cell which must have arisen by cell division because, in accordance with

Table 4. Proposed relationship of sporulation and cell division

Stage	Sporulation event	Relationship to cell division
I	Preseptation:DNA in axial filament form	Slow separation of DNA replicated in late log and early stationary phases
II	Septation	Asymmetric cell division results in two unequal sized cells each with a part of the DNA
III	Engulfment	Membrane synthesis out of step with cell wall formation (Stage IV) Envelope begins encystment of small cell
IV	Cortex formation	Peptidoglycan synthesis, equivalent to cell wall formation
V	Coat formation	Coat and cortex encyst the germ cell Coat protein is result of secondary metabolism
VI	Maturation	Net result of asymmetric cell division is one cell encysted in products of larger cell
VII	Lysis of larger cell and release of free spore	Lysis as a result of lack of further division

Virchow's postulate, a cell can only be formed by the division of a pre-existing cell. Second the morphological and compositional evidence available is consistent with this hypothesis which emphasizes the similarities of spore and vegetative cell composition and structure as well as accommodating the differences. The hypothesis incorporates the modern concepts of unbalanced growth, metabolic shifts, turnover and secondary metabolism in attempting to explain the dynamics of sporulation.

XI. Comparison of Endospore Formation with other Resting Cell Formation

As indicated previously the development of other resting forms, although having in common with endospore formation nutrient limitation as a key aspect of their formation, involves the transformation of the entire cell. Rather than being developed mainly inside another cell as in endospore formation, the cellular changes occur directly in the aqueous environment or inside specialized structures (as in myxospore formation) or on aerial hyphae (as in actinomycete conidia formation). The resulting resting forms have protective layers different from endospores although some similarity is noted between endospores and *Azotobacter* cysts. These layers surrounding the core in resting stages very probably contribute directly or indirectly to the degree of cryptobiosis (low or reduced metabolism), and, presumably, to the resistance demonstrated by the resting form. The marked degree of resistance and dormancy in endospores may be due to the complexity of their surface layers. Until recently the differences noted have been taken to mean that these differentiating systems have little in common. However, it can be expected that the broad outlines in the transition of one cell type to another, involving a distinct sequence of morphological and biochemical events produced by the synthesis of new mRNA and developmental proteins specified and controlled by genes would be similar in all resting cell formers. However, the details of the transition would be expected to reflect the different strategies which have evolved.

Although not as extensively studied as endospore formation both *Myxococcus* and *Azotobacter* developments show similarities to it (Figures 6, 7, 8 and 9). In all three systems, acetate metabolism occurs by the tricarboxylic acid cycle and the glyoxylate shunt, gluconeogenesis (for polysaccharide synthesis) is involved, nitrogen

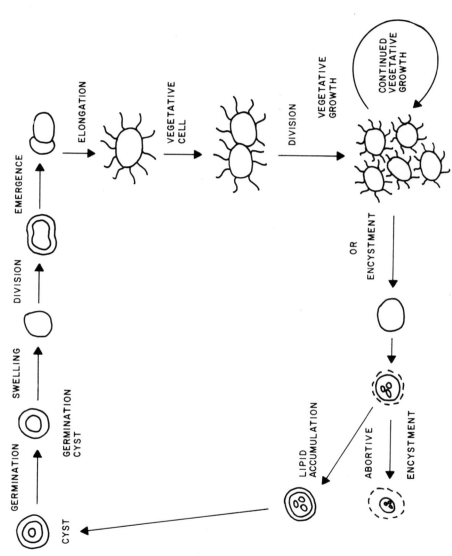

Figure 6. Cycle of germination, vegetative growth and encystment of *Azotobacter vinelandii* (modification of a figure provided by Professor Sadoff)

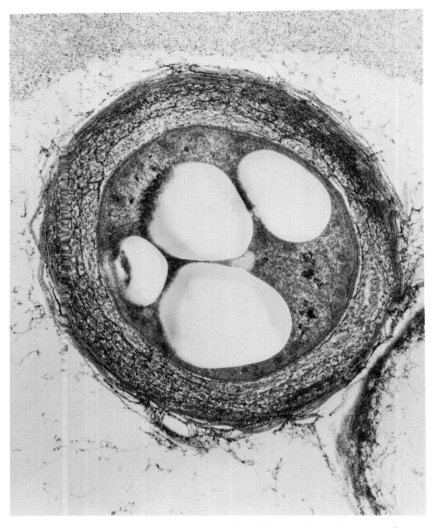

Figure 7. An electronphotomicrograph of a thin section of an *azotobacter vinelandii* cyst. The outermost layer is the exine and the next innermost layer is the intine. These layers surround the central body which contains granules of poly-beta-hydroxy butyrate (the large white egg-like bodies). Magnification is 53.3 K. (Photograph kindly supplied by Professor Sadoff.)

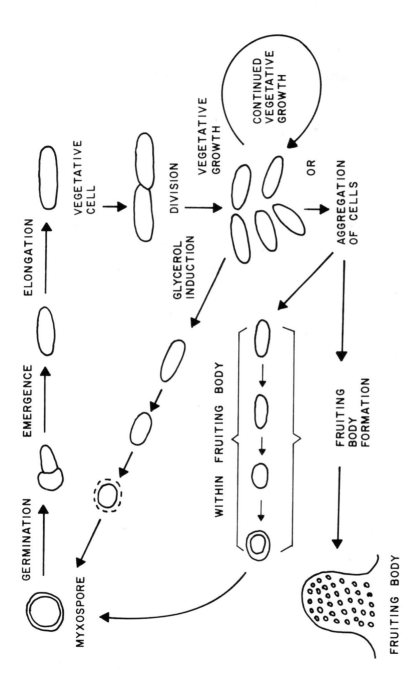

GERMINATION   EMERGENCE   ELONGATION

VEGETATIVE
CELL

DIVISION

VEGETATIVE
GROWTH

CONTINUED
VEGETATIVE
GROWTH

OR

GLYCEROL
INDUCTION

AGGREGATION
OF CELLS

MYXOSPORE

WITHIN FRUITING BODY

FRUITING
BODY
FORMATION

FRUITING BODY

Figure 8. Schematic diagram of the life cycle of *Myxococcus xanthus*. (Modification of a figure in Dworkin, M. (1973) In 'Microbial Differentiation', *23rd Symp. Soc. gen. Microbiology*.)

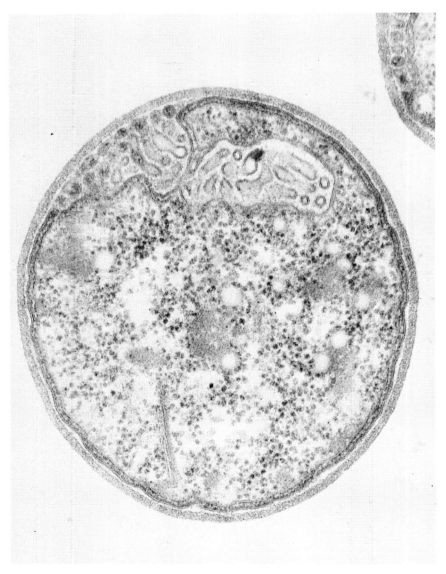

Figure 9.   An electronphotomicrograph of a thin section of a myxospore of *Myxococcus xanthus* induced by glycerol for two hours (photograph kindly supplied by Professor Herbert Voelz). Magnification is 58.5 K

deficiency is associated with initiation, and modifications of membrane synthesis occur. All three systems have been shown by hybridization competition experiments with RNA fractions from vegetative and sporulating or encysting cells to be transcribed predominantly by vegetative genes. Germination and outgrowth characteristics, although differing in time scale, are similar in that all show loss of refractility upon germination and extensive mRNA and protein synthesis prior to DNA replication. However, other differences in addition to those indicated previously are found. Although new rounds of chromosome replication are blocked, the developing cells complete the replication in progress in all three cases; however, the understanding of chromosome segregation and partition leading to compartmentalization of the genomes is not clear. *Myxococcus* sporulation is not preceded by a symmetric cell division as is *Azotobacter* encystment, or by an asymmetric division as in *Bacillus*. As a result the myxospores contain more DNA than their progenitor vegetative cells. The myxobacter are unusual also in that during the 'swarm' period which occurs in some instances, presumably due to a response to chemotactic substances, cell density is an important factor. It is believed that a high number of cells is required to elaborate enough exocellular enzymes to degrade the complex nutrients required for their development. In this regard they are different from *Bacillus*, for example, which depends to a greater extent on turnover within the cell during development.

Hence there are some similarities in the manner in which prokaryotes form resting cells but there are differences as well.

Some organisms produce more than one type of resting cell in 'normal' laboratory cultures. For example, methylotrophic bacteria form at least two types of cysts and exospores. It is also possible to manipulate cultures to obtain cyst-like bodies with low metabolism by shifting nitrogen starved outgrowing cells of *B. cereus* to a more enriched medium. Thus an organism considered only capable of forming one type of resting cell can under the proper conditions give rise to two differentiated types. Such experiments suggest that the phenotypic expression with regard to resting cell formation depends greatly on the environment. It may be that in those cultures where more than one resting form is observed, different nutritional states of the medium (particularly within colonies on solid medium) may be exerting an effect as to which resting form is developed. That a group of closely related organisms can contain members producing different types of spores (as is the case of exospores and endospores in actinomycetes) may indicate that the two types of sporulation mechanisms may have evolved in a similar manner from a common origin. It has been proposed that both types of sporulation are the result of a modified division. If one considers that 'arthrospores' of actinomycetes are morphologically similar in many ways to cysts, then all three 'major' resting bodies are found within the same group.

XII. Role of Resting Forms in Nature

It is assumed that an organism able to form, upon depletion of nutrients, a resistant resting form possessing low metabolism and able to return to an actively metabolizing state upon the availability of nutrients has a selective advantage over an organism not having a resting stage in its life cycle. The ability of endospores to persist for long periods of time, presumably due to their extreme dormancy properties, supports this view. Similarly the resistance of endospores to lytic enzymes of other organisms would seem to aid in the persistence of spores in various habitats. Longevity of other resting forms or resistance to lytic enzymes are not known. On the other hand, had there been a selective advantage it would be expected that resting cell formers would be the predominant members of a particular habitat. Except for desert soils where endospore forming bacteria prevail, this is not the case. Consistently, non-resting cell formers have been shown to be the dominant organisms in most habitats. Resistance to desiccation stands out as the one resistance property shared by all resting cells. The ability to produce and maintain an anhydrous state may give organisms able to form exospores, endospores or cysts survival advantages. One view suggests that resting cell formers evolved a different means of packaging the genome which would then be protected during unfavourable conditions. Alternatively, perhaps primitive cells protected DNA by first forming surface layers conferring resistance and other prokaryotes evolved from such cells.

XIII. Further Reading

Gerhardt, P., Costilow, R. N., and Sadoff, H. L. (1975). *Spores VI*, Washington, D.C.: American Society For Microbiology. (*Spores VII* is in press.)

Gould, G. W. and Hurst, A. (Eds) (1969). *The Bacterial Spore*. London: Academic Press. 724 pp. (contains much of what is known about endospores up to 1969 but a companion volume, *The Bacterial Spore II* covering more recent findings is in press).

Hitchins, A. D. and Slepecky, R. A. (1969). Bacterial Sporulation as a Modified Prokaryotic Cell Division. *Nature*, 223, 804–807.

Kalakoutskii, L. V. and Agre, N. S. (1976). Comparative Aspects of Development and Differentiation in Actinomycetes. *Bacteriol. Revs.*, 40, 469–524.

Sadoff, H. L. (1973). Comparative Aspects of Morphogenesis in Three Prokaryotic Genera. *Ann. Rev. Microbiol.*, 27, 133—153.

Sudo, S. Z. and Dworkin, M. (1973). Comparative Biology of Prokaryotic Resting Cells. In *Adv. in Microbial Physiol.* (Rose, A. H. and Tempest, D. W., Eds.), Vol. 9, pp. 153—224, New York: Academic Press.

Symbiosis in the Microbial World

D. C. SMITH

Melville Wills Professor of Botany, University of Bristol

I. Introduction

A. The Meaning of 'Symbiosis'

'Symbiosis' is a term which may have more than one meaning. The word was originally devised by De Bary in 1876 to describe any association between dissimilar organisms, and he expressly included parasitism. However, some biologists later excluded parasitism 'and restricted the term to cover only examples of 'mutualism', where each organism was presumed to derive 'benefit' from the association. Both of these usages present difficulties.

If 'symbiosis' is used in the original broad sense of De Bary, then it covers virtually all kinds of associations, so raising the need for classifying this large and complex assemblage. Recently, a comprehensive scheme has been proposed for classifying organismic associations according to a range of criteria such as: relative size; whether intracellular or extracellular; level of specificity; whether transient, prolonged or permanent; degree of dependence (i.e. facultative or obligate); degree of morphological integration, and the degree of harmful or beneficial effects. Unfortunately, although this is a good theoretical system, it is not a practical scheme which can be used easily by the general biologist.

The simpler classification into 'parasitic', 'commensal' and 'mutualistic' is therefore still widely used. While this classification easily accommodates the more extreme and obvious situations, there is a variety of intermediate associations which are less easy to categorize. Futhermore, in many associations commonly described as 'mutualistic', it is difficult to define or measure the 'benefit' that at least one of the partners is presumed to obtain. For example, a microbe may gain nutrients from its host, but at the same time its capacity for growth and reproduction may be severely curtailed, and it may even become eventually digested. Whether or not the microbe gains a net 'benefit' becomes a philosophical question, and it illustrates the difficulty confronting those who would define symbiosis as 'two organisms living together for their mutual benefit'.

Symbiosis can also be considered from an ecological standpoint. There are three main habitats in the world for microorganisms: land, water, and other living organisms. Those primarily adapted to the latter can be considered 'symbiotic' in the original broad sense of De Bary. If the habitat (= host organism) is 'harmed' (= put at a selective disadvantage) from colonization by microorganisms then the association is parasitic. If it 'benefits' (= gains a selective advantage), then it is usually described as 'mutualistic' even though, as outlined above, it may be arguable whether microorganisms gain an overall net benefit from the complex reactions of their living habitat towards them.

This article will be concerned with symbiotic associations considered to be mutualistic in this ecological sense, i.e. associations in which the host organism derives obvious selective advantage from the presence of microorganisms.

B. Nutrient Movement in Symbiosis: Necrotrophy and Biotrophy

The commonest selective advantage gained by the host is nutritional. There are two principal mechanisms of nutrient flow between organisms in symbiosis: necrotrophy and biotrophy. In necrotrophy, one organism first destroys part or all of the other before absorbing nutrients from dead tissues. In biotrophy, little or no destruction occurs, and nutrients move between living cells of the symbionts. Biotrophy permits one symbiont to exploit the other over a long period of time, and this is of particular advantage to hosts associating with microorganisms possessing major biochemical properties which they lack, such as nitrogen or carbon dioxide fixation. The success of biotrophy often requires the development of mechanisms for transferring nutrients between symbionts, and investigation of such mechanims has been a major area of experimental research into symbiosis. Although biotrophy is by far the commonest mode of nutrition in the associations to be discussed, necrotrophy may also occur.

C. Other Symbiotic Interactions

While the central feature of many associations is the flow of nutrients, a variety of other interactions may also occur. These may be grouped under the following headings:

1. Regulation of microbial growth

In all associations, the size to which the population of microbial symbionts can develop is strictly limited. This may be achieved partly by lysis, digestion or ejection of surplus microorganisms, but frequently, restriction of the rate of microbial growth is a major form of control.

2. Morphological changes

The morphology of both host and symbiont frequently alter in symbiosis. In the host, the gross morphology may be modified to improve or enlarge accommodation of the symbiont. The fine structure may also change, frequently so as to increase the area of contact with the symbiont; thus, absorptive surfaces may be convoluted to provide a greater area across which nutrient fluxes can occur. In the microbe, the general trend of modification is usually towards increased structural simplicity. Intracellular symbionts do not form motile structures such as flagellae; cell wall or sheathing structures are often reduced, and in certain cases, disappear altogether. Sometimes structures within the cell disappear, and in extreme cases the internal modification may be so great as to render the symbiont no longer identifiable with any free-living axonomic group.

3. Specificity mechanisms

In all associations, there is a high level of specificity between symbiont and host. Sometimes this is achieved by the direct transmission of symbiont from host generation to host generation, combined with adaptation of symbionts to an internal habitat in which no free-living microbe can successfully exist. Where hosts are newly reinfected by free-living forms of the symbiont in each generation, mechanisms are developed to ensure that only the symbiont and not other microbes colonize the host.

4. Interactions providing important selective advantages other than nutritional

There are a number of situations in which important selective advantages other than nutritional accrue to the host. Thus, algal symbionts of reef-forming corals not only provide nutrients, but also stimulate the massive calcium carbonate deposition which enables the reef to form (see Section II.B).

D. Principal Groups of Symbiotic Microorganisms

Symbiotic microorganisms can be classified according to the metabolic characteristic primarily exploited by the host. Five main groups may be recognized:

Types of symbiont	Characteristics primarily exploited by host
1. Autotrophic	Photosynthetic CO_2 fixation
2. Nitrogen fixers	N_2 fixation
3. Digestive tract symbionts	Breakdown of indigestible plant material
4. Mycorrhizal fungi	Efficient absorption of nutrients from soil
5. Miscellaneous	Various

This classification is rough and approximate for at least two main reasons. Firstly, a few symbionts may occur in more than one category; for example, some symbiotic blue-green algae can fix both carbon dioxide and nitrogen. Secondly, the nutritional role of a symbiont may extend well beyond the primary process exploited. For example, some algae symbiotic in animals may have the vital function of recycling waste nitrogen as well as supplying the host with products of photosynthesis.

II. Autotrophic Microbial Symbionts

There are two main groups of autotrophic microbial symbionts. Firstly, there are algae which associate with fungi to form lichens; the algae are always extracellular. Secondly, there are algae or chloroplasts forming associations with protists and lower invertebrates; the symbionts are mostly, but not always, intracellular. Both types can be of major ecological importance. Lichens cover very large areas of rock, tree, and ground, especially in arctic, subarctic and tundra regions. Coral reefs, all formed by coelenterates containing intracellular symbiotic algae, cover about 90 million square miles of the earth's surface.

Autotrophic or potentially autotrophic blue-green algae which associate with higher green plants will be considered with the nitrogen-fixing symbionts (Section III.C).

A. Lichens

With very few exceptions, each lichen consists of one species of fungus associated with one species of alga. The algae are either green or blue-green, and are unicellular or in simple short filaments. Twenty-seven different algal genera have been reported. The commonest, occurring in about 70 per cent of lichens, is *Trebouxia,* a unicellular member of the Chlorococcales which has never been found free-living. All other lichen algae belong to free-living genera (with *Trentepohlia* and *Nostoc* the commonest after *Trebouxia*), but except in a very few cases, they have been too poorly studied to establish whether the taxa in lichens exist free-living in nature.

Almost all lichen fungi are ascomycetes. They are not free-living, but are clearly related to a number of different free-living groups (including both apothecia and perithecia formers). There are a few basidiomycete lichen fungi. Over 18,000 species of lichens have been described, so they are a large group comprising about 25 per cent of the known species of fungi. Since both fungi and algae of lichens are taxonomically diverse, this large group of symbiotic associations must have originated on a number of separate occasions.

Two types of morphological structure occur. First, there are a few species in which algal and fungal filaments intermingle throughout the thallus; the algae (all blue-green) comprise a significant proportion of the bulk of the lichen. Secondly, there are the great majority of lichens in which the algae are restricted to a thin layer just beneath the surface, and the fungal hyphae differentiate into at least three distinct types of tissue: an upper cortex overlying the algal layer; an algal layer, with thin-walled hyphae closely encircling cells or short clumped filaments of algae; and an underlying medulla of larger and thicker-walled hyphae. Various additional tissues may be formed by different species. The algal symbionts form only a small proportion (less than 10 per cent) of the mass of the lichen.

During their long life-spans, most lichens are exposed to a range of environmental extremes, especially nutrient shortage, variations in temperature and frequent cycles of desiccation and rewetting. Indeed, the prolonged stability of the association seems to require these extremes. Addition of copious nutrients to lichens in the field causes the symbionts to disassociate, and in culture, isolated symbionts will only start to associate if the medium is very low in nutrients. The successful maintenance of healthy lichens in the laboratory can only be achieved if they are exposed to cycles of drying and wetting, and not a constant humidity level.

1. Physical nature of contact between the symbionts

Fungal hyphae are closely appressed to algal cells. Penetration of algal membranes by haustoria in healthy lichens is rare, and it is also uncommon to see dead algal cells. Algal and fungal cell walls are much reduced in thickness at the region of contact, but there is usually some kind of fibrous matrix material of uncertain composition and origin between the symbionts. Any materials moving between the symbionts have to cross an extracellular path through wall and matrix material.

2. Photosynthate movement between symbionts

A central feature of nutritional interactions between the symbionts is the flow of photosynthetically produced carbohydrate from the alga to its fungal host. Experimental investigations of photosynthetic ^{14}C fixation and movement in a wide range of lichens show five main features in common:

(i) Photosynthetically fixed carbon is released mostly as a single, simple carbohydrate. Blue-green symbionts release glucose, and green symbionts release a polyol — either ribitol, erythritol, or sorbitol. Export of fixed carbon through a single transport channel presumably gives the alga some measure of control and protects other essential metabolities from loss.

(ii) In the fungus, algal carbohydrates are immediately converted to fungal polyols — especially mannitol, often accompanied by arabitol. These polyols are normally not found in the alga, so that the transfer of photosynthetically fixed ^{14}C to the fungus can be studied in intact lichens simply by measuring the accumulation of label into fungal polyols.

(iii) The amount of carbohydrate released by the alga is substantial — probably up to 80 per cent of all the carbon fixed in photosynthesis.

(iv) Movement is rapid since photosynthetically fixed ^{14}C is detectable in fungal polyols within 2—5 minutes of fixation.

(v) As soon as the alga is isolated from the lichen, the massive carbohydrate release begins to disappear. Within 6—24 h, the released carbohydrate may no longer be detectable in extra-cellular products, and is either undetectable or present in only small amounts within algal cells.

A central question is therefore why existence in symbiosis induces the massive release of a specific carbohydrate from the alga. There are three possibilities, not necessarily exclusive.

Firstly, the fungus may secrete some compound which diffuses to the alga, and affects membrane properties so that the release of the specific carbohydrate is induced. However, exhaustive searches have so far failed to detect such compounds in lichens. Invertebrate hosts of symbiotic algae produce unidentified 'factors' causing release of specific compounds from the symbionts (Section III.B.1), but these have no effect on lichen algae.

Secondly, the membrane transport systems of the alga may be affected by a physical rather than a chemical influence of the fungus. Certainly, physical contact seems necessary for release from the algae to occur. It is possible that the membrane potential — important in determining the vectorial characteristics of some transport systems — may become altered by physical contact. In lichens with blue-green symbionts, digitonin inactivates the fungus, yet algal photosynthesis and release continues unabated (but ceases as soon as algae are isolated from physical contact with the fungus). Any affect on algal membrane potential would therefore have to be due to charges carried on cell walls rather than as a consequence of the metabolic activities of the fungus. The intransigent nature of lichen material has prevented exact and unequivocal measurements of membrane potentials of algae inside the thallus.

Thirdly, there may be some other aspect of existence in symbiosis which changes the internal metabolism of the alga so that a substantial surplus of the mobile carbohydrate is produced. This would then be released by normal transport mechanisms such as facilitated diffusion. It is certainly true of lichen algae — and indeed of all other symbiotic algae — that incorporation of fixed carbon into insoluble polysaccharides is substantially reduced in symbiosis. After isolation, decline in carbohydrate release is approximately paralleled by a rise of polysaccharide formation. So far, studies of isolated algae, whether symbiotic or not, yield no clue as to the kinds of factors which could repress polysaccharide synthesis and stimulate formation of a specific carbohydrate to such a high internal concentration that it is released in substantial quantities.

3. Simultaneous movement of fixed nitrogen and carbon in lichens with blue-green symbionts

In the light in lichens with blue-green symbionts, nitrogen is fixed by the heterocysts simultaneously with carbon by the vegetative cells. Most of the fixed nitrogen passes rapidly to the fungus, probably as ammonia. Hence, the fungus receives both ammonia and glucose from its symbiont; as with other fungi, ammonia uptake by lichen fungi is enhanced by glucose.

As soon as the algae are isolated, release of fixed nitrogen ceases even more rapidly than release of photosynthate. It is tempting to believe that, within the lichen, the same mechanism induces both photo-synthate and ammonia release. The primary enzyme of ammonia assimilation in the alga is glutamine synthetase. This enzyme becomes repressed in symbiosis, presumably explaining why substantial quantities of ammonia become available for release. However, even if parallel repression of key enzymes of carbohydrate metabolism could be demonstrated, the fundamental mechanism causing enzyme repression in symbiosis might still remain obscure.

4. Transport from fungus to alga

It is widely believed that the lichen fungus 'supplies the alga with minerals', but there is no experimental evidence that this occurs. Indeed, there is no *a priori* need for mineral nutrients to move between the symbionts. Since the algae are extracellular, minerals could presumably be acquired from external solutions permeating the intersymbiont space. Since growth is extremely slow, nutritional requirements are not high.

On the other hand, since lichen fungi are highly efficient at accumulating substantial amounts of solutes from dilute solutions, it would be surprising if some absorbed minerals do not pass in some way to the alga. One possible mechanism for this is suggested by two observations. Firstly, the great majority of lichens pass through frequent cycles of drying and wetting; and secondly, when air-dry lichens are plunged into water in the laboratory, they lose appreciable amounts of both organic and inorganic solutes during the first minute before membrane permeability barriers are re-established. In nature, it is rare for lichens to be suddenly immersed in water, but under the more common situation where a small area is suddenly saturated by a raindrop, there could be localized release and reabsorption of substances, so offering the possibility for passive movement of substances between the symbionts. This hypothesis has yet to be tested experimentally, but it could partly explain why lichens require cycles of wetting and drying rather than constant conditions for healthy growth.

5. *Culture of isolated symbionts and attempts at resynthesis*

Both algal and fungal symbionts of many lichens have been grown in culture. Isolation techniques are conventional, and the media relatively uncomplicated. Fungi show much better growth if biotin and thiamin are present, while algae prefer organic to inorganic media. Both grow more rapidly than when associated in a lichen, but more slowly than free-living relatives. The fungus shows none of the characteristic shape, tissue differentiation and colour of the lichen thallus.

Despite numerous attempts, especially in the first half of the century, the complete resynthesis of a taxonomically recognizable lichen has been unequivocally achieved on no more than two occasions. The symbionts show no tendency to associate when innoculated together on normal growth media. If almost all nutrients are withdrawn ('tap water agar'), then the initial stages of synthesis occur. Fungal hyphae tightly encircle the algae, and some tissue differentiation can be seen. Appreciable development only occurs if more natural substrates are used (rock, wood, soil) rather than agar, and if the cultures are subjected to slow periodic cycles of drying and wetting. Hence, synthesis only begins under nutrient-poor conditions disadvantageous to the growth of the isolated symbionts.

B. Associations of Autotrophs with Lower Invertebrates

A variety of lower invertebrates in marine or freshwater habitats possess autotrophic endosymbionts. The animal hosts may be protozoans, coelenterates, and certain groups of sponges, molluscs, and platyhelminths. The symbionts fall into four main groups: (1) dinoflagellates, (2) *Chlorella* spp., (3) chloroplasts of siphonaceous algae, and (4) other miscellaneous algae.

1. *Dinoflagellate symbionts*

By far the commonest symbionts of marine invertebrates are dinoflagellates. They are particularly prominent in tropical oceans, inhabiting the majority of coelenterates and being invariably present in reef-forming corals. In benthic animals, all symbionts so far isolated have been assigned to the single species *Gymnodinium microadriaticum*. In the less common pelagic forms (especially protozoa and turbellarians) they belong to a few species of the related genus *Amphidinium*.

The relationships between symbiont and host are best illustrated by studies of sea anemones, which are closely related to corals but much easier to investigate in laboratory experiments. As in virtually all symbiotic coelenterates, the dinoflagellate symbionts are confined to the digestive cells of the endodermis. It is relatively easy to culture the algae, and in liquid media they show all the typical characteristics of motile, free-living dinoflagellates: a rigid wall with a characteristic constriction around the middle of the cell, two unequal flagella — one anterior and the other running in a groove around the constricted waist — and chromosomes which remain permanently condensed. In symbiosis the cells appear substantially modified: the flagella do not develop, and a rigid wall is not formed so that the symbionts appear approximately spherical. However, the characteristic dinoflagellate nucleus remains. Each symbiont is bounded by a complex system of several membranes, but it is unclear whether the outermost part of the system is derived from host or symbiont.

The selective advantage of the symbionts to the host is illustrated by the observation that during starvation in the light, symbiotic anemones lose weight much less quickly than aposymbiotic. An obvious inference would be that the symbiotic algae provide a quantitatively important component of host nutrition. Since digestion of algae is hardly ever observed, products of photosynthesis presumably pass from intact algal cells to their host. This can be directly demonstrated by incubating anemones in the light in seawater containing $NaH^{14}CO_3$. Over half of the photosynthetically fixed ^{14}C accumulates in the animal tissues, much of it initially in proteins and the glycerol moiety of lipids.

Intact algae can be obtained by centrifugation of gently homogenized anemones. Such freshly isolated algae release a number of compounds, especially glycerol, to the medium. The total amount of photosynthate released by freshly isolated algae is less than half that released by algae in symbiosis. However, if an aqueous extract of animal tissue is added to freshly isolated algae, the release of glycerol is specifically stimulated. The belief that animal tissues contain release-stimulating factors is strengthened by the observation that no factors are detectable in aposymbiotic anemones, but if they become reinfected, then the 'factors' become apparent — as if the presence of symbionts induces their synthesis to detectable levels. These 'factors' have not been identified, although they are known to be thermolabile and effective on dinoflagellate symbionts from other phyla.

Under certain conditions, the release of alanine from freshly isolated algae is more prominent than glycerol, and is also specifically stimulated by 'factors'. This highlights a growing awareness that symbiotic dinoflagellates may be important in recycling nitrogen as well as in supplying photosynthate. As with almost all other kinds of algae, ammonia — frequently an end-product of animal metabolism — can be absorbed and converted to amino acids. The generally low levels of dissolved nitrogen in sea water would make any method of nutrient conservation and nitrogen recycling of particular advantage.

The pattern of relationships between the symbionts in sea anemones is similar for other animals with symbiotic dinoflagellates. The role of symbionts in nitrogen recycling may be particularly important in the coral-reef ecosystem. These develop in tropical waters where the nutrient supply is so low that they have been described as 'aquatic deserts'. Nitrogen shortage limits phytoplankton productivity in the open oceans, yet on the reef, photosynthesis per unit area may be 200-times higher, making it comparable to the most productive of terrestrial ecosystems. Two key factors reducing the nitrogen limitation of productivity are believed to be nitrogen fixation by free-living, blue-green algae, and nitrogen recycling with the dinoflagellate symbionts playing a major role.

Another vital function of the symbionts on coral reefs is that they greatly stimulate the rate of calcium carbonate deposition. A high rate is needed to enable the reef to replenish the losses caused by the contantly destructive actions of the sea. Using the radioactive isotope ^{45}Ca, it has been shown in a variety of corals that the rate of calcification is approximately ten times greater in the light than in the dark. This light-induced stimulation is completely abolished by DCMU which specifically inhibits photosynthesis without any detectable effect upon animal tissues. It is not clear how CO_2 fixation by symbionts in the endodermis stimulates calcium carbonate deposition by the ecto-dermis. Supplying the animal in the dark with glycerol and alanine — the photosynthetic products of the symbiont — does not increase calcification rates. In the aptly named 'Stag's Horn' coral, calcification rates are greatest at the tips, which have the lowest density of algae.

The principal dinoflagellate symbiont in the world, *Gymnodinium microadriaticum*, has so far not been found free-living, but this is probably because proper and intensive searches have not been made for it. Indeed, it is very difficult to believe that it will not be found free-living for at least two reasons. Firstly, symbiotic sea anemones in the laboratory can extrude packets of viable and healthy symbionts from their mouths, apparently as part of the mechanism of regulating their symbiont populations, and extruded algae can rapidly develop into motile forms. Secondly, in giant clams — where the mantle tissue is always densely packed with dinoflagellates — the eggs of the animal invariably hatch to produce uninfected larvae, so that reinfection is necessary at each generation. Since clams live on reefs, the most likely source of reinfection would be extrusion from other symbiotic reef organisms. Although infection is transmitted through the eggs of most coelenterates, there are some groups, notably Gorgonians, where reinfection is needed at each generation.

2. Chlorella symbionts

The commonest symbionts of freshwater invertebrates are *Chlorella* not yet assigned to a species but very similar to *C. vulgaris*. These symbioses are much less important ecologically than the marine associations involving dinoflagellates. On the other hand, two examples, green hydra and *Paramecium bursaria*, can be easily grown in mass culture, and are therefore particularly convenient for experimental analysis of symbiotic interactions.

A detailed consideration of green hydra will form a useful comparison with the work on sea anemones described above. As in the anemones, the symbionts are confined to the digestive cells of the endodermis. Each alga occurs within a vacuole, and on division, daughter algae segregate each into their own individual vacules. The algae are not much modified in morphology when compared to free-living relatives, but it is only very recently that they have been successfully cultured.

Symbiont-free (aposymbiotic) green hydra can be obtained by a variety of techniques. The growth rate of symbiotic and aposymbiotic hydra does not differ if they are fed daily, but with infrequent feeding, symbiotic animals grow more quickly, provided they are illuminated. Since the *Chlorella* symbionts are not digested, the evidence again suggests that photosynthesis is quantitatively important to host nutrition. When animals are gently homogenized, the algae remain undamaged because they have retained their cell walls, and so can be quantitatively separated from animal tissues. Using the radioactive isotope ^{14}C, the transfer of photosynthate from symbiont to host can therefore be studied with some precision.

Within two minutes, photosynthetically fixed ^{14}C appears in animal tissues, and continues to flow at a rate equivalent to 40–50 per cent of total fixation. Algae isolated from animal homogenates release fixed ^{14}C almost entirely as the disaccharide maltose. Animal tissues contain an active maltase enzyme: the glucose produced by its action is rapidly metabolized, so that photosynthetically fixed ^{14}C appears in a variety of lipid, protein, and carbohydrate moieties in the animal.

The maltose released by symbiotic *Chlorella* is synthesized from an internal pool of hexose phosphates. The site of synthesis is at or near the cell surface since no maltose is detectable within cells, and its production is extremely sensitive to the pH of the medium. Neither free-living *Chlorella* nor any other kind of green plant can synthesize free maltose, for it is normally produced only by starch breakdown. It is not known how a previously unknown synthetic reaction appears when an alga enters symbiosis. One possibility is that the action of a glucosyl transferase enzyme normally present at the cell surface has become modified. It might be analogous to the induction of lactose synthesis by mammary gland cells. A galactosyl transferase enzyme in one part of the cell combines with a milk protein, α-lactalbumin, synthesized in another part. The substrate specificity of the enzyme is changed so that it starts to synthesize lactose instead of transferring galactose to glycoproteins. The amino acid sequence of α-lactalbumin is indistinguishable from lysozyme, so it could have evolved from this very common enzyme. A model of this type might explain how autotrophic symbionts become modified by 'factors' when they enter animal cells. It is important to note, however, that no experimental investigation of this possibility in green hydra has been carried out.

Host cells can also regulate the population size of their symbionts. In regions of the animal where little or no host cell division normally occurs, there are 14—15 algae per digestive cell. If host cell division is initiated by inducing regeneration, the algal population per cell rises sharply by about 30 per cent, as if some restriction on algal division is removed as the host cell prepares to divide. In regions where host cell division occurs, the population of algae per cell is a mean of 20—21, but it drops back to about 15 when host cell division ceases (e.g. during starvation). Prolonged growth of animals in the dark causes the number

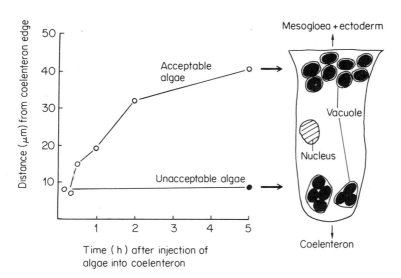

Figure 1. Migration of algae 'acceptable' as symbionts in the digestive cells of green hydra. 'Unacceptable' algae remain temporarily at the tip of the cell, become aggregated into large vacuoles, and are soon expelled

of algae per cell to drop well below 14, but they never completely disappear in most laboratory strains.

The host cell controls not only the number but also the position of algae in the cell. When aposymbiotic animals are reinfected by injecting dense suspensions of algae into the body cavity, the digestive cells rapidly take in about eight symbionts each. Over the next 5 hours, the algae are moved to their normal position which is at the far end of the cell away from the body cavity (Figure 1). The mechanism of movement is sensitive to inhibitors of microtubule polymerization such as vinblastine. Digestive cells can be induced to eject algae either slowly by incubation in DCMU or very rapidly by exposure to high light intensities. Again, ejection is prevented by inhibitors such as vinblastine.

Attempts to reinfect aposymbionts show that the animal cells are very specific as to the strains of *Chlorella* with which they will form an association. Only strains originally derived from *Hydra* are accepted; *Chlorella* symbionts from other animals are rejected, with the solitary exception of one particular strain isolated from *Paramecium bursaria*. 'Recognition' of symbionts by the host occurs after the algal cells are taken in. Acceptable algae are sequestered into individual vacuoles and transported to the distal end of the cell, while unacceptable algae remain in clumps at the proximal end and are soon ejected.

3. Chloroplast symbiosis

In certain kinds of sea slug, belonging to the family *Elysiidae* of the order Sacoglossa, cells lining the digestive tract are packed with photosynthetically active chloroplasts derived from some of the seaweeds upon which the animals feed.

The seaweeds providing symbiotic chloroplasts are always siphonaceous, i.e. they are constructed of vesicles or filaments in which cross-walls are infrequent, so that many nuclei and chloroplasts are in one compartment. The animals feed by puncturing a compartment and ingesting the contents, and presumably developed a preference for such seaweeds because more food could be obtained from a single puncture than from seaweeds constructed on a conventional cellular basis.

Chloroplasts isolated from siphonaceous seaweeds are the most robust type of organelle yet reported from eukaryotic organisms. They continue photosynthesis for at least a week when suspended in a simple mineral medium, and for up to 12 weeks after acquisition by sacoglossan digestive cells. They are not ruptured by the severe osmotic shock of transfer from sea water to distilled water, and the outer covering of the chloroplast, which is of unknown composition, is resistant to mild acid digestion. It is therefore not surprising that these organelles survive long periods in animal cells.

Siphonaceous chloroplasts have no greater genetic autonomy than those of higher plants, and in the absence of the plant cell nucleus, are unable either to divide or to synthesize certain key substances such as chlorophyll. Once isolated, their capacity for photosynthesis consequently declines, but with remarkable slowness.

The most detailed studies have been carried out with *Elysia viridis*, common around British coasts where there are large and well-established communities of its host seaweed, *Codium fragile*. The chlorophyll content per unit fresh weight is approximately the same in seaweed and in animals. The rate of photosynthesis per unit of chlorophyll in the animals is 60 per cent of that in the plant. Animals starved in the light lose weight less rapidly than those fed in the dark, illustrating the quantitative importance of photosynthesis to animal nutrition.

During photosynthesis, chloroplasts rapidly release 40—50 per cent of all the carbon they fix to the animals, probably in the form of sugar or sugar phosphates, and neutral amino acids. These are rapidly metabolized, and fixed carbon accumulates in a range of animal products, especially mucus. Isolated chloroplasts release relatively little fixed carbon, but aqueous extracts of animal tissues contain unidentified thermolabile 'factors' which stimulate release of specific compounds. These 'factors' are also active on symbiotic dinoflagellates. In *Elysia*, 'factors' are only detectable in the chloroplast-containing tissues. They are absent from related sacoglossans which do not contain chloroplasts.

Apart from elysioid sacoglossans, there are scattered reports in the literature of certain protozoans with 'chloroplast-like' symbionts. The best known example is *Mesodinium rubrum*, a flagellate which contains chloroplasts of a cryptomonad alga, accompanied by a number of other organelles of algal origin.

4. Miscellaneous associations

There are a few reports of algae other than those described above entering into symbiosis with animals. In some cases the descriptions are inadequate to permit firm conclusions on the identity of the alga, or even as to the permanence of the associations.

However, there is one well-studied animal of considerable interest, *Convoluta roscoffensis*. This is a small, acoelous Platyhelminth growing intertidally on sandy beaches of the North Brittany coast and the Channel Islands. Each animal contains the unicellular alga *Platymonas convolutae*, packed between the subepidermal cells. The alga is a normal free-living component of the phytoplankton, and has four anterior flagella. In symbiosis, it loses its flagella and cell wall, and becomes a membrane-bound cell of flexible shape fitting between host cells.

Eggs of *Convoluta* hatch to produce white, uninfected larvae. These feed voraciously, but only develop into mature animals if infected by *Platymonas* during feeding. Mature animals cease to ingest particulate matter, so that they rely almost entirely on algal photosynthesis for their nutritional needs, apart from such dissolved substances as they can absorb from the sea.

One of the major end products of animal metabolism is uric acid. *Platymonas* in culture will grow on uric acid as a sole nitrogen source. Within the animal, the alga receives uric acid from its host, breaks it down to ammonia, and then synthesizes amino acids, especially glutamine, glycine, and serine, using carbon skeletons derived from photosynthesis. These neutral amino acids are then released back to the animal. This tight recycling of nitrogen is reminiscent of the situation on coral reefs.

Young larvae can be experimentally infected with algae closely related to *Platymonas convolutae*, although the association takes longer to establish. If such animals are then exposed to the natural symbiont, it will infect and entirely replace the foreign symbionts over a period of 5—7 days.

III. Nitrogen Fixing Symbionts

Biological fixation of nitrogen is a property exclusively restricted to prokaryotes. There is very little combined nitrogen in the sedimentary rocks of the earth's crust, so that most of that in living organisms is originally derived from gaseous nitrogen, with biological fixation a more important source than non-biological. Even today, it is estimated that more nitrogen is fixed in the world each year by micro-organisms than by industrial processes. Because there is a continuous loss of nitrogen back to the atmosphere — especially by microorganisms carrying out denitrification in anaerobic habitats such as waterlogged soils — there is a continuous need for fixation.

More nitrogen is believed to be fixed by symbiotic microorganisms than non-symbiotic in the world. Although such global estimates have to be treated with great caution, it illustrates that, quantitatively, symbiotic nitrogen fixation is much more important than symbiotic carbon fixation. A possible explanation for this may lie in the nature of the enzyme responsible for nitrogen fixation, nitrogenase. This enzyme is very sensitive to oxygen, yet requires substantial amounts of energy (much more readily produced by aerobic than anaerobic processes). It is also readily inhibited by the end product of fixation, ammonia. If a microorganism can form a symbiotic association such that the host can supply energy, create a suitably anaerobic local environment for the microbe, and remove the products of fixation quickly, then fixation can proceed more rapidly than in non-symbiotic situations. By contrast, there is no evidence that carbon fixation is more rapid in symbiotic than non-symbiotic algae.

There are three main kinds of nitrogen-fixing symbionts: bacteria, actinomycetes, and blue-green algae.

A. Bacteria

1. Rhizobia and root nodules

Rhizobium occurs in nodules on the roots of many legumes, but has also been found in one non-leguminous plant, *Trema aspera*, a member of the *Ulmaceae*.

Rhizobium occurs free-living, but cannot fix nitrogen when it is existing as a natural member of the soil population. Indeed, it is only recently that is has been induced to fix nitrogen in culture. The sequence of events in the infection of roots and formation of an active, nitrogen-fixing nodule reveals a complex series of interactions between symbiont and host.

Initially, free-living *rhizobia* aggregate around the root hairs, presumably in response to an unidentified attractant in root exudates. The root hair often curls, and it has been suggested (though not proved) that this is because the bacteria secrete the plant hormone auxin. The host root then secretes extracellular polygalacturonase, which weakens the external plant cell wall adjacent to the bacteria, and it invaginates to form a structure called the infection thread. Bacteria become enclosed in the infection thread, which moves inward through the root cortex; at the same time, cortical cell division is stimulated to form the bulk of the eventual nodule tissue. The wall of the thread is cellulose, and entirely of host plant origin. The infection thread may branch and spread through the developing nodule tissue, the branches terminating in host cells which have previously become polyploid (usually tetraploid if the host is diploid). By means not entirely clear, the bacteria become deposited in the host cell where they become enclosed in vesicles or host membranes. They may undergo several divisions, but then swell to about 40 times the volume of the original bacterium to become 'bacteroids', often with simple branches to give 'Y' or 'X' shapes. Finally, the host tissue synthesizes a pigment, leghaemoglobin, which colours the nodule pink. Only if this is formed can the nodules fix nitrogen. The function of leghaemoglobin is probably to combine with molecular oxygen so that nitrogenase is not inhibited, but the combined oxygen becomes available for oxidative metabolism in the host cytoplasm near the bacteroids.

Because no convenient radioactive isotope of nitrogen is available, and because of the morphological complexity of the nodule, detailed studies of the transfer of fixed nitrogen from symbiont to host have so far proved impossible. However, host plant photosynthate is clearly essential for maximal nitrogen fixation. In soyabean, the supply of photosynthate is the major limiting factor in nitrogen fixation under natural conditions. If the atmosphere is artificially enriched with CO_2, net photosynthesis is stimulated and nitrogen fixation is increased fivefold — partly by increased nodule mass, but partly also by increased

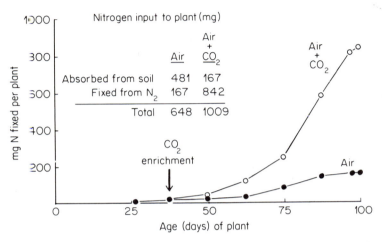

Figure 2. Effect on nitrogen fixation by soya beans of enriching air with CO_2 (800–1200 p.p.m.). Note that uptake of combined nitrogen from the soil is reduced as fixation is increased in plants growing in enriched air. Reproduced by permission of L. Muscatine.

activity of nodules. Photosynthate presumably provides both energy and carbon skeletons for the incorporation of fixed nitrogen. By analogy with free-living, nitrogen-fixing bacteria, ammonia produced by fixation is believed to be immediately assimilated within the bacteroid by glutamine synthetase to form glutamine, which is then released to the host plant. The host nodule tissue contains 'transfer cells' which are normally associated with rapid fluxes of solutes across tissues.

2. *Klebsiella and leaf nodules*

Some tropical plants of the genera *Psychotria*, *Pavetta*, and *Ardisia* have leaf nodules packed with the facultative nitrogen fixing anaerobe, *Klebsiella*. The bacteria occur extracellularly in subepidermal leaf protuberances. The contribution of these nodules to host nutrition has not been studied.

3. *'Associative' symbioses*

Under certain conditions, much higher rates of nitrogen fixation occur in the rhizosphere of some plants than in the rest of the soil. For example, in some habitats, roots of maize and other tropical grasses are closely associated with nitrogen fixing-bacteria of the genus *Spirillum*, and it explains how these crops can grow without added fertilizer in poor soils. Neither the morphological nor the biochemical aspects of the interaction have been studied, but the bacteria presumably depend on a supply of organic substrates from root exudates in a manner analagous to the dependence of *Rhizobium* on host photosynthate. The amount of organic matter released by roots may be appreciable, especially if the continuous sloughing of root-caps is taken into account.

4. Bacteria associated with animals

There is a variety of reports of nitrogen-fixing activity in the intestinal flora of both ruminant and non-ruminant mammals, and *Klebsiella* is known to be involved in the nitrogen fixation found in human and swine faecal matter. However, there is no evidence that nitrogen fixers are obligate components of the flora, or that they make an important contribution to the nitrogen nutrition of their hosts. Although the highly anaerobic atmosphere of the rumen would favour nitrogenase activity, the amount of gaseous nitrogen present is very small.

On the other hand, the recent discovery of substantial nitrogen-fixation activity in termite guts may be significant, since the diet of these animals is often very low in combined nitrogen. As with the rumen, the termite gut is anaerobic, but gaseous nitrogen is present. The organism responsible for fixation has not yet been identified.

B. Actinomycetes and Non-legume Root Nodules

Apart from legumes, plants from 14 genera of angiosperms scattered through eight families may develop nitrogen-fixing nodules on their roots. Some of them are important components of natural ecosystems on poor soils. For example, *Casuarina* spp. are particularly important on coastal soil and dune systems throughout tropical Asia, Australia, and the Pacific Islands.

The nodules are often distinctive in appearance, and vary in size from a few millimetres to over 10 cm in diameter. Generally, they form as modifications of lateral roots. The organism responsible for nitrogen fixation closely resembles an actinomycete, though the crucial test of reinfection with isolates has never been carried out unequivocally. Experimental infections of roots can be carried out using crushed nodules from other plants.

Within the nodules, the microbes are restricted to the cortex, and are packed in vesicles in large cells which are devoid of contents. No haemoglob n-like pigments occur.

C. Blue-gree algae

Nitrogen fixation in most blue-green algae occurs in heterocysts, empty-looking cells with refractive walls which are larger than their neighbouring vegetative cells. Heterocysts contain nitrogenase, but lack photosystem II, the oxygen-yielding reaction of photosynthesis. They possess a functional photosystem I, which presumably supplies ATP essential for nitrogenase activity. Carbon skeletons and perhaps additional energy are supplied to the heterocysts by adjoining vegetative cells.

Blue-green algae enter into symbiosis with representatives of all main plant groups, as well as certain protozoa, sponges, ascidians, and an echiuroid worm. They are either intracellular or extracellular. Intracellular forms may be either short filaments or single cells — often of reduced morphology and termed 'cyanelles'.

Approximately 8 per cent of lichen species contain blue-green algae, and as described in Section II.A.3, most of the fixed nitrogen rapidly passes to the fungus as ammonia. In most of these lichens, blue-greens are the only algal symbionts, but there is a small group in which the main symbiont is green, and the blue-greens are restricted to special structures, termed cephalodia, usually on the surface of the plant. Where blue-greens are the only algal symbionts, the filaments contain a much lower proportion of heterocysts (3.3 per cent of cells) than algae in the free-living state (about 15 per cent), but rates of nitrogen fixation are almost comparable, implying either very efficient heterocyst fixation or some vegetative cell fixation. In cephalodia, the proportion of cells which are heterocysts increases to 40 or 50 per cent with a fivefold increase in fixation.

Heterocyst frequency is also high in all associations involving green plants, but the frequency drops sharply if the symbionts are isolated into culture. Special structures are developed by the host to house the symbionts (usually either *Nostoc* or *Anabaena*). In the bryophytes *Anthoceros* and *Blasia*, and in the water fern *Azolla,* the symbionts occur in special cavities. In *Blasia*, filamentous outgrowths from the host into the cavities are developed, as if to facilitate the interchanges of material.

About one third of the species in the gymnosperm order Cycadales have root nodules, coralloid masses superficially resembling some of those on non-leguminous angiosperms, such as Alder. Nodules near the surface of the soil are packed with blue-green algae, and have fixation rates comparable to Alder nodules. *Gunnera*, a member of the *Haloragidaceae*, is the only angiosperm with blue-green symbionts.

These are borne in wart-like swellings at the base of the leaves or stem.

In *Azolla, Blasia,* and *Gunnera,* the symbionts have substantially reduced photosystem II activity, and also have reduced CO_2 fixation. These reductions are greater than can be accounted for simply by increase in heterocyst frequency. In *Blasia* and *Anthoceros,* there is a substantial transfer of photosynthetically fixed carbon from host to symbiont, and a movement of fixed nitrogen in the form of ammonia in the opposite direction.

Hence, in association with photosynthetic green plants, the CO_2 fixing capacity of the symbiont is reduced, while the N_2 fixing capacity is increased. In lichens this occurs in cephalodia, but in those where blue-green algae are the only symbionts, CO_2 fixing capacity remains high.

IV. Digestive Tract Symbionts

A. Introduction

The digestive tract of animals usually contains a rich microflora of organisms specifically adapted to this specialized habitat. Hosts are assumed to benefit since animals artificially freed of their microflora rarely thrive as well as the naturally infected controls. The reasons for this are not always clear, and may be a complex of minor factors. However, there is one type of situation where the advantage to the host

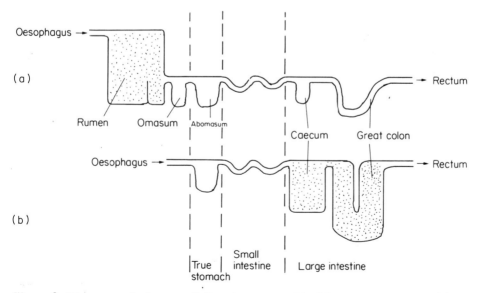

Figure 3. Diagrammatic layout of digestive tracts of herbivorous vertebrates: (a) a cow, in which main microbial fermentation is pregastric and in a special organ, the rumen, and (b) a horse, in which microbial fermentation is post-gastric. Main regions of fermentation are stippled

is of obvious and major importance: herbivorous animals which rely on their internal microflora to break down otherwise indigestible plant material.

In terrestrial ecosystems, carbon fixed in photosynthesis by green plants accumulates mainly in cellulose and lignin. Ability to digest these compounds is therefore essential to herbivores. It is paradoxical that vertebrates, which provide the dominant herbivores in most terrestrial ecosystems, do not possess cellulase or xylanase. Instead their alimentary tracts have become successfully modified to create one or more large fermentation chambers to house microorganisms which can break down cellulose and lignin. Unlike the symbioses described in preceding sections where a host organism associated with one specific microbe, herbivorous animals form an association with a very complex and diverse population of microbes.

The fermentation chamber may either precede the animal's own digestive action or follow it. Where it precedes, it may be in a compartmented stomach (e.g. camel, hippopotamus) or in a separate, clearly defined organ, the rumen. Where the chamber follows, it may be the caecum, colon, or large intestine which is modified, but wherever the chamber is located, the same three principles for its operation apply. Firstly, fermentation is anaerobic, so that ingested plant material is only partially broken down to products (usually volatile fatty acids, especially acetic, propionic, and butyric acids) which then enter the blood stream to be metabolized by the host. Secondly, the fermentation chamber has to be adequately large; in the cow, the rumen may be up to 100 l in capacity; in a non-ruminant such as a horse, fermentation occurs in the caecum and great colon whose combined volume may be up to 120 l (the stomach is only 10 l). Thirdly, since microbial breakdown of solid substrates is relatively slow, the flow of food through the chamber has to be regulated so that material can spend a relatively long time (8–12 h) there.

B. Rumen Digestion

Rumen digestion has received particularly close study because ruminants include animals of major domestic importance such as sheep and cattle. The rumen contains a dense culture of protozoa ($10^5 - 10^6$ per ml) and bacteria ($10^{10} - 10^{11}$ per ml), which is continuously mixed by contractions of the wall at 1–2 minute intervals. There is a large input of saliva (equivalent to 1–3 times the volume of the rumen per day) containing 100–140 mM bicarbonate and 10–50 mM phosphate. This acts as an alkaline buffer to neutralize the fatty acids produced by fermentation and keep the pH favourable (approximately 6.5) for further microbial action. The saliva is removed in the omasum, the next chamber in the alimentary tract after the rumen, and is then recirculated. Much of the body water of a cow is in saliva.

The rumen atmosphere is highly anaerobic, and consists of CO_2 (60–70 per cent), CH_4 (30–40 per cent) together with traces of H_2 and H_2S. Most rumen microorganisms are extreme obligate anaerobes, and it is only relatively recently that satisfactory techniques for studying them have been developed. Various groups with different functions can be recognized, though a complete picture of their complex interrelationships *in vivo* is not yet available. As might be expected, various cellulose-digesting bacteria occur, but surprisingly, they form only 1–2 per cent of the rumen population although cellulose typically comprises about 30 per cent of the food intake. Bacteria capable of metabolizing a variety of other plant materials, such as xylans, lipids, pectins, and hemicelluloses, also occur. Many of these bacteria can also ferment the sugars resulting from polysaccharide breakdown to produce the volatile fatty acids which pass to the host.

There remain at least two other major groups of rumen micro-organisms whose function is less clear: methanogenic bacteria and protozoa. Methanogenic bacteria are present in all ruminants, and they produce methane by combining carbon dioxide and the free hydrogen produced by various biochemical processes. The amount of methane produced is probably considerable, bearing in mind that a cow may release over 1000 l of gases a day by belching. There is some experimental evidence that a rise in the amount of free hydrogen in the rumen may inhibit bacteria involved in hydrogen-yielding processes.

Rumen protozoa are a highly specialized group consisting chiefly of ciliates. Some can produce cellulose, and others convert sugar to starch, but their overall numbers do not indicate that they are quantitatively important. They may ingest bacteria, and it has been suggested that they play a role in regulating bacterial turnover. However, since ruminants can be reared free of protozoa, they are not essential.

All of the protein in the food entering the rumen is rapidly converted to ammonia. This may either be absorbed by the microbes to be converted to their own amino acids and proteins, or it may pass into the blood stream and be converted to urea in the liver. The urea may reenter the rumen to be broken down to ammonia, or excreted. The rumen continuously produces a substantial surplus of microbes, and these pass to the abomasum (true stomach) where they are digested and the host recovers usable protein and amino acids.

Further fermentation may occur after the abomasum in the distal portion of the small intestine and caecum, where cellulose which escaped rumen digestion is broken down. The products of this fermentation are small compared to those from the rumen.

C. Non-ruminants

In general, the processes of microbial fermentation in non-ruminants are broadly similar to ruminants, except that, presumably, there is no facility for host digestion of surplus microbes if the fermentation chamber is after the true stomach. On the other hand, there could be increased autolysis of microbes to yield amino acids.

D. Insects

Some invertebrate herbivores such as molluscs can produce cellulase, but the largest group of animals, the insects, cannot. Insects feeding primarily or exclusively on wood and cellulose rely on fermentation by microbes housed in various types of chamber or distensions of specific regions of the alimentary tract. In wood-destroying termites, the fermentation chamber is a relatively large paunch in the hind-gut which harbours a dense population of flagellate protozoans as well as bacteria, spirochaetes, and rickettsias. In the intermoult period, the biomass of protozoa is one third of that of the whole animal. Unlike the situation in ruminants, the principal cellulolytic agents are protozoans. In other respects, however, recent studies indicate some remarkable similarities with the rumen. The termite hind gut is anaerobic, and the protozoans break down glucose to H_2, CO_2, acetic, and other fatty acids. Methane production also occurs, and at a rate greater than ruminants if compared on a body weight basis.

V. Mycorrhizas

'Mycorrhiza' is the term applied to a wide range of associations between diverse fungi and the roots, rhizomes and thalli of higher plants. Only one species of fungus usually forms a mycorrhizal association with any particular root. The fungal mycelia are intimately associated with the host root and also ramify extensively into the soil, functioning as an extension of the root system and playing an important role in absorbing nutrients.

In the symbiotic associations described previously, the microbial symbionts all had key biochemical properties not possessed by the host. In mycorrhizas (other than those of the *Orchidaceae* and possibly the Ericales), the symbiont extends rather than complements the properties of the host. Four main kinds of mycorrhizas are recognized: ectotrophic (or 'sheathing'), vesicular—arbuscular, orchidaceous, and ericaceous.

A. Ectotrophic and Vesicular—Arbuscular Mycorrhizas

These two types are similar in function though different in

morphology. Ectotrophic mycorrhizas mainly occur on North Temperate forest trees, including conifers, and beech, oak, birch, and their relatives. The feeding roots become completely sheathed in a thick mantle of compact fungal tissue. A branching mycelium enters the cortex of the root, but does not normally penetrate cells. The fungus is also directly connected to an extensive mycelium in the soil. The host root becomes somewhat modified in degree of branching, cell-shape, and meristematic activity so that infected roots are recognizably different in appearance from uninfected, and root hair development is suppressed. The fungal components of ectotrophic mycorrhizas are mainly basidiomycetes such as agarics and boleti, etc. — indeed, some of the fruiting bodies commonly found in woodlands are mycorrhiza formers.

Vesicular—arbuscular is undoubtedly the commonest type of mycorrhiza, being found in almost all families of flowering plants as well as in many ferns. It also occurs on trees in the South Temperate and Tropical regions. The fungus does not form an external sheath, but ramifies through the root cortex. There is usually extensive intracellular penetration of unspecialized parenchymatous tissue of the cortex, with coils of hyphae often forming in cells. Branches of the main hyphal system may also penetrate cells to produce dichotomously branched haustoria — the 'arbuscules'. Intercalary or apical swellings occur on the main hyphae — the 'vesicles'. The infection is named after the abundant occurrence of vesicles and arbuscules in the host tissue. Only one family of fungi, the *Endogonaceae* (Phycomycetes), provide the symbionts for these associations.

Mycorrhizal infection is not universal for either ectotrophic or vesicular—arbuscular types. Plants on soils rich in mineral nutrients may be quite uninfected, but on poor soils (the more usual situation in natural conditions) infection is normal, especially if roots have high carbohydrate levels.

A variety of experiments have compared infected and uninfected plants. On poor soils, infected plants show greater vigour, and have a higher mineral content than the uninfected. On rich soils, the difference is much less striking, and may even be not apparent. If poor soils are labelled with trace amounts of $^{32}PO_4$, infected plants acquire much more label than uninfected, because the mycorrhizal fungus is able to tap a much greater volume of soil than the roots of uninfected plants, even though the latter have root hairs. Although uptake into the mycelium is substantial and rapid, release from fungus to root is slower, and the system functions as if the fungal component can act as a reservoir of minerals for the host plant.

There is a reverse flow of photosynthetically fixed carbon from host to fungus, and this may be why infection is favoured by roots with a high carbohydrate content. Photosynthate movement has been extensively studied in the ectotrophic mycorrhiza of forest trees, where it is

Table 1. Examples of experimental results showing effects of mycorrhizal infection on host plant growth

Type of infection	Control soil		Soil with added phosphate	
	Uninfected	Mycorrhizal	Uninfected	Mycorrhizal
Ectotrophic				
Pinus strobus seedlings after 1 year's growth in Prairie soil				
Dry wt (mg)	303	405	—	—
Nitrogen content (% dry wt)	0.85	1.24	—	—
Phosphorus content (% dry wt)	0.074	0.196	—	—
Potassium content (% dry wt)	0.425	0.744	—	—
Vesicular–arbuscular				
Maize plants on poor soil at harvest. 'Mycorrhizal' inoculated with *Endogone* as seedlings				
Length of ear (cm)	14.0	22.0	19.5	22.7
No. grains/ear	31	354	279	321
1000 grain wt (g)	2.4	19.8	23.7	20.9
Vesicular–arbuscular plus Rhizobium				
Stylosanthes guyanensis in poor soil (pH 6.5) after 10 weeks' growth. All plants previously inoculated with *Rhizobium*, 'mycorrhizal' also with *Endogone*.				
Sterile (irradiated) soil				
Dry wt (mg)	93	580	310	610
Total P per plant (µg)	77	1077	383	1791
No. *Rhizobium* nodules per plant	0	34	7	37
Unsterile soil				
Dry wt (mg)	—	—	337	343
Total P per plant (µg)	—	—	592	748
No. *Rhizobium* nodules per plant	—	—	35	31
Mycorrhizal (i.e. including infection from indigenous fungi in the unsterile soil)	—	—	55%	70%

substantial: in some Swedish forests, the amount of fixed carbon passing to the fungus is equivalent to 10 per cent of total timber production. The carbon moves mostly as sucrose, and on entering the fungus is converted to mannitol, trehalose, and glycogen — fungal products which the host cannot utilize. Unlike most non-symbiotic basidiomycetes, the fungi of ectotrophic mycorrhizas cannot break down cellulose and lignin.

Many legumes associated with *Rhizobium* may also be infected with vesicular—arbuscular mycorrhizas. Such plants show greater vigour, more nodulation and more nitrogen fixation than those without mycorrhizal infection, especially on poor soil.

B. Orchid Mycorrhizas

These mycorrhizas are distinct in that during at least the early development of the host, organic compounds move from the fungus to the developing plant. Orchids produce large numbers of very tiny seeds, which may contain only 8—100 cells and with very little food reserve. Unless they become infected with a fungus soon after germination, they usually do not develop further. During infection of the embryo, the fungus penetrates and forms coils inside cortical cells. The hyphae later swell, become disorganized and are then digested by the host; the cycle of infection and digestion may be repeated in the same cell. Tissues other than the cortex seem resistant. The fungus is always a basidiomycete, e.g. *Fomes, Corticium, Armillaria.*

The developing embryo spends an appreciable period underground, which in some tropical species may be as long as 1 or 2 years. During this time the host presumably relies heavily on the fungus for nutrients; experiments have shown that the fungus can break down cellulose and that some of the products pass to the host. Hence, carbon compounds flow in the opposite direction to the other mycorrhizas described above.

Some, such as the bird's nest orchid, remain non-photosynthetic even in adult life, but most become green and autotrophic. The fungal infection persists in the adult state, but the interaction with the host at this stage has never been investigated.

C. Mycorrhizas of the Ericales

Mycorrhizal infections of the Ericales (heather and related plants) have some resemblance to ectotrophic forms, and some to orchidaceous. A distinct external sheath of mycelium is formed on the host root which may be almost as well developed as in ectotrophic mycorrhizas. However, unlike these, there is extensive intracellular penetration of cortical cells with the fungi sometimes even forming coils within the cells. As in orchidaceous types, there may be repeated digestion of the hyphae by the host cell. The fungal symbiont is always a septate form which has never been conclusively identified.

The functioning of these mycorrhizas is not at all well understood. Many species of Ericales grow on soils deficient in minerals, especially nitrogen. In the early part of this century it was claimed that the fungus could fix nitrogen, but this is now known to be incorrect. On the other hand, it has been demonstrated that infected plants accumulate more ammonia from the soil than do uninfected, presumably because, like ectotrophic mycorrhizas, they can tap a larger volume of soil.

There are some non-photosynthetic members of the Ericales, such as *Monotropa*. This plant shares a common mycorrhizal fungus with trees like spruce and pine, and it has been shown that sugars will pass from the tree to *Monotropa* along the fungal mycelium. In this case the relationship of fungus to *Monotropa* is analagous to orchid mycorrhizas in supplying organic substances to the host plant.

VI. Miscellaneous Associations

The diverse collection of remaining associations may be broadly classified into three groups: (a) associations with microbes of known identity where the main selective advantage to the host is from microbial properties other than those discussed previously; (b) associations with microbes of known identity where a selective advantage to the host is demonstrable but not explicable; and (c) associations with microbes of unknown identity.

A. Associations in which Selective Advantage to Host Derives from Known Microbial Properties

Two examples will be mentioned: the complex association of bacteria and spirochaetes with the protozoan *Myxotricha paradoxa*, and the amoeba *Pelomyxa palustris*.

Myxotricha paradoxa is a flagellate protozoan which itself is a symbiont in the anaerobic hindgut of the primitive wood-eating Australian termite, *Mastotermes*. In addition to the large anterior flagella, *Myxotricha* also appears to be clothed in cilia which beat in synchrony. Close examination shows that the 'cilia' are in fact motile spirochaetes attached to the cell surface. These undoubtedly affect the movement of *Myxotricha* since it progresses steadily forward in a single direction, in marked contrast to other flagellates which continuously change direction. At the point of attachment of each spirochaete, there is a second symbiont, a bacterium which appears as if it is clamping the spirochaete to the surface, though its exact function is not known. Further bacterial symbionts occur within the host cytoplasm, whose function again remains uninvestigated. *Myxotricha* was originally considered unique, but other 'motility symbioses' between spirochaetes and protozoans in termite guts have now been described.

Pelomyxa palustris is a giant amoeba which does not possess mitochondria, but has a population of intracellular bacteria which appear to function as mitochondria. This is a rare example of a symbiont replacing an organelle.

B. Associations with a Demonstrable but Inexplicable Selective Advantage to Host

There is a variety of situations in which removal of a microbial symbiont results in decreased host vigour, yet the reason remains obscure. As noted previously, some digestive tract symbionts fall into this category. Frequently, microbial ability to synthesize essential micronutrients, such as vitamins and sterols, is suggested as a possible cause, but it is technically very difficult to produce direct experimental evidence of transfer of such compounds to the host *in vivo*, and other types of interaction between host and symbiont cannot be ruled out.

A remarkable example of the evolution of this kind of symbiosis is presented by a particular laboratory strain of *Amoeba proteus*. This strain became infected by bacteria which were originally very harmful to the host cells. Within 1000 generations of the original infection, the amoebae now grow normally but have become obligately dependent upon their symbionts. The nucleus of an infected amoeba can form a viable cell with the cytoplasm of an uninfected strain only when live symbionts are present. The symbionts are now not culturable, but can exist in the cytoplasm of most strains of *A. proteus*. Clearly, in the infected strain, there has been some loss of host function which has been taken over by the symbiont.

C. Associations with Microbes of Unknown Identity

Particularly in intracellular symbiosis, there is a marked tendency for microbial morphology to become progressively reduced during evolution of the association. The extreme case would be where all that remains of the original symbiont is some of its DNA, either in the form of plasmids or even attached to host nuclear DNA. Indeed, it may no longer be possible to prove that such DNA originated from a foreign symbiont rather than the host's own DNA. Here we will be concerned with less extreme situations where it is still clear that an intracellular object was originally a symbiont, but where morphological reduction has obscured its identity. Two examples will be discussed: gamma particles in *Blastocladiella,* and Blochman bodies in insects.

Blastocladiella emersoni is a water mould forming flagellate zoospores. Apparently identical zoospores can produce either tiny vegetative structures, thick-walled resistant large thalli, or orange or colourless thin-walled thalli. The type of differentiation is correlated with the number of 'gamma' particles in the zoospore. These are horse-shoe shaped particles containing DNA, RNA, and protein, including at least one enzyme, chitin synthetase; however, they lack ribosomes. The particles are photosensitive and undergo a complex morphogenetic cycle involving reversible loss of the horse-shoe shape. Observations of the changes in these bodies and of correlated developments in the host suggest that they are the functional equivalent of Golgi bodies in other eukaryotes, and are particularly responsible for the genesis of host cyst walls. Their apparent involvement in the determination of zoospore developmental pathways is not understood, but indicates how deeply integrated into host cell functioning they have become.

Perhaps the symbionts of greatest importance in this 'unidentifiable' category are the 'Blochman bodies' of insects. With more than 700,000 species, insects are the largest class of living organisms. About 10 per cent of species have symbiotic microbes: these may be identifiable forms in the digestive tract as described in Section IV.D or they may be the unidentifiable 'Blochman bodies' housed in special cells called mycetocytes. An insect may have either or both types.

Mycetocytes may occur as some of the cells surrounding the gut, or they may be associated with Malpighian tubules, or they may aggregate into special organs, mycetomes, in the body cavity. The symbionts are membrane bound and contain DNA. They are the same size as mitochondria but lack internal cristae. Sometimes two morphologically distinct types of Blochman body occur in the same mycetome. While such polymorphism is suggestive of Ricksettias and mycoplasmas, the DNA of Blochman bodies is shorter and smaller (approximately 2×10^6 daltons, compared with 1×10^9 for mycoplasmas and up to 2.5×10^9 for ricksettias). Indeed, there have been suggestions that they may constitute a new group of prokaryotes.

Some Blochman bodies can be suppressed by antibiotics and sulfa drugs, and insects treated in this way are substantially less vigorous than controls. Blochman bodies are transmitted from generation to generation through the egg cytoplasm. It is not known how the egg becomes infected. In some insects, special host cells appear to carry the symbionts to transitional embryonic mycetomes during development, before they become relocated in the fat bodies of the adult.

VII. Conclusions

Many biologists would widen the definition of symbiosis beyond that used in this article, and would include situations where no demonstrable advantage to the host need occur, with the symbiont simply finding an ecological niche in the host without much effect on the latter. Even in the restricted sense of this article, it might be argued that symbiosis is a term covering such a diverse and disjunct collection of separate phenomena that unifying concepts or themes cannot be discerned. From an ecological standpoint, however, it is just as valid to consider that symbiotic microbes are a single group as it is, say, marine microbes.

Indeed, the capacity of a living host to make dramatic responses to infection imposes some unique characteristics upon symbiotic microbes. Further, while a living organism contains all the nutrients needed by a microbe, some of them are often much less easily available than in a non-living habitat, and in so far as intracellular symbionts are concerned, the host cell's own metabolic processes are often potent competitors for nutrient pools.

In an ecological context, a host organism may therefore be considered an extreme habitat, with intracellular location more extreme than extracellular. Non-living extreme habitats (e.g. semideserts) are characteristically colonized by relatively few species, and this is also the case with living habitats. Of the identifiable intracellular symbiotic and 'mutualistic' microbes, only three types are at all common and widespread: *Gymnodinium microadriaticum* and *Chlorella* amongst the algae, and *Rhizobium* amongst bacteria. Even including siphonaceous chloroplasts, the unidentified actinomycete symbionts of non-legume nodules and Blochman bodies of insects, the list is not long. Extracellular symbionts are more numerous, but nevertheless either dominate as the only symbiont in their habitat (e.g. the algae of lichens or the fungus of a mycorrhiza) or become part of so highly specialized a population of symbionts that casually infecting microbes are excluded (e.g. digestive tract symbionts).

Unlike non-living habitats, the rapid response of living habitats to infection imposes substantial additional selective pressures, so that symbiotic microbes may evolve more rapidly. The change of the bacterial symbionts of *Amoeba proteus* from parasitic to mutualistic in less than 1000 generations is testimony to this, as is the evolution of microbes from the identifiable to the unidentifiable condition. Hosts, too, can evolve, and the evolution of composite organs of host and symbiont tissue is common (e.g. legume nodule, mycetome, lichen thallus, rumen, etc.). Considering the problems of synchronizing the growth and differentiation of two separate organisms, the frequency of such composite organs is remarkable.

Finally there is the question of the relevance of symbiotic phenomena to theories that eukaryotic organelles such as chloroplasts and mitochondria arose from symbiotic colonization of ancestral prokaryotes. The ability of symbiotic microbes to become deeply integrated into host cell functioning certainly shows that such a theory is possible. The replacement of mitochondria by bacteria in *Pelomyxa*, and Golgi bodies partially by 'gamma-particles' in *Blastocladiella* make the attraction of the theory seductive, but they do not prove it.

VIII. Further Reading

General

Jennings, D. H. and Lee, D. L. (1975). *Symbiosis. 29th Symp. Soc. exp. Biol.*, Cambridge: University Press

Autotrophic Symbionts

Smith, D. C. (1974) Transport from symbiotic algae and symbiotic chloroplasts to host cells. In *Transport at the Cellular Level, 28th Symp. Soc. exp. Biol.*, M. A. Sleigh and D. H. Jennings (Eds.) pp. 485–520. Cambridge University Press.
Smith, D. C. (1978) Autotrophic endosymbionts of invertebrates. In *Companion to Microbiology*, A. Bull and P. Meadow (Eds.) London: Longmans.
Farrar, J. F. (1978) Symbiosis between Algae and Fungi. In *Handbook of Nutrition and Food*, M. Rechcigl, Ed.: CRC Press.

Nitrogen Fixers

Burns, R. C. and Hardy, R. W. F. (1976) *Nitrogen Fixation in Bacteria and Higher Plants.* Berlin and New York: Springer.
Nutman, P. S. (Ed.) (1976). *Symbiotic Nitrogen Fixation in Plants.* International Biologica Programme Monograph No. 7. Cambridge: University Press.

Digestive Tract Symbionts

Hobson, P. N. (1976) *The Microflora of the Rumen.* Patterns of Progress. Co. Durham, England: Meadowfield Press.
Hungate, R. E. (1975) The Rumen Microbial Ecosystem. *Ann. Rev. Ecology and Systematics*, 6, 39–66.

Mycorrhizas

Harley, J. L (1969) *The Biology of Mycorrhizas*, 2nd edition, London: Leonard Hill.
Marks, G. C. and Kozlowski, T. T. (Eds.) (1973). *Ectomycorrhizas.* New York and London: Academic Press.
Sanders, F. E., Mosse, B., Tinker, P. B. (Eds.) (1975) *Endomycorrhizas*. New York and London: Academic Press.

Bacterial Nutrition

ROGER WHITTENBURY

Department of Biological Sciences, University of Warwick, Coventry, CV4 7AL

I. Preamble

At first sight bacterial nutrition would seem to be a dull topic; indeed it is frequently treated as such — long lists of compounds and elements being divided into groups and subgroups according to the categories of microbes that do or do not thrive on them. Such a treatment, unfortunately too common, only gives the shadow of the story and ignores the dynamic aspects of bacterial nutrition which have evolved as the complex ecosystem embracing all living organisms has developed over aeons of time.

Equally important in the story of bacterial nutrition is the manner in which different species of bacteria are able to compete with each other for the same nutrients. Consequently, this discussion will be in two parts: (1) the categorization of bacteria according to their ability to use different compounds and elements as nutrients — *heterotrophy* and *autotrophy*, and (2) the ways and means whereby bacteria are able to acquire nutrients and so survive in competition for existence with other microbes — *adaptation for survival*.

II. Heterotrophy and Autotrophy

All living organisms require nutrients to grow, function, and multiply. Some require many nutrients, some only a few; some are very flexible in the range of substances they can use as nutrients, some are rigidly inflexible in that they can use only one carbon source and a limited range of inorganic compounds for growth. Certain nutrients are essential for a bacterium to grow, others are useful (in that they spare energy used in synthesis) but dispensable. They serve as energy sources, electron sources (in the case of photosynthetic bacteria using light as an energy source) and/or as building blocks from which the macromolecules of the cell and its products are made. They may be required by microbes in large amounts (macronutrients) or in small amounts (micronutrients); requirements differ from species to species.

Although all microbes are composed, in the main, of the same elements (the major ones are C, H, O, N, S, P, and Mg), they vary considerably in their ability to acquire these elements from the compounds which contain them. This variation, besides being reflected in which compounds are used and not used by the different types of bacteria, is also reflected in the variety of mechanisms by which bacteria take up these compounds and transform them into energy and substances used in cell synthesis. This latter activity is a major subject in itself and will not be dealt with in any detail here.

All the organic compounds formed in nature are utilizable by microbes, as are many of the inorganic compounds (both in a reduced and oxidized state) formed in biogeochemical cycles. That some substances accumulate in vast reserves (coal, oil, natural gas, and peat) does not imply their immunity to microbial attack: environmental conditions are responsible here; once in a favourable environment which usually means one containing oxygen, these substances are rapidly utilized and their oxidized products (carbon dioxide and minerals) returned to the biosphere for recycling through other organisms. The only organic materials known to accumulate and to be impervious to microbial attack are man-made materials such as some plastics and pesticides.

However, it is possible that even some of these intractable compounds will eventually be made amenable to microbial attack by a combination of genetic engineering techniques and mutation (evolution) of enzymes in certain bacteria. For instance, Professor Patricia Clarke and her colleagues at University College, London, have demonstrated the feasibility of 'training' *Pseudomonas putida* to use organic compounds — in this case certain amides — which do not occur naturally.

The initial step in the nutritional categorization of bacteria is the division of them into two camps — those which are *heterotrophs* and those which are *autotrophs*.

A. Heterotrophs

Heterotrophs (*chemo-organotrophs* and *organotrophs* are alternative terms) are those bacteria which require organic compounds for growth. They may require them as macronutrients or, in extreme cases, they may only require one as a micronutrient. This is obviously an unsatisfactory start to the nutritional categorization of microbes in that such an extremely wide range of nutrient behaviour is encompassed within the one bracket. As will be seen in the subsequent discussion about autotrophy, a definition of heterotrophy based on whether or not an organic compound is utilized as a nutrient is becoming too imprecise to be really meaningful. However, common sense tends to prevail in these situations and most microbiologists accept that a heterotrophs and it is possible to subdivide them into groups on the basis prime carbon source, transforming it into cellular components: such an organism is *Escherichia coli*. The great majority of bacterial species are heterotrophs and it is possible to subdivide them into groups on the basis of their nutritional requirements, but these groups cut across both species and genus barriers in many cases, serving only to illustrate nutritional diversity of microbes. Some genera are founded on heterotrophic nutritional characteristics (e.g. the lactic acid bacteria such as the genera *Pediococcus*, *Streptococcus*, *Lactobacillus*, *Leuconostoc*, and *Aerococcus*), others are not — so it is not possible to treat nutritional behaviour grouping as a logical extension of bacterial classification.

Nevertheless, it is probably useful at this point to describe three major categories of heterotrophs — *oxidative, fermentative,* and *facultative* — if only because it is necessary to point out that the traditional groupings of organisms into aerobic, anaerobic and facultatively anaerobic categories is now seen to blur similarities on the one hand and to maintain dissimilar organisms within one group on the other hand.

1. Oxidative Bacteria

These are mainly those microbes able to use both O_2 and NO_3^- (anaerobically) or only O_2 as an electron acceptor during oxidative phosphorylation. Some strict anaerobes are in this category, e.g. *Desulfovibrio desulfuricans* and the endospore-forming *Desulfotomaculum* species which use SO_4^{2-} as an external acceptor in a cytochrome mediated oxidation, some methane producing bacteria which use CO_2 as an external electron acceptor, and possibly some Gram-negatives that use fumarate as an electron acceptor when they have formed cytochromes on the provision of haem compounds.

The oxidative bacteria, by far the largest group of heterotrophs, embrace microbes of varying capacity; some are able to use in excess of 100 organic compounds as combined carbon and energy sources (e.g. some *Pseudomonas* species and *Thiobacillus* A_2 — which can also grow autotrophically), some can use only one organic compound as an energy source (e.g. *Coxiella burnetii* can only oxidize glutamate). The pseudomonads able to use a great number of compounds introduce another facet into this story — the plasmid. Genetic ability to utilize some of these compounds (e.g. *n*-alkane oxidation) has been found to reside in plasmids which may move from strain to strain or species to species; consequently, an individual strain of a species may be found to fluctuate in its powers of organic compound utilization, depending on whether it gains or loses plasmids within its environment. An outcome of this fact is that nutritional versatility may be more aptly defined as being a population characteristic rather than a fixed characteristic of a particular strain of the species.

Within the oxidative group it is possible to create nutritional subcategories, for instance the parasites of eukaryotic cells (e.g. *Rickettsia* species), the parasites of bacterial cells (e.g. *Bdellovibrio* species, which bore through the cell wall of Gram-negative bacteria and grow and multiply within the periplasmic region), the *Myxobacter* species (which lyse Gram-negative bacteria and grow on the lysed contents), the symbionts, such as *Rhizobium* and some cyanobacteria species, and the co-oxidizing bacteria such as *Methylomonas methanica*. Some of these particular modes of nutrition will be examined later in a discussion on how nutrients are obtained.

2. Fermentative Bacteria

These organisms are those which only obtain their energy via substrate level phosphorylation (e.g. as in the fermentation of glucose to acetic and butyric acids, CO_2 and H_2 by *Clostridium butyricum*). Most microbes of this sort are strict anaerobes, with the lactic acid bacteria (e.g. *Pediococcus, Lactobacillus, Leuconostoc, Streptococcus,* and *Aerococcus*) being an exception in that many species can grow as well in the presence of oxygen as they do in its absence.

Again, as with the oxidative bacteria, this group contains a wide variety of nutritional types. Some bacterial species, such as the lactic acid bacteria, require a large range of growth factors (i.e. amino acids, purines, pyrimidines, and vitamins), including small peptides, and only use the major carbon source, such as glucose, as an energy source. Other species require less assistance of this type to grow, but a general tendency among obligate fermentative bacteria is a requirement for growth factors. This is not surprising really in that the tricarboxylic acid cycle does not operate in these organisms as it commonly does in oxidative bacteria, hence a lack of ability to synthesize certain carbon skeletons for biosynthesis. *Escherichia coli*, when growing fermentatively, exhibits a requirement for growth factors not required when it grows oxidatively; this can be partially explained by the failure of 2-oxoglutarate dehydrogenase to be induced under anaerobic conditions with the consequent incomplete functioning of the tricarboxylic acid cycle.

Many fermentative bacteria utilize carbohydrates, some do not (e.g. *Peptococcus* species use only amino acids, *Veillonella* spp. use only lactic and succinic acids, while some clostridia obtain energy only from a coupled oxidation—reduction reaction involving a pair of amino acids. One amino acid serves as the electron donor, the other as the electron acceptor—the so-called Stickland reaction).

3. Facultative Bacteria

These microbes are able to grow both by oxidative and fermentative mechanisms. Under aerobic conditions both mechanisms may operate, the latter being followed by the former in an expression of the *diauxie* phenomenon. An example of aerobic diauxie is the utilization of glucose [when present in the medium at a concentration of over 2.0 per cent (w/v)] by *Pseudomonas* species. Glucose is first utilized via the Entner—Doudoroff pathway mechanism and the resultant pyruvate accumulates until the glucose level is considerably reduced. Once this has occurred, repression of the oxidation mechanism is reversed, cytochromes and other relevant enzymes and electron carriers are synthesized, and the pyruvate is oxidized via the tricarboxylic acid cycle. Again, this example reinforces the earlier viewpoint that oxidation and fermentation should not be regarded as being respectively aerobic and anaerobic activities, although most oxidative bacteria are aerobes and most fermentative bacteria are anaerobes.

B. Autotrophs

Very few bacterial species are capable of autotrophic growth, yet, in contrast to heterotrophy, a great deal of argument and discussion about the concept of autotrophy has appeared in print. Clearly, the idea that some bacteria can survive and grow solely on inorganic compounds has proved intriguing to microbiologists since the time Winogradsky first described the phenomenon. At present the definition commonly accepted as embracing autotrophic bacteria is that they oxidize inorganic compounds for energy (or as a source of electrons for photophosphorylation) and obtain their carbon from CO_2 fixed via the Benson—Calvin pathway. Two major groups fit this description (Table 1 and 2) the *chemolithoautotrophs* and the *photolithoautotrophs*. Often these two groups are respectively referred to as

Table 1. Chemolithoautotrophs (can use carbon dioxide as sole carbon source and inorganic compounds as energy source)

Energy substrate	Examples	2-Oxoglutarate dehydrogenase present (+) or absent (−)	Obligate (O)/ facultative (F) growth potential
Hydrogen (aerobic)	*Paracoccus denitrificans*	+/−[a]	F
Sulphur and	*Thiobacillus thiooxidans*	−	O
reduced sulphur	*T. thioparus*	−	O
compounds	*T. neapolitanus*	−	O
	T. novellus	+	F
	T. acidophilus	?	F
	T. A2	+/−[a]	F
Ferrous iron	*Thiobacillus ferrooxidans*	+/−[a]	O/F?
	Gallionella sp.		?
Ammonia	*Nitrosolobus multiformis*	−	O
Nitrite	*Nitrobacter agilis* (Winogradsky)	+	F
Hydrogen (anaerobic)	*Methanobacterium thermoautotrophicum*	?	O
	Methanobacterium formicicum	?	F

[a] Inducible in those strains growing on organic compounds but negative when growing autotrophically.

Table 2. The photosynthetic bacteria (photolithoautotrophs)

Organism	Comments on nutrition
1. The Cyanobacteria	All possess photosystems I and II and, therefore, are able to use H_2O as electron source and are aerobic when growing photosynthetically. Growth can occur in the dark at the expense of storage compounds; most notably N_2 can be fixed by some strains in the dark — a property restricted to this group of photosynthetic organisms. Some species are able to use a reduced S compound as an electron donor (analogous then to purple S bacteria). Some grow heterotrophically in the dark as part of a symbiont relationship with cycads (the cyanobacteria grow on root surfaces well down in the soil at the expense of nutrients released from the roots and are presumed, in return, to provide the plant with N compounds).
2. The *Rhodospirillaceae*	These organisms (e.g. *Rhodopseudomonas* and *Rhodospirillum* species) possess only a photosystem of the Type I type and cannot use H_2O as an electron donor. All fix N_2 photosynthetically. Organic compounds and H_2 are used by all species as electron donors (they are, therefore, facultative autotrophs in that they can grow both photoautotrophically and photoheterotrophically). They can also grow in the dark as chemoheterotrophs both aerobically and anaerobically. Photosynthesis, however, is a strictly anaerobic process. Recently some strains have been shown to use H_2S as an electron donor when present at very low concentrations.
3. The *Chromatiaceae* and the *Chlorobiaceae*	These organisms (e.g. *Chromatium vinosum* and *Chlorobium thiosulphatum*) are strict anaerobes which possess only a photosystem of the Type 1 type. H_2S and other partially oxidized S compounds are the major electron sources. Some species can grow heterotrophically (e.g. on pyruvate) in the dark. All fix N_2 photosynthetically. It has been claimed that a reversed tricarboxylic acid cycle mechanism of CO_2 fixation operates in place of the Benson—Calvin pathway in some of the *Chlorobiaceae*.

chemo- and *photolithotrophs* but such a description refers only to an ability to use inorganic compounds as a source of energy or electrons — which is also a property of some heterotrophs with an obligate requirement for organic compounds as their major carbon source.

All photosynthetic bacteria are capable of autotrophic growth and most of them can also grow heterotrophically. Exceptions are some cyanobacteria (formerly called blue-green algae and sometimes called blue-green bacteria), and some sulphur-purple and sulphur-green photosynthetic bacteria (families *Chromatiaceae* and *Chlorobiaceae*). Detailed comments on these organisms are given in Table 2. A commonly held view is that the cyanobacteria are a group of organisms apart from other photosynthetic bacteria (see Table 2). However, there are properties of some cyanobacteria which bridge this gulf; the H_2S-utilizing powers of some strains growing anaerobically with, presumably, only a photosystem type I mechanism functioning, and the anaerobic photosynthesis of the differentiated heterocysts which lose photosystem II and apparently function mainly as nitrogen fixers in the service of adjacent undifferentiated cells in some filamentous cyanobacteria.

Examples of bacteria presently considered to be chemolithoautotrophs (bacteria growing aerobically or anaerobically in the dark solely on inorganic compounds) are listed in Table 1. Although carbon dioxide is used as a sole carbon source by these organisms there is no evidence that the methane producing bacteria fix CO_2 via the Benson—Calvin pathway, a property accepted nowadays to be a property of autotrophs, though not originally, of course, as this pathway had not been described. Inconsistencies of this sort which have not been clarified in reviews on the topic of autotrophy have bedevilled the subject for some time as will become evident in the following discussion.

The first question to consider is — why are some autotrophs obligate? A subsidiary part of the same question, frequently coupled with this one is: is there such a thing as an obligate autotroph? This latter element of the question can be dealt with first as it really is a matter of nutritional semantics. A literal interpretation of obligate autotrophy is that no organic compound can serve as a carbon source to an obligate autotroph. This has proved not to be the case: all such organisms examined have been found able to utilize substances such as acetate as a subsidiary carbon source, *provided* that CO_2 is being utilized as the *prime* carbon source. No organism has been found with the 'submarine' type property (as it has been termed) of total exclusion of all organic substrates; consequently it is reasonable to accept obligate autotrophs as being organisms capable of using amino acids, etc., in addition to CO_2.

As to why some organisms are obligate, four major explanations have been proposed from time to time:

(a) Metabolic 'lesions'; the absence of a key enzyme central to heterotrophic growth.

(b) A regulatory dependence on CO_2-based metabolism; balanced synthesis from a primary carbon source other than CO_2 is not attainable.

(c) Failure to grow on organic matter because of substrate toxicity or because of production of autoinhibitory substances from organic substrates.

(d) Inability to obtain sufficient energy from the oxidation of organic compounds.

The first and third of these proposed explanations represent distinct hypotheses and will be considered in some detail below. The second suggestion is a subsidiary consideration of these two. No information is available about the fourth idea.

1. Metabolic Lesions

Two in particular have been postulated. One, the absence of NADH oxidase coupled with a lack of ability to link NADH oxidation to synthesis of ATP, is now discounted. A second is the absence of 2-oxoglutarate dehydrogenase. A consequence of this lesion is that the tricarboxylic pathway cannot function as an energy generating pathway but, instead, takes on a purely biosynthetic role, operating as a 'horse-shoe'. The right-hand 'half' operates as normally, the left-hand 'half' operates in reverse.

Even though this particular 'lesion' occurs in thiobacilli, nitrifying bacteria and cyanobacteria, there seems no obvious reason why it should lead to obligate autotrophy — especially as some heterotrophic bacteria lack a functional tricarboxylic acid cycle (e.g. *Escherichia coli* growing anaerobically, the lactic acid bacteria and *Acetobacter suboxydans*). Obviously if acetate were the alternative energy source to the inorganic substrate, then the organisms would have no choice other than that of growing obligately by autotrophic mechanisms. However, this does not account for a failure to use organic substrates, such as glucose, which lead to substrate level phosphorylation or oxidative phosphorylation during NADH oxidation. Alternatively, if the incomplete tricarboxylic acid cycle (and the absence of a glyoxalate cycle) meant that certain growth factors were not furnished — except when CO_2 was being fixed then, theoretically, they could be added to the medium as supplements and the organisms could dispense with CO_2 as a carbon source. This has not proved to be the case — even though such supplements are taken up by the organisms. Therefore it appears that trying to pinpoint a single metabolic 'lesion' as an all embracing reason for obligate autotrophy is too simplistic an approach. It may well be more realistic to think that such lesions probably act in concert with unique control mechanisms or complement metabolic problems related to relative invariability of major enzymes and/or metabolite autotoxicity, which in itself is the subject of the second hypothesis — to be enlarged upon below.

2. The Autoxicity Theory

This notion originated with Winogradsky — though obviously not in its present form which is derived from recent knowledge. The idea now is that autotrophs may oxidize glucose and other organic substrates — and grow as a result — but that these substrates rapidly become non-metabolizable because of the generation of toxic intermediary metabolites, such as keto acids, 2-oxoglutarate (a consequence of the lesion in the tricarboxylic acid cycle), and p-hydroxyphenylpyruvate, as has been found to happen with some thiobacilli metabolizing glucose.

Attempts have been made to overcome this toxicity problem — and thus demonstrate growth on glucose by thiobacilli — by using dialysis culture techniques. Claims have been made that such techniques were successful, but the evidence was not conclusive. A common inhibitory substance may not be produced by all autotrophs from organic substrates. Two other classes of inhibiting metabolite may be formed: 1) those overproduced because of different regulatory processes prevailing during growth on organic substrates (many amino acids at very low concentrations are toxic to autotrophs); and (2) those accumulated by pathways operating at an abnormal level as a result of excess substrates being available over and above that usually resulting from autotrophic metabolism.

Despite these theories and a considerable input of research into the question, no clear idea has yet emerged to explain 'obligate auto-trophy'.

3. An Examination of the Conventional View of Autotrophy

If the concept of autotrophy is to mean anything at all of value, that is to say if it is to describe all organisms which are now seen to share an important nutritional mechanism central to their growth, then this should be evident in the definition. This does not seem to be so as becomes clear when the two halves of the present definition are analysed, i.e. (a) the use of inorganic substrates as energy sources, and (b) the use of CO_2 as a sole carbon source via the Benson–Calvin pathway.

(a) *The energy source.* The first point to make is that there is *no* shared mechanism of inorganic chemical oxidation amongst autotrophs. The different substrates (NO_2^-, NH_4^+, reduced S compounds, H_2, Fe^{2+}) are all oxidized by different enzyme complexes and pathways; for instance, NH_4^+ requires a mixed function oxygenase to incorporate $\frac{1}{2}O_2$ into the compound to form hydroxylamine en route to NO_2^-; H_2 reduces NAD^+ and the product, NADH, is oxidized via the electron transport chain; NO_2^- and reduced sulphur compounds release electrons on oxidation at potentials too positive to reduce NAD^+, hence NADH for biosynthesis is provided via ATP-dependent reversed electron transport. The second point is that inorganic chemicals are oxidized by organisms considered to be conventional heterotrophs, e.g. *Desulfotomaculum* species oxidize hydrogen, various pseudomonads oxidize thiosulphate to tetrathionate, strains of actinomycetes oxidize NH_4^+ to nitrite. Some of these oxidations clearly result in energy production for the heterotrophs concerned, e.g. *Desulfotomaculum* species generate ATP as a result of SO_4^{2-} reduction to H_2S with H_2. The conclusion, therefore, is that reduced inorganic compound oxidation is not a single process, and that it is not a magical property restricted to autotrophs. That it is not a key feature of autotrophy is exemplified by the metabolism of *Pseudomonas oxalaticus* on formate, and *Rhodopseudomonas* species and *Paracoccus denitrificans* on methanol. These substrates are oxidized and support growth of these organisms but do not serve as a direct source of organic compounds for biosynthesis. Oxidation proceeds through to CO_2 which is then fixed via the Benson—Calvin pathway. Clearly these organisms pose a problem about the validity of a definition of autotrophy which is exclusive about the inorganic nature of the energy source, especially in the case of *Pseudomonas oxalaticus* and formate. Although formate is undoubtedly organic it is used as though it were H_2 and CO_2 and, as a consequence, seems to point to the notion that the organic or inorganic nature of the energy source is of secondary importance and not central to the definition of autotrophy.

(b) *The carbon source.* All autotrophs fix CO_2 via the Benson—Calvin pathway. Other prime modes of carbon assimilation have been proposed from time to time (e.g. the reversed tricarboxylic acid pathway in *Chlorobium* species) but, to date, the ribulose bisphosphate pathway is the one proven and indispensable mechanism of carbon assimilation. Secondary carbon sources, as already mentioned, can be assimilated, but only when CO_2 is the prime carbon source.

Therefore, of the two halves of the definition of autotrophy, only that concerned with CO_2 fixation seems to be indispensable. But before

Table 3. Categories of organisms according to the organic or inorganic nature of their energy and carbon sources, and facultative or obligate ability

Category	Nature of energy and carbon sources	Examples	Nutritional type
(a)	Inorganic energy and inorganic carbon sources	*Thiobacillus thioparus*	Obligately autotrophic
(b)	Inorganic energy and inorganic carbon, inorganic energy + organic carbon, organic energy + organic carbon	(a) Most rapid growth on inorganic energy + organic carbon: *Thiobacillus intermedius* *Nitrobacter winogradskyi*	Facultatively autotrophic
		(b) Most rapid growth on organic energy + organic carbon: *Paracoccus denitrificans* *Pseudomonas saccharophila*	Heterotrophic
(c)	Inorganic energy + organic carbon, organic energy + organic carbon	*Thiobacillus perometabolis*	Heterotrophic
(d)	Organic energy + organic carbon, but autotrophic mode of metabolism	*Pseudomonas oxalaticus* (formate) *Paracoccus* (*Micrococcus*) *denitrificans* (methanol) *Thiobacillus* A2 (formate)	Autotrophic, although uses organic energy source
(e)	Organic energy + organic carbon but mode of metabolism akin to autotrophy	*Methanomonas methanica* (methane) *Pseudomonas methylotropha* (methanol)	
(f)	Organic energy + organic carbon using 'serine' pathway	*Methylosinus trichosporium* (methane) *Hyphomicrobium vulgare* (methanol)	
(g)	Inorganic energy + inorganic carbon, but inorganic carbon (carbon dioxide) serves initially as an oxidant	*Methanobacterium* sp. (anaerobic formation of methane from hydrogen and carbon dioxide)	Autotrophic (because totally inorganic nutrition) but may technically be heterotrophic

settling for this nutritional definition, the question ought to be considered as to whether it is too exclusive in that there may exist other organisms which seem closely similar to autotrophs in principle if not in exact mechanism. In other words, is it worth widening the scope of autotrophy?

A categorization of bacteria according to the organism and/or inorganic nature of their energy and carbon sources (Table 3) sets the scene for this debate.

Both autotrophic forms [categories (a), (b), (d), and possibly (g)] and heterotrophic forms [categories (c), (e), (f), and possibly (g)] are incorporated into the Table. The apparent anomaly of category (d) has already been discussed, while category (g) remains unresolved because of the lack of evidence as to the pathway(s) concerned with the assimilation of carbon from CO_2.

Heterotrophic categories (e) and (f), but primarily (e), embrace the organisms (the methane and the methanol oxidizing bacteria) which might be included within a broader definition of autotrophy. An immediately obvious (perhaps) reason for this is that they are able to synthesize all their components from a C_1 compound, e.g. CH_3OH or CH_4. The relevant nutritional properties of these organisms are discussed below.

4. C_1 Compound Utilizers

Many microbes come into this category. Some, such as formate- and/or methanol-utilizing species of *Bacillus, Caulobacter, Asticcacaulis*, and the 'mushroom-shaped bacterium' still remain unexplored as far as their C_1 metabolism is concerned. However, a great deal has been learned about the carbon assimilation pathways of the methane oxidizing bacteria and the pseudomonads, hyphomicrobia and yeasts which oxidize methanol, the N-methyl compounds and other C_1 compounds. Professor Quayle and his colleagues at Sheffield have been the major contributors to the unravelling of the pathways of carbon assimilation in these cases.

Two major pathways have emerged (Figures 1 and 2), the 'ribulose monophosphate pathway' and the 'serine pathway'. Both have variants but the essential features are that formaldehyde formed in the oxidation of the C_1 compounds serves as the prime carbon source for biosynthesis, and that carbon dioxide, acetate, and other carbon compounds serve as secondary carbon sources.

The 'serine pathway' clearly has no obvious similarity to the Benson—Calvin pathway and will not be discussed further. The 'ribulose

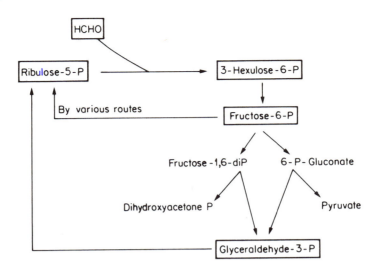

Figure 1. Ribulose monophosphate cycle of formaldehyde fixation incorporating the two variant routes of glyceraldehyde-3-P formation

monophosphate' pathway — starting with a $C_1 + C_5$ condensation (formaldehyde and ribulose monophosphate) — is clearly analogous to the $C_1 + C_5$ condensation process which initiates the Benson—Calvin cycle. It is found in hyphomicrobia, pseudomonads, and yeasts, but was first detected in the obligate methane oxidizers and it is these last organisms that will be compared in more detail with the autotrophs to bring out other similarities which add to the case for considering organisms with the 'ribulose monophosphate' pathways to be part of the autotrophic complex of microbes.

There are two distinct groups of methane oxidizers as judged by biochemical, physiological, morphological, and ultrastructural studies. One group assimilates formaldehyde resulting from methane oxidation via the 'ribulose monophosphate pathway' (Figure 1) and the other via the 'serine pathway' (Figure 2). All these organisms require carbon dioxide as an essential carbon source in addition to formaldehyde.

Similarities of the 'ribulose monophosphate' methane oxidizers to 'conventional' autotrophs are quite marked in a number of respects. For example all the strains of this group tested lack 2-oxoglutarate dehydrogenase; consequently the incomplete tricarboxylic acid cycle acts as a biosynthetic unit, providing carbon skeletons and amino acids but not energy, as in 'conventional' autotrophs and in anaerobically growing *Escherichia coli*. This 'missing' enzyme does not seem to be

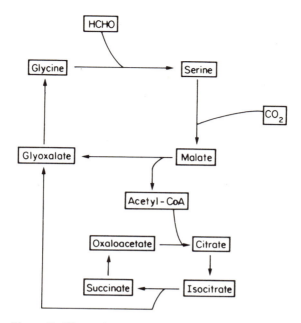

Figure 2. The serine pathway (isocitrate lyase variant)

inducible in these organisms, which are obligate methane utilizers and, logically, would not be expected to require a complete cycle. In the aerobic microbes, as mentioned earlier, this form of incomplete tricarboxylic acid cycle seems to be restricted to obligate autotrophs, facultative autotrophs growing autotrophically, and the 'ribulose monophosphate pathway' C_1 compound utilizers when utilizing C_1 compounds.

In addition, all methane oxidizers have a complex fine structure the most prominent feature being the membranous system which fills most of the cell. This membranous unit appears to be an invagination of the cytoplasmic membrane and is organized in various patterns depending upon the type of bacterium. Similar membrane arrangements have been found in ammonia and nitrite oxidizing bacteria, photosynthetic bacteria, and methane-producing anaerobic bacteria.

The interesting point here is that certain autotrophs and the methane-oxidizing bacteria share this membranous development. What the function of these membranes is remains unproven. However, it is possible to make an educated guess by considering the physiology of the organisms concerned.

Photosynthetic bacteria have been shown to have a membrane and chlorophyll content in an inverse ratio to the intensity of illumination at which they are grown, i.e. at low light (light-limiting) intensities, chlorophyll and membrane content is high, whereas at saturating light intensities chlorophyll and membrane content are considerably reduced. This implies that the organism is able to adjust, within limits, its rate of ATP synthesis per cell by varying these two components according to the photon intensity, but more importantly, points to the idea that increased membrane content over that of the cytoplasmic membrane is a way of overcoming kinetic problems of one sort or another related to poor and/or very low concentrations of an energy source or electron donor.

The nitrite oxidizers are probably among the least favoured of bacteria in respect of the high energy requirement for their biosynthesis coupled with low energy yield resulting from the oxidation of the substrate nitrite to nitrate. When NADH and ATP requirements for carbon dioxide fixation and the synthesis of NADH at the expense of ATP and reduced cytochrome c and other demands are set against the energy yield from nitrite, such an organism would seem to need a vast number of nitrite oxidizing sites in order to compensate for the low energy yielding oxidations by high turnover rates. As nitrite oxidizing and associated ATP synthesizing systems are located in the membrane, a proliferation of such membranes would be an advantage to such bacteria as compared with a nitrite oxidizing cell with a simple cytoplasmic membrane. Turnover rates in the same cell with multiple membranes could be increased twentyfold or more as judged by the membrane content of nitrite oxidizers in addition to the cytoplasmic membranes. Even so, nitrite oxidizers are slow growers (generation times of more than 20 h) and nitrate accumulation is pronounced before any turbidity is observed in cultures.

Ammonia oxidizers pose similar problems to the nitrite oxidizers as regards energy production and consumption for biosynthesis (e.g. carbon dioxide fixation and NADH synthesis). Even though ammonia oxidation releases more energy per mole than does nitrite oxidation, an additional energy-demanding reaction is necessary in the hydroxylation of ammonia to its presumed first intermediate, hydroxylamine. Assuming this to be a conventional mono-oxygenase type of reaction, NADH is expended in this reaction without any return in the form of ATP synthesis.

Perhaps the ammonia oxidizers (of all autotrophs) and the 'ribulose monophosphate pathway' methane oxidizers have the closest *apparent* relationship, because in addition to having in common a complex fine structure, an incomplete tricarboxylic acid cycle, the ability to assimilate C_1 compounds by mechanisms similar in principle (a $C_5 + C_1$

condensation), methane oxidizers are also able to oxidize ammonia to nitrite — as do the ammonia oxidizers. Although it is not yet clear whether 'useful' energy is released in this oxidation, other 'co-oxidations' do release useful energy as will be discussed later. The methane mono-oxygenase enzyme complex appears to be the same one which oxidizes ammonia.

Obviously, these arguments may also illustrate convergence rather than common ancestry of methane oxidizers and ammonia oxidizers. However, the methane oxidizers and the other particular autotrophs referred to show basic similarities of principle, if not of the exact mechanisms, of metabolism with the conventional autotrophs and it would be unrealistic to ignore this fact in reassessing the concept of autotrophy. Perhaps this can be summarized, in one example. Methane oxidizers can grow on hydrogen and formaldehyde in continuous flow systems using the 'ribulose monophosphate pathway' for the assimilation of formaldehyde; hydrogenomonads growing on hydrogen and carbon dioxide using the 'ribulose bisphosphate pathway'* to assimilate carbon dioxide are obtaining their energy and carbon in a very similar way to these methane oxidizers. Both must share many biosynthetic mechanisms in common. This latter point, of course, can be made for all microbes in relation to each other and also serves to warn against the temptation to overemphasize differences between groups of microbes. A tentative conclusion then is that in redefining autotrophy, these particular C_1 compound utilizers should be within that concept.

Taking the 'ribulose monophosphate pathway' feature as being central to this argument, then certain methanol utilizing bacteria would also be brought under this new umbrella of autotrophy. Finally, the 'serine pathway' organisms. Similar arguments (based on both analogy and identity of nutritional pathways) cannot be advanced to bolster the claims of these organisms to be considered as autotrophs. However, they can synthesize all their carbon compounds from a C_1 compound (both the obligate methane utilizers and the methanol utilizing bacteria, such as *Pseudomonas* AM1). It has, therefore, been proposed that they all be considered autotrophs within a new definition, if only because such a grouping will encourage comparative studies of all these organisms. This may lead to new insights into their nutrition, which is the only valid excuse for tinkering with an established concept.

As a tail-piece — two recent pieces of information support the idea of keeping 'autotrophy' under continual review: the discovery of phosphoribulokinase and ribulose bisphosphate carboxylase activity in the obligate methane oxidizer *Methylococcus capsulatus* (strain Bath) and

*'Bisphosphate' is the most up-to-date terminology for this chemical; it is often called 'diphosphate' in earlier literature.

Table 4. Classification of organisms as autotrophs based on carbon assimilation occurring via a $C_1 + C_5$ condensation

Cell carbon via either 'ribulose bisphosphate pathway' (RBP) or 'ribulose monophosphate pathway' (RMP)		Energy source
Obligate		
Nitrosomonas europaea	(RBP)	NH_4^+
Thiobacillus neapolitanus	(RBP)	$S_2O_3^{2-}$
Methanomonas methanica	(RMP)	CH_4
Methylococcus capsulatus	(RMP and RBP)	
Facultative		
(a) Non-NAD$^+$ reducing oxidations:		
Nitrobacter winogradskyi	(RBP)	NO_2^-
Thiobacillus novellus	(RBP)	$S_2O_3^{2-}$
(b) NAD reducing oxidations:		
Pseudomonas saccharophila	(RBP)	H_2
Pseudomonas oxalaticus	(RBP)	$H.COOH$

in the strictly anaerobic sulphate reducing bacterium *Desulfovibrio desulfuricans*.

To summarize this area of discussion, a nutritional classification of organisms as autotrophs based on carbon assimilation via either the

Table 5. Autotrophy — embracing all organisms which assimilate C_1 compounds as their sole carbon source

	Carbon source	
'Ribulose bisphosphate pathway'	CO_2	*Nitrosomonas europaea* (growing on NH_4^+)
'Ribulose monophosphate pathway'[a] and a CO_2 fixing mechanism (unknown)	$HCHO + CO_2$	*Methylomonas methanica* (growing on CH_4 or CH_3OH)
'Serine pathway'[a] and possibly one or more heterotrophic CO_2 fixing mechanisms as well as CO_2 fixing activity within serine pathway	$HCHO + CO_2$	*Methylosinus trichosporium* (growing on CH_4 or CH_3OH)
Unknown pathways in anaerobic CO_2 reducing microbes	CO_2	(a) *Methanobacterium* spp. (growing on H_2) (b) *Clostridium aceticum* (growing on H_2)

[a]There is more than one variant of the main pathway

'ribulose monophosphate pathway' or the 'ribulose bisphosphate pathway' is given in Table 4, and a nutritional classification of organisms as autotrophs based on their ability to assimilate their carbon solely from a C_1 compound is given in Table 5.

5. A Definition of Autotrophy

Over the years, the definition of autotrophy has veered between two standpoints:

(1) That autotrophs are organisms which grow on inorganic nutrients, no specification of pathways of C_1 compound assimilation being made. Following more detailed biochemical studies on the organisms in such a category, anomalies have presented themselves, e.g. hydrogen utilizers which fix carbon dioxide are included in the same category as such metabolically dissimilar microbes as anaerobic methane producers, but are separated from organisms of overall metabolic similarity such as the formate oxidizing *Pseudomonas oxalaticus*.

(2) That autotrophs are organisms which fix carbon dioxide as their prime carbon source, via the Benson—Calvin pathway, and obtain energy from the oxidation of inorganic chemical compounds. This more recent definition suffers as it is too narrow and excludes obviously similar organisms either on grounds of the nature of the energy source and/or different (but rather similar) pathways of carbon assimilation.

A conclusion is that all microorganisms which are able to assimilate C_1 compounds as their prime carbon source for cellular biosynthesis, irrespective of their energy source, should be included with a new definition of autotrophy. Such a definition covering both photo- and chemotrophic organisms could be: *Autotrophs are microorganisms which can synthesise all their cellular constituents from one or more C_1 compounds*. There are at least three categories of autotrophic microbes fitting this definition: (1) those possessing the 'ribulose bisphosphate pathway'; (2) those possessing the 'ribulose monophosphate pathway'; (3) those possessing the 'serine pathway'. Leniency in the application of this definition to metabolic cripples (e.g. vitamin requirers) is obviously good common sense, as a requirement for one or two growth factors reflects minor metabolic and biosynthetic inadequacies — it does not impinge upon the importance of the central assimilation and biosynthesis routes.

In this regard, those organisms classed as being photoheterotrophs because of obligate vitamin requirements should be regarded as photoautotrophs, as should *Desulfovibrio desulfuricans* previously regarded as a heterotroph because of its dependence on an unidentified growth factor(s) but now known to be able to fix CO_2 autotrophically.

III. Adaptation for Survival

A discussion on nutrition cannot be reduced simply to a series of comments on what nutrients are used by bacteria — as was said at the beginning this is only a part of the nutritional story. There is a vast array of bacterial species, many of which ostensibly use the same nutrients. The question is how do they survive in competition with each other. Broadly, the answer is because they have adapted themselves to grow and survive in a bewildering variety of niches — or environments — and have also adopted modes of life which give them a particular advantage — or protected environment — not afforded to a competitor which might be a more efficient utilizer of that substrate.

In so far as the environment is concerned, niches have developed where many physical factors contribute to their individual uniquenesss, i.e. pH value, temperature, free-water availability, gas partial pressures, light, concentration of nutrients, and so on. This has led, for example, to the development of osmophilic, thermophilic, acid tolerant, and micro-aerophilic species. Some have only one of these attributes, some have more. In this way species have evolved which compete successfully for a nutrient simply because these additional characteristics give a competitive advantage. However combined with these factors, or instead, are other modes of nutritional competitiveness such as consortium growth, interdependent growth in specialized ecosystems, symbiotic associations (all really related), parasitism, ability to co-oxidize substrates, ability to adopt a major mode of metabolism in the presence of a particular nutritional supplement, the ability to acquire additional genetic powers to enable otherwise unavailable substrates to be used (e.g. plasmids which were referred to earlier and will not be discussed further here), ability to move towards nutrients (tactic response), ability to adhere to surfaces in aquatic environments and so gain a nutritional advantage, and, finally, enhanced nutritional competitiveness as a result of adopting a differentiated multicellular form. Other examples could be quoted, but these will serve to illustrate the plethora of ways in which bacteria have evolved in order to survive and succeed within the microbial world.

A. Consortium Growth

Very recently a new nutritional category has been defined — the consortium. It describes an interdependent nutritional relationship between two or more microbes. The idea first arose from the discovery that the methane producing organism, '*Methanobacterium omelianskii*' was not a single-membered culture in the ethanol/CO_2 medium in which it was grown. Two organisms were present — one oxidized ethanol to acetate and H_2, the second (the methane producer) formed methane from H_2 and CO_2. The interdependence was as follows: the ethanol utilizer was sensitive to the concentration of H_2 and ceased to grow unless the level was kept very low, which it was by the methane oxidizer which consumed the hydrogen. The methane oxidizer, in turn, was obviously dependent upon the ethanol utilizer for a supply of H_2. Environmentally, this has proved to be an important nutritional concept; a number of consortia have been described (some contain up to six members) allegedly exhibiting a variety of cross-dependent relationships (detoxification or substrate and growth factor require-ments). 'Synergism', 'mutualism' and 'cross-feeding' can also be regarded as extensions of the consortium principle. Such a phenomenon illus-trates how crude a technique is batch culture which, far from demon-strating the presence of organisms making an essential contribution to the cycling of a substrate, brings to light organisms able to tolerate the artificially high concentrations of substances in a batch culture. Only continuous flow culture techniques will overcome this problem, in that substrate levels can be maintained at those concentrations in which they occur in the environment. New information on organisms of nutritional significance in different environmental conditions is now coming to light as a result of this approach. Unfortunately, 'Consort-ium' systems can be a convenient excuse for sloppy work and poor isolation techniques; nothing is easier than to explain away a 'mixed' culture by sophisticated chatter about consortia!

B. Interdependent Growth

Interdependent growth might be loosely regarded as being an ex-tension of the consortium idea, in that a number of microbial types coexist in a population by using the products of other microbes as their substrates. The idea differs mainly from the consortium notion in that not all of the members are totally dependent upon each other for growth. Two main categories can be defined, the linear category and the circular category.

1. The Linear Category

One example is in the rumen. Here in a specialized anaerobic environment a succession of fermentations takes place; basically they are (1) the breakdown of cellulose, pectins and hemicelluloses by bac-teria with the necessary extracellular enzymes; (2) the fermentation of the products by a second group of bacteria to alcohols, acids, H_2, and CO_2; (3) the breakdown of these alcohols and acids by a further group of bacteria to the fatty acids, acetate, propionate, butyrate, etc.; and finally the conversion of CO_2 and H_2 to CH_4 by the methane-producing bacteria. Many protozoa are also involved in this process, but the complexity of the rumen fermentation need not be invoked here to illustrate the interdependence being discussed. A second example is really a short length of food chain: bacteria (e.g. pseudomonads) utilize glucose and the resultant pseudomonad population is parasitized by species of *Bdellvibrio* which bore into the bacterial cells and grow and multiply within the periplasmic region on contents leaking through the pseudomonad cytoplasmic membrane. Perhaps *Bdellovibrio* can be considered analogous to a carnivore eating herbivores.

2. *The Circular Category*

In essence the whole of the ecosystem is an example of the circular category, but within it is a series of smaller ecosystems. An artificial example of a circular microbial ecosystem is the Winogradsky column driven by the one energy input, light (Figure 3). Products and sub-

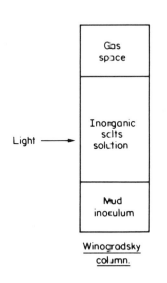

Light ⟶

Gas space

Inorganic salts solution

Mud inoculum

Winogradsky column.

Figure 3. Cycling of nutrients in a Winogradsky column

The system. A glass column containing a mud inoculum (in which cellulose and $CaSO_4$ have been mixed), a nutrient salts solution and a gas space above the column — sealed and exposed to sunlight and maintained at room temperature.
Major microbial events
(1) Oxygen rapidly removed by aerobic heterotrophs utilizing dissolved organic compounds released from mud-system becomes ANAEROBIC.
(2) Cellulose hydrolysed by clostridia and products fermented. H_2, CO_2, alcohols, keto- and fatty acids are major end products.
(3) A series of events now take place more or less simultaneously.
(a) $H_2 + CO_2 \rightarrow CH_4$ which accumulates in gas space (methane producers).
(b) H_2, alcohols and keto acids $+ SO_4^2 \rightarrow H_2S$ (the SO_4^2 reducers such as *Delsulfovibrio* and *Desulfotomaculum* species).
(c) H_2, organic compounds $+ CO_2 \rightarrow$ cellular products via photosynthetic action of *Rhodospirillaceae*.
(d) H_2S, S and other partially oxidized compounds $+ CO_2 \rightarrow$ cellular products and SO_4^2 via photosynthesis action of *Chromatiaceae*, *Chlorobiaceae* and some *Rhodospirillaceae* and cyanobacteria.
(e) Cyanobacteria able to tolerate H_2S fix CO_2 and photolyse H_2O and evolve O_2.
(f) Lysed cell products utilized by heterotrophs and autotrophs.
Cycles. A steady state evolves which represents a balance of all microbial activities, especially if oxygen is formed by cyanobacteria. An anaerobic to aerobic gradient is set up in the column. Oxidized products are reduced (as detailed above) and reduced products oxidized by photosynthetic bacteria and be aerobes. In other words the different types of microbe broadly feed each other and substrates are cycled — ideally. Obviously over production of a nutrient (e.g. H_2S) will bring all activity to a halt.
Visual appearance of columns. Most noticeable are the colours of the photosynthetic bacteria — the body of the column may become pink, mauve, green or brown depending upon the dominating species. Rings of colour are seen in the mud layer. Green bacteria — the most H_2S tolerant and low light intensity tolerant — usually form the lower most bands in the mud. Total blackening (FeS) of the column reflects excess H_2S-production.

strates are recycled (one organism's products being another's substrate) thus keeping a balanced population under set conditions. Again this is really an extension of the consortium idea.

C. Symbiosis

Symbiosis is probably yet another facet of the consortium principle — but remains a phenomenon apart because it usually involves a bacterium (in the present context) and a eukaryote which may be single or multicelled. Examples (there are many: see earlier in this book) include the *Rhizobium*—legume relationship, the bacteria — termite relationship, and two involving cyanobacteria, the cyanobacterium—cycad relationship (see Table 2) and composites such as the lichens; in this particular instance a fungus—cyanobacterium relationship.

In these relationships each symbiont benefits nutritionally from its partner — or that is the assumption. As far as the *Rhizobium*—legume relationship is concerned, it is clear what the plant obtains — fixed nitrogen from N_2-fixing activities of the *Rhizobium* bacteroid in the nodule formed on the plant roots: only prokaryotes have the ability to fix nitrogen and so any relationship of this sort is clearly of enormous benefit to the plant in a fixed-nitrogen-hungry ecosystem. What the bacterium gains, however, is unclear. In the bacteroid form it does not grow in the nodule; consequently, at the most, it can only be claimed that the symbiotic relationship provides a period of protected existence. Perhaps this particular type of relationship should be more realistically viewed as being a slave—master relationship whereby the legume has in effect converted the bacterium into an organelle — 'protecting' it from oxygen toxicity (thus allowing nitrogenase activity to occur) and 'feeding' it with reduced compounds to provide the necessary ATP and NADH to reduce N_2.

The concept of 'captive' microbes being exploited in this manner is not novel; many people believe that chloroplasts and mitochondria, the organelles that provide energy for eukaryotes, have evolved from prokaryotes which were at one time free-living before being 'modified' to suit their host's purposes. The case for this interpretation of the origin of chloroplasts and mitochondria has yet to be settled, if it ever can be settled.

D. Parasitism

Parasitism is where a bacterium is dependent upon a host cell — usually a eukaryotic cell — for its nutrition. Such a process benefits the bacterium, not the host.

E. Co-oxidation

A number of organisms are able to oxidize compounds but not to grow on them as a single source of energy. In other words the oxidation is a secondary activity to their growth on a utilizable substrate. This phenomenon — co-oxidation or co-metabolism — is still something of a mystery. It can be useful in a mixed population in that the co-oxidizing organism can make a substrate available to other microbes by oxidizing it to a compound utilizable by them. But what the co-oxidizing organism gains from this apparently altruistic activity is not known — unless of course a consortium situation exists whereby the other microbes release metabolities essential to the growth of the co-oxidant.

A recent study of this phenomenon in my laboratory has shown, as might be expected, that altruism is not an explanation; in the case of methane oxidizers, co-oxidation is of direct nutritional benefit and proof that this is so is worth spelling out in some detail.

Methane oxidizers can oxidize a wide range of compounds which do not support their growth when methane — their prime carbon and energy source — is absent. These compounds fall into two categories: (1) those which have initially to be oxygenated by methane monooxygenase (e.g. ethane, propane, and butane) and (2) those which are directly oxidized (e.g. ethanol and propanol).

The failure of these compounds to support growth in the absence of methane was assumed to be due to an obligate requirement for formaldehyde (formed from methane or methanol) as a carbon source; formaldehyde is not a product of the oxidation of the co-oxidizable compounds. However, it seemed reasonable to suppose that ATP and reduced pyridine nucleotides might well result from co-oxidation and that in this way co-oxidation could contribute energy if not carbon to growth. The problem was how to demonstrate that obligate energy was a product An unambiguous answer was arrived at by taking advantage of the N_2 fixing ability of *Methylococcus capsulatus*. To fix N_2, both ATP and reduced pyridine nucleotides are necessary and, of course, have to be furnished by oxidation of an energy source. Therefore, if nitrogen fixation by cell suspensions of this organism grown on methane and N_2 could be demonstrated as occurring as a direct response to co-oxidation — then it could be unequivocally accepted that the co-oxidized compounds furnished energy to the organism. The test was done with assays which employ analogues of N_2, e.g. if acetylene, one of the analogues used, is reduced to ethylene, then the organism is understood to be able to fix N_2. This test was positive when applied to this organism furnished with directly oxidizable compounds such as ethanol and H_2, but was negative in the case of compounds initially oxygenated by mono-oxygenase activity. The negative result proved to be a consequence of mono-oxygenase inhibition by acetylene — therefore another N_2 analogue, N_2O, was employed instead; its successful reduction to N_2 indicated that the compounds initially requiring oxygenation were also a source of utilizable energy to the methane oxidizer.

Some of these compounds, such as ethane and ethanol, may also be used as a supplementary carbon source by CH_4 oxidizers in that the oxidized product, acetate, is assimilable and usable in biosynthetic activities — provided, of course, that formaldehyde carbon is also available from methane oxidation. In this particular instance, therefore, ethane and ethanol serve both as an energy and carbon source *when methane is also being utilized.* Co-oxidation in this particular case is not, therefore, some aberration of metabolic behaviour but of direct nutritional benefit to the co-oxidizing organism.

F. Non-energy Yielding Nutrients

All bacteria require for growth a range of elements and inorganic compounds apart from those which serve as energy sources. These substances are essential components of cell macromolecules (e.g. ATP, enzymes, and nucleic acids). Not all organisms have exactly the same requirements — for instance, they may not all form nitrogenase for which molybdenum is essential — but, of course, many elements and inorganic compounds, such as magnesium and phosphate, are required by all microbes.

Many bacteria can provide themselves with all the carbon compounds they require by biosynthetic mechanisms stemming from the one carbon source. However, a considerable number have biosynthetic deficiencies in that they require preformed organic compounds in their medium. These are taken up and incorporated into the cellular system. These requirements may be limited to one or two vitamins, as in the case of strains of the *Rhodopseudomonas palustris* — a photosynthetic bacterium — or may be extensive as in the case of many species of lactic acid bacteria which may require many amino acids, vitamins, purines, pyrimidines and even small peptides.

Finally there are nutrients which are not essential for the growth of the bacteria that can use them, but which improve both growth rate and cell yield per mole of energy-yielding substrate by sparing energy expenditure on synthesis of them from simple compounds — or by allowing a more economic pathway of metabolism to operate. Awareness of the possibility that some organisms might benefit from the latter process can be very important when considering how such organisms behave in their natural environment — which frequently is not mirrored in the laboratory and can lead to a total misconception of the real nature of the microbe being considered. Merely because an organism grows well in the medium provided should not be taken as meaning that this necessarily reflects that it has expressed its genetic potential.

An excellent example of this state of affairs is exemplified by some lactic acid bacteria. These bacteria are unique in that the great majority of them can grow in the presence of oxygen — yet do so by employing fermentative mechanisms employed in anaerobic conditions. For instance, *Streptococcus faecalis* utilizes glucose via the Embden–Meyerhoff pathway both aerobically and anaerobically; substrate level phosphorylation is the source of ATP. All other types of bacteria growing aerobically obtain most, if not all, of their ATP via oxidative phosphorylation mechanisms involving a cytochrome mediated electron transport chain with oxygen being the indespensable electron acceptor. There has grown up, therefore, the myth that lactic acid bacteria are in an evolutionary position midway between aerobes and strict anaerobes and have 'learned' to tolerate oxygen.

Recent evidence points to this notion as being mostly nonsense and that lactic acid bacteria are probably best regarded as being retro-

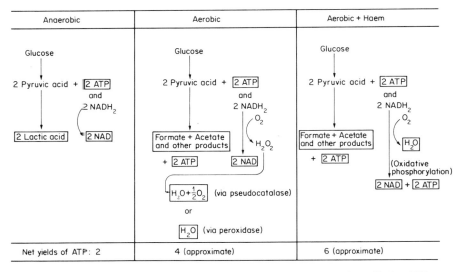

Figure 4. Glucose as an energy substrate for *Streptococcus faecalis* in different environments. Boxes enclose final products

gressive aerobes, that is organisms that have lost certain properties of biosynthesis so making it impossible for them to function as aerobes *unless an essential growth supplement is included in their medium*. This supplement is haematin. It is not an exotic substance, being commonly present in the natural environment of many of these organisms (e.g. plant juices, meat surfaces, blood, and tissues of animals and humans). Oxygen, even in the absence of haem, can act as nutrient to some lactic acid bacteria (e.g. *Streptococcus faecalis*) in that cell yields per mole of substrate can be considerably improved over that obtained in the absence of oxygen (i.e. anaerobically). Effect of oxygen alone and then in the presence of haematin is discussed in more details below (see Figure 4).

1. Oxygen in the Absence of Haematin

Anaerobically, *Streptococcus faecalis* ferments 1 mole of glucose to 2 moles of lactic acid with a net yield of 2 moles of ATP. In the presence of oxygen (Figure 4) NADH is not oxidized in the reduction of pyruvic acid to lactic acid (as must occur anaerobically if reduced pyridine nucleotides are to be recycled) but is oxidized by oxygen via a flavoprotein oxidase – to yield H_2O_2 which is subsequently per-oxidized to H_2O or destroyed by pseudocatalase as follows,

$$H_2O_2 \rightarrow H_2O + O_2$$

The pyruvate (2 moles) spared from being reduced to lactic acid is further metabolized to yield an additional 2 moles of ATP. That is to say, aerobically (using oxygen as an electron acceptor) *Streptococcus faecalis* can obtain 4 moles of ATP as opposed to the 2 moles obtained anaerobically. Again it needs emphasizing that ATP formed under aerobic conditions by this organism in the absence of haem is by substrate level mechanisms, that is by fermentation processes usually associated only with anaerobic growth.

2. Oxygen in the Presence of Haematin

When haematin is added as a supplement to media in which *Streptococcus faecalis* is growing aerobically, further increases in ATP yield per mole of substrate are obtained over those described earlier. This has proved to be a direct consequence of this organism being able to use preformed haematin (which it cannot synthesize) in the biosynthesis of cytochromes and catalase (which becomes an additional enzyme in the armoury of enzymes able to dispose of H_2O_2 formed as a product of

superoxide dismutase activity and flavoprotein oxidase activity). The
formation of cytochromes now allows *Streptococcus faecalis* to synthesize ATP by oxidative phosphorylation processes – in addition to
substrate phosphorylation mechanisms; the result is 6 or more net
moles of ATP being formed per mole of glucose (for instance) instead of
the 4 net moles (net because, of course, ATP is utilized in phosphoryl-
ating reactions during growth) formed in the presence of oxygen alone.
Most aerobes produce a much better yield of ATP aerobically than
Streptococcus faecalis, but this is so because they have a fully
functional tricarboxylic acid cycle which *Streptococcus faecalis* does
not.

At this juncture, it should be pointed out that not all lactic acid
bacteria have this ability to use haematin in forming an oxidative
phosphorylation system. Lactic acid bacteria show considerable variation of ability in their ranks to use oxygen and haematin, e.g. some
can form catalase from haematin but not cytochrome. On the other
hand some strictly anaerobic bacteria when provided with haematin
appear now able to carry out oxidative phosphorylation reactions with
fumarate serving as an electron acceptor (instead of oxygen). Clearly,
therefore, the present concept of aerobic and anaerobic metabolism will
have to be modified to account for these recent discoveries.

G. Tactic Responses to Light and Nutrients

An ability to move towards a desirable nutrient (or away from a
potentially toxic substance) is obviously a useful attribute. This
directed movement (flagella driven or creeping motility) in response to
signals generated within the organism, is called taxis. Chemotaxis –
response to dissolved chemicals – and phototaxis – response to light –
are two main classes of taxis. In the latter instance, motile photosynthetic bacteria, probably by negative taxis to unfavourable wavelengths of light, are able to accumulate within those bands of the
spectrum in which their pigments (carotenoids and bacteriochlorophylls)
absorb. In practice this phototactic ability is expressed in the depths of
a water source (lake, pond etc.). Obviously electron source, light, and
oxygen tension all affect the final outcome of a photosynthetic
bacterium's position within a body of water – as does its ability to
synthesize additional amounts of carotenoids in response to restricted
wavelength bands of light which are available at different depths. A
number of tactic responses are involved here and the outcome is a
compromise, but is obviously of direct importance to the nutrition of
the bacterium.

An absolute requirement for tactic properties to survive is expressed by *Thiovulum majus* (a large — 25 nm in diameter or more — chemoautotrophic bacterium) which oxidizes H_2S for energy with O_2 as the electron acceptor in an environment with a pH value around neutral. Controlled movement by the organism is essential to position it at the interface of the H_2S/O_2 zone in the environment. Either side of this zone, one or the other of the two nutrients does not exist (the H_2S is rapidly autoxidized at neutral pH values). As H_2S is generated anaerobically (by sulphate reducing bacteria) and oxygen enters at the surface, the two substances form concentration gradients which diminish as they meet. Consequently the tactic responses of *Thiovulum majus* (probably reacting to both H_2S and O_2) are essential to its survival as the interface zone of H_2S and O_2 probably rises and falls in response to changes affecting the production of H_2S and the solution of O_2. In a tank where conditions are suitable for the growth of this organism, the hundreds of millions of organisms appear as a wreathlike white veil which undulates slowly up and down as the H_2S/O_2 interface changes position.

H. Morphogenetic Response to Levels of Dissolved Nutrients

A particular nutrient (it varies with species) can, in some unknown manner, profoundly affect the morphogenesis of a microbe — only a few species are known to react in this way. What advantage such morphogenetic changes confer on the bacteria concerned is not clear — except, it seems, in the case of certain prosthecate species. These are microbes which can form appendages (integral parts of the cell) which emerge from all over the cell's surface to give — in extreme cases — an appearance of a many-armed starfish. The advantage of such a change can be deduced from the 'cause', in this case appendage formation is in response to increasing dilution of nutrient concentrations — the reverse is the case when nutrient concentration is increased — that is to say no appendages are formed and, therefore, as the bacterium grows and divides less appendages are found per bacterium as they are 'diluted' out by successive rounds of division. Therefore, the advantage would seem to be that appendages result in a dramatic increase of surface to volume ratio and, consequently, a greater ability to acquire nutrients from a dilute medium.

This particular form of adaptation of bacteria to dilute nutrient habitats may be reflected in other ways. A study of this phenomenon is in its infancy; spirilla and spirochaetes may be naturally adapted by their morphology to a dilute nutrient existence. The massive cytoplasmic membranous invaginations of methane oxidizers, some methane producers and other autotrophs including the photosynthetic bacteria, appear to offer a greatly increased area of metabolic sites compared with that normally present on a single cytoplasmic membrane. Nutritional characteristics of these bacteria support such an idea, especially in the case of the photosynthetic bacteria in which membrane and chlorophyll content varies inversely with the light intensity — the lower the intensity the more chlorophyll and membrane content per organism.

I. Adhesion to Surfaces and Colony Formation

1. Adhesion

Many microbes have the capacity to stick themselves firmly to a surface in their environment. Mechanisms of 'sticking' are many, e.g. by fimbriae, by excreted polysaccharides, and by special 'hold-fasts' located at a particular point on the surface of the bacterium. Indeed some bacteria have evolved morphologically to be specially suited to anchoring themselves to surfaces (e.g. the stalked *Caulobacter* species which have a sticky hold-fast at the distal end of the stalk). Some budding bacteria have 'sticky' poles at one end of the cell which facilitates permanent anchorage as the cell grows and multiplies only from the opposite end. The advantage conferred on such bacteria is assumed to be a nutritional one in that in aquatic environments nutrients adsorb to surfaces. As the adsorbed nutrients are captured by those microbes located on their surfaces, a nutrient gradient from the surrounding water to the adsorbing surfaces is set up, thus favouring, nutritionally, those bacteria on the surface over those suspended in the water itself — or so it is presumed. This element of the nutritional story is still in its infancy as regards firm information.

2. Colony formation

In the aquatic environment the progeny of many species of bacteria remain linked, either in a matrix of slime, by being retained in a sheath or envelope (e.g. *Sphaerotilus natans*), by 'hold-fasts', or by remaining joined to each other (e.g. *Rhodomicrobium vannielii*). The immediate nutritional advantage conferred is that once an organism has found a rich source of nutrients, it is able to proliferate more successfully as a species by retaining its descendents within that nutrient-rich sphere. The alternative possibility — not developing into a system resistant to dispersal — would be the possible loss of most of the progeny, by uncontrolled distribution of them through the aquatic environment to nutrient-poor areas.

What seems to be a sophisticated mechanism for enhancing species survival, over and above the systems for keeping progeny of microbes captive in one area, is practised by *Rhodomicrobium vannielii*. This photosynthetic microbe grows as a branching colony in appropriate nutrient-balanced environments, but when certain nutrients or light become limiting, it forms either *swarm cells* (motile cells separated via a binary fission process from the branching colony), which swim away from the colony and do not proceed to grow and multiply (into a branching colony) until adequate environmental conditions are found, or *exospores* (up to four per bacterium) which can remain quiescent until they end up in a favourable nutrient environment where they germinate and grow.

J. Nutritional Advantage Conferred on a Multicellular Prokaryote

As the last component of this nutritional story, a clear example will be described of how the ability to exist as a multicellular prokaryote can confer a nutritional advantage over all similar species which are single-celled. The organism concerned is *Anabaena cylindrica*, a filamentous cyanobacterium (blue-green alga) able to use water as an electron source (as it possesses photosystems I and II) and fix nitrogen in an aerobic environment. It normally exists as a chain of cells which have the property of intercellular dependence, that is adjacent cells are in communication with others and are able to influence each other's catabolic and metabolic activity, and, more importantly, be the recipient of nutrients from each other.

Selected cells in the filament have the ability to differentiate into specialized cells of varying function. The most important of these cells — from the point of view of this discussion — are those able to differentiate into heterocysts. These particular differentiated cells have the power to fix nitrogen as a result of both structural and enzymic modifications. Principally these are the loss of the ability of photolyse water (disappearance of photosystem II) — so ceasing to produce oxygen within the cell which would inhibit the nitrogenase activity now induced — and the gain of the ability to receive reduced organic substrates from adjacent undifferentiated cells. All this results in the gaining of an ability to fix nitrogen in an aerobic environment — the resultant fixed nitrogen being passed on in some way to the adjacent cells which are photosynthesizing normally. This organism, once differentiated in this manner, is unique amongst microbes in its ability to photolyse water and fix nitrogen simultaneously. As a consequence these types of cyanobacteria, made up of cells with different functions, are the first to colonize effectively and grow without limitation in new inorganic environments (e.g. volcanic islands) as their nutrient requirements, CO_2, H_2O, and N_2, inorganic salts and elements, give them a nutritional advantage over all other microbes in such circumstances, and of course, over an individual cell of the same species.

IV. Further Reading

Autotrophy

Kelly, D. P. (1971). Autotrophy: concepts of lithotrophic bacteria and their organic metabolism. *Annual Reviews of Microbiology*, 25, 177–210.
Rittenberg, S. C. (1972). The obligate autotroph — the demise of a concept. *Antonie van Leeuwenhoek*, 38, 457 – 478.
Whittenbury, R. and Kelly, D. P. (1977). Autotrophy: A Conceptual Phoenix. In *Microbial Energetics*, Ed. B. A. Haddock and W. A. Hamilton, pp. 121–149. *Symposium of the Society for General Microbiology*, 27, London: Cambridge University Press.
Pfennig, N. (1967). Photosynthetic Bacteria. *Annual Reviews of Microbiology*, 21, 286–313.

Methylotrophy

Quayle, J. R. (1972). The metabolism of one-carbon compounds by micro-organisms. *Advances in Microbial Physiology*, 7, 119–203.
Anthony, C. (1975). The biochemistry of methylotrophic micro-organisms. *Science Progress*, Oxford, 62, 167–206.

Prosthecate Bacteria

Whittenbury, R. and Dow, C. S. (1977). Morphogenesis and Differentiation in *Rhodomicrobium vannielii* and other budding and prosthecate bacteria. *Bacteriol. Reviews*, 41, (in press).

Subject Index

Bold type indicates Chapter number

SI/15

symbiosis, 16, 23
Syncephalastrum racemosum, 3, 16
Synchitrium endobioticum, 3, 13, 19
synergism, 16, 21
syphilis, 13, 15, 18
systole, 5, 22

Taphrina, 3, 9
T. deformans, 3, 4, 23, 24, 25
Taphrinales, 3, 25
taxa, definition of, 9, 29
taxometrics, 9, 2
taxon-radius models, 10, 14, 20, 24
taxonomic,
    distance, 9, 14
    hierarchy, 9, 3
    map, 9, 14, 21
    models, 9, 15
    rank, 9, 3
    structure, 9, 17
taxonomy, 9, 2
taxospecies, 9, 4, 28
Technitella thompsoni, 5, 8
teichoic acid, 6, 19, 20, 25, 26; 7, 25,
    26; 12, 10
teichuronic acid, 6, 20, 25; 7, 25, 26
teletroch, 5, 26
Telosporea, 5, 5
telosporidian, 5, 20
temperate phage, 12, 27
termites, 2, 26; 5, 21, 29; 15, 24
tetanus toxin, 6, 21; 13, 16
tetracycline, 6, 28
Tetradinium, 1, 27, 28
Tetrahymena, 5, 3, 4, 6, 12, 13
T. pyriformis, 5, 26, 28
T-even bacteriophages (myovidiae), 4,
    28, 29
Textularia, 5, 8
Thamnidium elegans, 3, 16
Thecamoeba verrucosa, 5, 22
Thermoactinomyces, 14, 2, 4, 5
Thielavia, 3, 4
Thiobacillus, 14, 4
T. acidophilus, 16, 6
T. ferrooxidans, 16, 6
T. intermedius, 16, 12
T. neapolitanus, 16, 6, 18
T. novellus, 16, 6, 18
T. perometabolis, 16, 12
T. thioparus, 16, 6, 12
T. thiooxidans, 2, 22; 16, 6
Thiovulvum majus, 16, 29

thrush, 13, 28
thylakoids, 1, 17
thymine, 6, 5; 11, 4, 18
Tintinnopsis, 5, 7
toxin, 13, 16
    diphtheria, botulinum, and tetanus,
        13, 16
toxoid, 13, 16
Toxoplasma gondii, 13, 31
Toxoplasmea, 5, 5
transduction, 1, 17; 12, 31
transferrin, 13, 11
transient-state phenomena, 7, 25
Trema aspera, 15, 17
Tremella, 3, 31
Treponema, 2, 8, 27
T. pallidum, 13, 13, 18
Triactinomyxon, 5, 3
tricarboxylic acid cycle, 8, 18, 24
Trichodina, 5, 15, 16
Trichomonas, 5, 3
Trichophyton, 13, 29
Trichoscyphella willkommii, 3, 6
Trinema, 5, 11
t-RNA, 6, 5; 11, 21, 22, 23, 26
    -synthetase, 11, 20
Trypanosoma, 5, 3, 4, 5, 13, 21, 26
T. cruzi, 13, 31
T. gondii, 13, 31
trypanosomes, 13, 30
tryptophan, 8, 29, 30
tryrothricin, 14, 19
Tuberculariaceae, 3, 18
tuberculosis, 13, 5, 15, 18
tubulin, 1, 29
Tulasnellales, 3, 29
turbidostat, 7, 11, 14, 15
Twort, F. W., 12, 2
tymovirus, 4, 17, 18
typhoid,
    bacillus, 13, 5, 10
    fever 13, 4, 17
typing phages, 10, 12

Ulmaceae, 15, 17
undulating membranes, 5, 14
Unitunimycetidae, 3, 4, 25
uracil, 6, 5, 11, 5, 18
urease, 8, 6
Uredinales, 3, 7, 28
Uronema, 5, 14
Ustilaginales, 3, 7, 11, 28
Ustilago violacea, 3, 29

This book is to be returned on
or before the date stamped below

UNIVERSITY OF PLYMOUTH

ACADEMIC SERVICES
PLYMOUTH LIBRARY
Tel: (0752) 232323
This book is subject to recall if required by another reader
Books may be renewed by phone
CHARGES WILL BE MADE FOR OVERDUE BOOKS